LONDON MATHEMATICAL SOCIETY LECTURE NOTE SERIES

Managing Editor: Professor N.J. Hitchin, Mathematical Institute,
University of Oxford, 24–29 St Giles, Oxford OX1 3LB, United Kingdom

The titles below, and earlier volumes in the series, are available from booksellers
or from Cambridge University Press at www.cambridge.org

London Mathematical Society Lecture Notes Series 315

Structured Ring Spectra

Edited by

Andrew Baker
University of Glasgow

Birgit Richter
Universität Bonn

CAMBRIDGE
UNIVERSITY PRESS

CAMBRIDGE
UNIVERSITY PRESS

University Printing House, Cambridge CB2 8BS, United Kingdom

One Liberty Plaza, 20th Floor, New York, NY 10006, USA

477 Williamstown Road, Port Melbourne, VIC 3207, Australia

314-321, 3rd Floor, Plot 3, Splendor Forum, Jasola District Centre, New Delhi - 110025, India

103 Penang Road, #05-06/07, Visioncrest Commercial, Singapore 238467

Cambridge University Press is part of the University of Cambridge.

It furthers the University's mission by disseminating knowledge in the pursuit of education, learning and research at the highest international levels of excellence.

www.cambridge.org
Information on this title: www.cambridge.org/9780521603058

© Cambridge University Press 2004

First published 2004

A catalogue record for this publication is available from the British Library

ISBN 978-0-521-60305-8 Paperback

Contents

Preface

This book arose out of the *Workshop on structured ring spectra and their applications* held in Glasgow in January 2002. Although it is not intended to be a proceedings of this conference, nevertheless the articles reflect the subject matter of the conference and the papers of Elmendorf, Robinson and Schwede have their origins in series of overview talks which these authors gave in Glasgow. All the papers published here have been refereed.

We would like to thank the London Mathematical Society, Edinburgh Mathematical Society and Glasgow Mathematical Journal Trust Fund for their financial support for the Workshop.

Since the middle of the 1990's there has been a renewed interest in structured ring spectra and several new models for the homotopy category of spectra of Boardman or Adams have been constructed, for example the category of S-modules constructed by Elmendorf, Kriz, Mandell and May [8], the category of symmetric spectra of Hovey, Shipley and Smith [11], the category of Γ-spaces constructed by Lydakis [13], the category of orthogonal spectra defined in [15] and [16], and many more. All of these categories possess a smash product which is strictly associative, commutative and unital, and therefore it makes sense to talk about monoids and commutative monoids, i.e., associative and commutative ring spectra.

Before these constructions have been found, one merely had a smash product of spectra which fulfilled associativity, commutativity only up to homotopy and therefore multiplicative structures on spectra were always given up to homotopy as well. Nevertheless, the techniques of Boardman and Vogt [4] and May, Quinn, Ray and Tornehave [14] allowed for various stricter notions of stricter ring structures, namely ring spectra which are homotopy associative (resp. commutative) up to higher coherences, the so-called A_∞ (resp. E_∞) ring spectra. The coherences were encoded via *operads* in the May approach and via *categories of operators in standard form* in the work of Boardman and Vogt.

With the help of these methods it was possible to identify important spectra as E_∞ ring spectra. For instance, the complex cobordism spectrum MU and other bordism spectra are E_∞ spectra [14, IV §2], as are real and complex connective K-theory ko and ku [14, VIII §2]. But for instance, it is currently unknown whether the Brown-Peterson spectrum BP at a prime possesses an E_∞ structure.

The development of a strict smash product allows additional methods for proving that certain spectra have multiplicative structures and all examples of A_∞ and E_∞ spectra get a strict model by the comparison results of [8, II 4.5, II 4.6]. But many new examples can be gained by working in these strict monoidal categories. For instance, as Bousfield localizations preserve algebra structures on spectra [8, VIII 2.2], the commutative structures on ko

3

and ku can be used to prove that real and complex K-theory, KO and KU are commutative S-algebras [8, VIII 4.3].

In addition to the construction of algebra spectra as above there are other approaches to imposing associative or commutative algebra structures on a given spectrum: There are several obstruction theories where certain obstruction groups are identified whose vanishing implies the existence of an associative or commutative multiplication on a spectrum. The earliest work in this direction is contained in the paper by Alan Robinson [18] from 1989, which developed an obstruction theory for the detection of A_∞ spectra and applied it to prove that the Morava K-theories $K(n)$ at odd primes have uncountable many A_∞-structures. Later in [1] this method was used to show that completed versions of the Johnson-Wilson spectra possess A_∞ structures. Other obstruction theories for associative algebra structures are contained in the work of Hopkins and Miller [17], Goerss [9] and Lazarev [12].

Later obstruction theories for commutative structures were developed by Basterra [2], Goerss and Hopkins [10], and Robinson [19]. A common feature of all of these obstruction theories is the fact that the obstruction groups live in some generalized versions of André-Quillen cohomology. There will be a short overview about the different types of these cohomology theories for commutative algebras and comparison results between them in [3].

For example the Hopkins-Miller approach led to an A_∞ structure on each Lubin Tate spectrum E_n; the work of Goerss-Hopkins (compare the paper "Moduli Spaces of Commutative Ring Spectra" [10] in this volume) shows that this structure can be refined to an E_∞ multiplication.

The papers in this book deal with two different general topics. On the one hand the papers by Elmendorf, Schwede, and Joachim deal with foundational matters and with the construction of E_∞ structures.

- Tony Elmendorf's article "The development of structured ring spectra" [5] gives background material to the evolution of the symmetric monoidal category of spectra of S-modules of [8]. His second paper "Compromises forced by Lewis's theorem" [6] describes problems one necessarily meets when constructing such a symmetric monoidal category which models the stable homotopy category. There is a list of natural properties such a category might reasonably be required to have and Lewis's theorem shows that these are inconsistent. The third paper [7] in this series represents joint work of Tony Elmendorf and Mike Mandell. That paper sketches how one might use permutative categories to model connective spectra. They construct a symmetric monoidal product on permutative categories and give evidence why this should model the smash-product of spectra. A by-product is a short proof for the fact that bipermutative categories give rise to commutative spectra by forming the associated K-theory spectrum.

- Stefan Schwede's paper "Morita theory in abelian, derived and stable model categories" discusses an enlarged version of Morita equivalence. The subject of the paper [21] is to transfer the well-known notion for Morita equivalence of rings to the derived category of a ring and further to modules over a ring spectrum.
- In "Higher coherences for equivariant K-theory" Michael Joachim gives an E_∞ model for G-equivariant K-theory, for G a compact Lie group and he proves that the G-equivariant analog of the Atiyah-Bott-Shapiro orientation from Spin^c-bordism to K-theory has an E_∞-model as well.

The second part of this book is concerned with obstruction theories for algebra structures on spectra.

- We start with a survey called "(Co-)Homology theories for commutative (S-)algebras" where we describe some of the cohomology theories for commutative algebras and discuss some comparison results for these.
- Alan Robinson gives an overview over his obstruction theories for A_∞ and E_∞ structures on spectra in his paper "Classical obstructions and S-algebras". Besides this he proves, that there are no obstructions to upgrading a homotopy unit to a strict unit for an A_∞ (resp. E_∞) spectrum.
- The paper by Paul Goerss and Mike Hopkins "Moduli spaces of commutative ring spectra" is a summary of the obstruction theory using methods of Dwyer, Kan and Stover; in particular it discusses the example of the Lubin-Tate spectra.
- Andrey Lazarev's paper "Cohomology theories for highly structured ring spectra" describes the approach to obstruction theory using topological André-Quillen homology, which was defined by Basterra, and its counterpart in the theory for associative algebra structures, called *topological derivations*.

Andrew Baker Birgit Richter

References

[1] A. Baker, A_∞ *structures on some spectra related to Morava K-theory*, Quart. J. Math. Oxf. **42**, 2 (1991), 403–419.

[2] M. Basterra, *André-Quillen cohomology of commutative S-algebras*, J. Pure Appl. Algebra **144**, 2 (1999), 111–143.

[3] M. Basterra & B. Richter, *Homology theories for commutative (S-)algebras*, this volume

[4] J.M. Boardman & R.M. Vogt, *Homotopy invariant algebraic structures on topological spaces*, Lecture Notes in Mathematics, **347** Springer-Verlag, Berlin-New York (1973) x+257 pp.

[5] A.D. Elmendorf, *The development of structured ring spectra*, this volume

[6] A.D. Elmendorf, *Compromises forced by Lewis's theorem*, this volume

[7] A.D. Elmendorf & M.A. Mandell, *Permutative categories as a model of connective stable homotopy theory*, this volume

[8] A.D. Elmendorf, I. Kriz, M.A. Mandell & J.P. May, *Rings, modules, and algebras in stable homotopy theory*, with an appendix by M. Cole, Mathematical Surveys and Monographs, **47**, AMS, Providence, RI (1997)

[9] P.G. Goerss, *Associative MU-algebras*, (http://www.math.nwu.edu/~pgoerss/)

[10] P.G. Goerss & M.J. Hopkins, *Moduli Spaces of Commutative Ring Spectra*, this volume

[11] M. Hovey, B. Shipley & J. Smith, *Symmetric spectra*, J. Amer. Math. Soc. **13**, 1 (2000), 149–208

[12] A. Lazarev, *Cohomology theories for highly structured ring spectra*, this volume

[13] M. Lydakis, *Smash products and Γ-spaces*, Math. Proc. Cambridge Philos. Soc. **126**, 2 (1999), 311–328.

[14] J.P. May, E_∞ *ring spaces and* E_∞ *ring spectra*, with contributions by F. Quinn, N. Ray and J. Tornehave, Lecture Notes in Mathematics **577** (1977).

[15] M. A. Mandell, J. P. May, S. Schwede and B. Shipley, *Model categories of diagram spectra*, Proc. London Math. Soc. **82** (2001), 441–512.

[16] M.A. Mandell, J.P. May, *Equivariant orthogonal spectra and S-modules*, Mem. Amer. Math. Soc. **159**, 755 (2002), x+108 pp.

[17] C. Rezk, *Notes on the Hopkins-Miller theorem*, Contemp. Math., **220** (1998) 313–366.

[18] A. Robinson, *Obstruction theory and the strict associativity of Morava K-theories*, in 'Advances in Homotopy Theory', eds. S. Salamon, B. Steer, W. Sutherland, London Mathematical Society Lecture Note Series **139** (1989) 143–52

[19] A. Robinson, *Gamma homology, Lie representations and* E_∞ *multiplications*, Invent. Math. **152**, 2 (2003), 331–348

[20] A. Robinson, *Classical obstructions and S-algebras*, this volume

[21] S. Schwede, *Morita theory in abelian, derived and stable model categories*, this volume

THE DEVELOPMENT OF STRUCTURED RING SPECTRA

A. D. ELMENDORF

ABSTRACT. The problem of giving a succinct description of multiplicative structure on spectra was recognized almost as soon as the idea of a spectrum was formulated. This paper aims to describe the major features of the historical precursors to the S-module approach of [2]. In particular, we consider the purely homotopical notion of a ring spectrum, May's concepts of external smash product and its internalization, the Lewis-May twisted half-smash product, and this product's use in formulating May and Quinn's notion of an E_∞ ring spectrum. We then describe how three essentially trivial (but crucial) observations led to the idea of an $\mathbb{L}$-spectrum, and soon thereafter to S-modules. We conclude by describing the good formal and homotopical properties of the category of S-modules.

The aim of this paper is to give some historical background to the first of the modern treatments of structured ring spectra: the S-module approach of [2]. There have been subsequent models developed as well; I'd like to mention in particular the symmetric spectra originally developed by Jeff Smith [3] and the orthogonal spectra of Mandell and May ([6] and [7]). There has been quite a lot of work done relating these various approaches, but this paper is concerned with the S-module approach only.

The invention of spectra, in the sense of algebraic topology, is usually credited to Lima in the late 1950's, although the first definition in print appears to be Spanier's [10]. There were a number of sources for the idea, of which I'd like to mention three: stable maps and the Spanier-Whitehead category, cohomology and Eilenberg-Mac Lane spaces, and cobordism. All of these involve sequences of spaces $A_0, A_1, \ldots$ and maps

$$A_i \to \Omega A_{i+1}$$

(or equivalently, $\Sigma A_i \to A_{i+1}$.) Two of the problems that were recognized early on were

(1) What is the "right" notion of morphism; some only "start to exist" after n stages, and

(2) what is the correct way to formulate multiplicative structure?

Boardman gave what were quickly recognized as the right answers **after** passage to homotopy – he constructed a symmetric monoidal closed triangulated category, now universally called the stable category, whose study has

---

2000 *Mathematics Subject Classification.* 55P42.
Key words and phrases. ring spectra, smash product.

a large literature of its own. Although Boardman never published his construction, an account was given by Vogt [11]. Adams gave a treatment in his Chicago notes [1], and May also gave a construction in a series of papers and books; see particularly [8] and [9]. All these constructions shared a common problem: the smash product construction in the underlying category of spectra was not associative until passage to homotopy. As a consequence, there were no strict ring spectra, and no "good" categories of module spectra. One problem in particular will illustrate this: if R is a ring spectrum in the weak sense, i.e., it descends to a ring object in the stable category, and if M and N are R-modules in the same weak sense, and further we have a map $M \to N$ of R-modules, then the cofiber of this map need not even be an R-module. Although topologists were able to use these weak notions to good effect nonetheless, the situation was clearly less than completely satisfactory.

Progress came first from Peter May, who began by resolving problem 1; see [8]. His solution was to restrict attention to spectra for which the structure maps $A_i \to \Omega A_{i+1}$ are homeomorphisms. Having done so, he showed that all spectra are weakly equivalent to ones of this restrictive form, and further, the "naive" sort of morphism, consisting of sequences of maps making obvious diagrams commute, suffice for this sort of spectrum. (The modern point of view is that he restricts his attention to fibrant objects.) The next step was to remove the indexation on natural numbers, by developing what he called coordinate-free spectra ; see [9], although some details were later deleted in the equivariant version due to Lewis and May [4] (and ironically enough, restored in the definition of $\mathbb{L}$-spectra, below.) These are defined by first picking a **universe** $\mathcal{U}$: a real inner product space isomorphic to $\mathbb{R}^\infty$, topologized using the colimit topology from the sequence

$$\{0\} \subset \mathbb{R}^1 \subset \mathbb{R}^2 \subset \cdots,$$

with this topology used to topologize spaces of linear isometries that will arise shortly. The index set for a spectrum over the universe $\mathcal{U}$ is the set of finite dimensional subspaces of $\mathcal{U}$. In detail, a spectrum E assigns a space EV to each finite dimensional subspace $V < \mathcal{U}$, and whenever $W \perp V$, there is a structure homeomorphism

$$EV \xrightarrow{\;\cong\;} \Omega^W E(V \oplus W),$$

subject to an associativity diagram. Here $\Omega^W X$ is the function space $F(S^W, X)$, and S^W is the one-point compactification of W. Morphisms from E to E' consist of maps $EV \to E'V$ making the obvious squares commute. We get the category $\mathcal{S}\mathcal{U}$ of spectra over $\mathcal{U}$. We note for later use that $\mathcal{S}\mathcal{U}$ is both complete and cocomplete, meaning it has all limits and colimits; this is not related to the coordinate-free nature of the spectra in $\mathcal{S}\mathcal{U}$.

A key point about spectra with structure maps consisting of homeomorphisms is that the constituent spaces need only be given for a cofinal set of indices: the rest can be filled in by looping spaces given for larger indices. This allows us to define the **external smash product**, which is a functor $\mathcal{S}\mathcal{U} \times \mathcal{S}\mathcal{U}' \to \mathcal{S}(\mathcal{U} \oplus \mathcal{U}')$ for any pair of universes $\mathcal{U}$ and $\mathcal{U}'$. From the previous remark, we need only consider subspaces of $\mathcal{U} \oplus \mathcal{U}'$ of the form $V \oplus V'$ for $V < \mathcal{U}$ and $V' < \mathcal{U}'$, and for these, we make a preliminary definition of

$$(E \wedge E')(V \oplus V') := EV \wedge E'V'.$$

For structure maps, we use the composite

$$\Sigma^{W \oplus W'} EV \wedge E'V' \cong \Sigma^W EV \wedge \Sigma^{W'} E'V' \to E(V \oplus W) \wedge E'(V' \oplus W')$$
$$= (E \wedge E')(V \oplus V' \oplus W \oplus W').$$

The main philosophical point here is that there is no issue with permuting indices, since all of the indexing subspaces are orthogonal, so the direct sum is completely independent of the order of the summands. There is one technical annoyance to be confronted: the adjoint structure maps

$$(E \wedge E')(V \oplus V') \to \Omega^{W \oplus W'}(E \wedge E')(V \oplus V' \oplus W \oplus W')$$

are not homeomorphisms any more. However, there is a "spectrification" functor that corrects this situation, and we do get a good smash product $E \wedge E'$ indexed on $\mathcal{U} \oplus \mathcal{U}'$, where by "good" I mean that we get a symmetric monoidal structure on the category

$$\coprod_{n \geq 0} \mathcal{S}(\mathcal{U}^n).$$

Unfortunately, the homotopy category we get is wrong: instead of the stable category, we get a coproduct of one copy of the unstable homotopy category (for $n = 0$) and infinitely many of the stable category. May's solution for this is to pick an element $f \in \mathcal{I}(\mathcal{U}^2, \mathcal{U})$ (the space of linear isometries from $\mathcal{U}^2$ to $\mathcal{U}$), and "push down" $E \wedge E'$ using f: we get a spectrum $f_*(E \wedge E') \in \mathcal{S}\mathcal{U}$. (The push down process proceeds by defining $(f_* D)(V) = \Sigma^{V - ff^{-1}V} D(f^{-1}V)$, and then spectrifying.) Unfortunately, this destroys the associativity and commutativity of the external smash product, since no such choice of f is associative and commutative. May does show that this gives the right construction in homotopy.

The technical heart of the solution via S-modules is the twisted half-smash product introduced by Lewis, May, and Steinberger [4]. This is a functor with input two universes $\mathcal{U}$ and $\mathcal{U}'$, an unbased space A with a structure map $A \to \mathcal{I}(\mathcal{U}, \mathcal{U}')$, and a spectrum E over the first universe $\mathcal{U}$. The output is a spectrum $A \ltimes E$ over $\mathcal{U}'$. This construction has the following important formal properties:

(1) Given $A \to \mathcal{I}(\mathcal{U}, \mathcal{U}')$ and $B \to \mathcal{I}(\mathcal{U}', \mathcal{U}'')$, then there is a canonical isomorphism

$$B \ltimes (A \ltimes E) \cong (B \times A) \ltimes E.$$

(2) Given $A_1 \to \mathcal{I}(\mathcal{U}_1, \mathcal{U}_1')$ and $A_2 \to \mathcal{I}(\mathcal{U}_2, \mathcal{U}_2')$, then there is a canonical isomorphism

$$(A_1 \ltimes E_1) \wedge (A_2 \ltimes E_2) \cong (A_1 \times A_2) \ltimes (E_1 \wedge E_2).$$

(3) If $f \in \mathcal{I}(\mathcal{U}, \mathcal{U}')$, then $\{f\} \ltimes E \cong f_* E$. In particular, $\{\mathrm{id}_\mathcal{U}\} \ltimes E \cong E$.

In addition, the most important homotopical property of the twisted half-smash product is the following: given a homotopy equivalence $A_1 \to A_2$ (**not** a weak equivalence) and a structure map $A_2 \to \mathcal{I}(\mathcal{U}, \mathcal{U}')$, then the induced map

$$A_1 \ltimes E \to A_2 \ltimes E$$

is a homotopy equivalence when E is "tame"; this hypothesis is satisfied if E is a CW-spectrum. In general, we don't know if this map is even a weak equivalence. The point is that the homotopy equivalence between A_1 and A_2 need not be over $\mathcal{I}(\mathcal{U}, \mathcal{U}')$. For further details, see [2], especially the Appendix by Michael Cole. This homotopical property implies that the inclusion map $\{f\} \subset \mathcal{I}(\mathcal{U}^2, \mathcal{U})$ induces a homotopy equivalence

$$f_*(E \wedge E') \simeq \mathcal{I}(\mathcal{U}^2, \mathcal{U}) \ltimes (E \wedge E')$$

for tame spectra E and E'.

We are now in a position to describe the original notion of a structured ring spectrum, called an E_∞ ring spectrum. First, for notation, we let $\mathcal{L}(n) = \mathcal{I}(\mathcal{U}^n, \mathcal{U})$; this is the nth space in the **linear isometries operad** using the universe $\mathcal{U}$. The intuition is that given a spectrum E, the spectrum $\mathcal{L}(n) \ltimes E^{\wedge n}$ encodes all possible n-fold smash powers of E, and has the correct homotopy type (at least when E is tame.)

Definition 1. *An E_∞ ring spectrum R over $\mathcal{U}$ is a spectrum over $\mathcal{U}$ together with structure maps*

$$\xi_n : \mathcal{L}(n) \ltimes R^{\wedge n} \to R.$$

These must be "coherent" in the sense that several diagrams must commute; the most important (and largest) is the following, in which the map γ is the structure map for the operad $\mathcal{L}$:

$$
\begin{array}{ccc}
\mathcal{L}(n) \ltimes ((\mathcal{L}(j_1) \ltimes R^{\wedge j_1}) \wedge \cdots \wedge (\mathcal{L}(j_n) \ltimes R^{\wedge j_n})) & & \\
\Big\downarrow{\scriptstyle \cong} \qquad \qquad \quad \searrow^{1 \ltimes (\xi_{j_1} \wedge \cdots \wedge \xi_{j_n})} & & \\
(\mathcal{L}(n) \times \mathcal{L}(j_1) \times \cdots \times \mathcal{L}(j_n)) \ltimes R^{\wedge(j_1 + \cdots + j_n)} & & \mathcal{L}(n) \ltimes R^{\wedge n} \\
\Big\downarrow{\scriptstyle \gamma \ltimes id} & & \Big\downarrow{\scriptstyle \xi_n} \\
\mathcal{L}(j_1 + \cdots + j_n) \ltimes R^{\wedge(j_1 + \cdots + j_n)} & \xrightarrow{\;\;\xi_{j_1 + \cdots + j_n}\;\;} & R.
\end{array}
$$

Further, the maps must be commutative, in the sense that ξ_n descends to a map from the orbit spectrum $\mathcal{L}(n) \ltimes_{\Sigma_n} R^{\wedge n}$.

This definition can be given in an alternative, somewhat more formal way:

Definition 2. *Given a spectrum E over* $\mathcal{U}$, *let*

$$\mathbb{C}E = \bigvee_{n \geq 0} \mathcal{L}(n) \ltimes_{\Sigma_n} E^{\wedge n}.$$

Then $\mathbb{C}$ *is a monad in* $\mathcal{SU}$, *and an* E_∞ *ring spectrum is the same thing as a* $\mathbb{C}$*-algebra.*

And there things stood for 15 or 20 years.

It wasn't until 1993 that a combination of three essentially trivial observations led to a breakthrough with the development of $\mathbb{L}$-spectra and, from them, S-modules; see [2] for full details. The first observation is that $\mathcal{L}(2) \ltimes (E \wedge E')$ is a **canonical** smash product for E and E', encoding all possible choices of f_*, and further it has the correct homotopy type. The stumbling block is that it's not associative. The key to correcting this defect comes from the second observation, which is that $\mathbb{C}$ has a tiny submonad $\mathbb{L}$, defined as

$$\mathbb{L}E := \mathcal{L}(1) \ltimes E.$$

An **$\mathbb{L}$-spectrum** is simply an algebra over the monad $\mathbb{L}$. Since $\mathbb{L}$ is a submonad of $\mathbb{C}$, it follows automatically that every E_∞ ring spectrum is an $\mathbb{L}$-spectrum. Further, although this took a bit of work, $\mathbb{L}$-spectra form a perfectly good model of the stable category.

The third observation, due to Mike Hopkins, tells us how to put the first two observations together in order to construct an associative smash product. It is:

Lemma 3. (Hopkins' Lemma) *Consider the left action of* $\mathcal{L}(1)$ *on* $\mathcal{L}(j)$ *for any* j *and the right action of* $\mathcal{L}(1) \times \mathcal{L}(1)$ *on* $\mathcal{L}(2)$, *both by means of composition. Then if* $i \geq 1$ *and* $j \geq 1$, *the structure map* γ *of the operad* $\mathcal{L}$ *induces an isomorphism*

$$\mathcal{L}(2) \times_{\mathcal{L}(1) \times \mathcal{L}(1)} \mathcal{L}(i) \times \mathcal{L}(j) \cong \mathcal{L}(i+j).$$

Proof. By choosing isomorphisms $\mathcal{U}^i \cong \mathcal{U}$ and $\mathcal{U}^j \cong \mathcal{U}$, the coequalizer splits. □

This allows us to make a key definition.

Definition 4. *Given* $\mathbb{L}$*-spectra M and N, their smash product is given by*

$$M \wedge_{\mathbb{L}} N := \mathcal{L}(2) \ltimes_{\mathcal{L}(1) \times \mathcal{L}(1)} (M \wedge N).$$

Here the unsubscripted smash product is the external smash product described above.

Proposition 5. *The smash product of* $\mathbb{L}$*-spectra is coherently associative and commutative.*

Proof. The essential point is the associativity, and this follows by using Hopkins' Lemma to show that both ways of associating are canonically isomorphic to

$$\mathcal{L}(3) \ltimes_{\mathcal{L}(1)^3} (M_1 \wedge M_2 \wedge M_3).$$

$\square$

A small problem is that this smash product of $\mathbb{L}$-spectra is not quite unital; instead, there is a canonical weak equivalence

$$\lambda : S \wedge_{\mathbb{L}} M \to M$$

for any $\mathbb{L}$-spectrum M, but this is sufficient to formulate most concepts. In particular, we can define a strictly commutative $\mathbb{L}$-ring spectrum as an $\mathbb{L}$-spectrum A together with a unit map $\eta : S \to A$ and an associative, commutative, and unital map

$$\mu : A \wedge_{\mathbb{L}} A \to A.$$

It is now an easy proposition that this recovers exactly the definition of E_∞ ring spectrum!

However, we don't have to be satisfied with the weak notion of units present with $\mathbb{L}$-spectra, because of a stroke of good luck: it turns out that the unit map for the sphere spectrum, $\lambda : S \wedge_{\mathbb{L}} S \to S$, is an isomorphism. This is because of the "accident" that Hopkins' lemma is true when $i = j = 0$, although the proof is different (and the lemma fails when one index is 0 and the other is not.) It follows immediately that $\lambda : S \wedge_{\mathbb{L}} M \to M$ is an isomorphism precisely when M is of the form $S \wedge_{\mathbb{L}} M'$, and it is these M's that we call *S*-**modules**. A bit of extra work gives us the following:

Proposition 6. *The smash product of $\mathbb{L}$-spectra is symmetric monoidal on the full subcategory of S-modules, and this subcategory models the stable category with its smash product.*

We write the category of S-modules as $\mathcal{M}_S$.

We are now in a good position to mimic all the formal apparatus of commutative algebra, once a few more details are settled; it is to these we now turn. First, we would like to have a function spectrum construction adjoint to the smash product, just as one has in categories of (ordinary) modules. This relies ultimately on the fact that the twisted half-smash product has a right adjoint, called the twisted function spectrum, written $F[A, E')$, with input $A \to \mathcal{I}(\mathcal{U}, \mathcal{U}')$ and a spectrum E' over $\mathcal{U}'$, and output a spectrum over $\mathcal{U}$. For details, see [2]. The end result is all we could expect: there is a function spectrum construction on S-modules, written $F_S(M, N)$ for which

$$\mathcal{M}_S(M \wedge_{\mathbb{L}} N, P) \cong \mathcal{M}_S(M, F_S(N, P)).$$

Our second piece of unfinished business is to show that $\mathcal{M}_S$ has all the limits and colimits we could possibly want.

Proposition 7. *The category of S-modules is complete and cocomplete.*

Proof. First, May's category $\mathcal{SU}$ is complete and cocomplete, and the category of $\mathbb{L}$-spectra is a category of algebras over it. Therefore, by [5], section VI.2, exercise 2, the category of $\mathbb{L}$-spectra is complete. Further, $\mathbb{L}$ has a right adjoint $\mathbb{L}^{\#}$, given by $\mathbb{L}^{\#}E := F[\mathcal{L}(1), E)$. $\mathbb{L}^{\#}$ is consequently a comonad and the category of algebras over $\mathbb{L}$ can be identified with the category of coalgebras over $\mathbb{L}^{\#}$. By the dual exercise, $\mathbb{L}$-spectra form a cocomplete category. Next, we examine the functor $S \wedge_{\mathbb{L}} \_ : \mathbb{L}$-spectra $\to \mathcal{M}_S$, and find that it has both a left and a right adjoint, with the right adjoint being the inclusion of $\mathcal{M}_S$ into $\mathbb{L}$-spectra. Consequently, colimits in $\mathcal{M}_S$ are created in $\mathbb{L}$-spectra, and limits exist and are gotten by applying $S \wedge_{\mathbb{L}} \_$ to the limit in $\mathbb{L}$-spectra. $\square$

Now we can import the entire formal apparatus of commutative algebra into stable homotopy theory: rings, commutative rings, algebras, left and right modules, tensor products, and function objects, with all the expected properties. As an example, we define a commutative S-algebra to be simply a commutative monoid in the symmetric monoidal category of S-modules, and we quickly see that all of them are E_∞ ring spectra, and further, given an E_∞ ring spectrum A, then $S \wedge_{\mathbb{L}} A$ is a commutative S-algebra. Because of the isomorphism $S \wedge_{\mathbb{L}} S \cong S$, this accounts for all commutative S-algebras, and since the unit map $\lambda : S \wedge_{\mathbb{L}} A \to A$ is always a weak equivalence, and easily seen to be a map of E_∞ ring spectra, we recover all the homotopical properties of E_∞ ring spectra by considering only commutative S-algebras.

I'd like to close by mentioning additional structure that is present: the categories of S-modules, S-algebras, commutative S-algebras, and the categories of algebras and modules over a given S-algebra are all topological model categories in which all objects are fibrant. In some sense this is the most exciting part of the new developments with structured ring spectra, since it allows us to talk about homotopy categories that were not even in the picture previously. These new homotopy categories have already inspired a considerable body of work, some of which appears elsewhere in this volume, but clearly much more is still to be done. Let's look forward to the exploration of these brave new worlds!

REFERENCES

[1] J. F. Adams, Stable homotopy and generalised homology, Chicago Lectures in Mathematics, University of Chicago Press,Chicago, IL, 1974

[2] A. D. Elmendorf, I. Kriz, M. A. Mandell, and J. P. May, Rings, modules, and algebras in stable homotopy theory, with an appendix by M. Cole, Mathematical Surveys and Monographs, vol. 47,American Mathematical Society, Providence, RI, 1997

[3] M. Hovey, B. Shipley, and J. Smith, Symmetric spectra, J. Amer. Math. Soc. vol 13, 2000, 149–208

[4] L. G. Lewis, Jr., J. P. May, and M. Steinberger, Equivariant stable homotopy theory. With contributions by J. E. McClure, Lecture Notes in Mathematics vol. 1213, Springer-Verlag, Berlin, 1986

[5] Saunders Mac Lane, Categories for the working mathematician, Springer-Verlag, New York, Berlin, 1971

[6] M. A. Mandell, J. P. May, S. Schwede, and B. Shipley, Model categories of diagram spectra, Proc. London Math. Soc. (3) vol 82, 2001, 441–512

[7] M. A. Mandell and J. P. May, Equivariant orthogonal spectra and S-modules, Mem. Amer. Math. Soc. vol 159 (2002) no 755

[8] J. P. May, Categories of spectra and infinite loop spaces, 1969 Category Theory, Homology Theory and their Applications, III (Batelle Institute Conference, Seattle, WA 1968), 448–479

[9] J. P. May, E_∞ ring spaces and E_∞ ring spectra, with contributions by Frank Quinn, Nigel Ray, and Jørgen Tornehave, Lecture Notes in Mathematics vol. 577, Springer-Verlag, Berlin-New York, 1977

[10] E. H. Spanier, Function spaces and duality, Ann. of Math. (2) vol 70, 1959, 338–378

[11] Rainer Vogt, Boardman's stable homotopy category, Lecture Notes Series No. 21, Matematisk Institut, Aarhus Universitet, Aarhus, 1970

DEPARTMENT OF MATHEMATICS, PURDUE UNIVERSITY CALUMET, HAMMOND, IN 46323

E-mail address: aelmendo@math.purdue.edu

COMPROMISES FORCED BY LEWIS'S THEOREM

A. D. ELMENDORF

ABSTRACT. In 1991, Gaunce Lewis published a theorem showing that a quite minimal list of desiderata for an "ideal" category of spectra was inconsistent; see [4]. This result requires any category modeling stable homotopy theory to make some compromises in its formal structure. This short paper describes the compromises present in $\mathcal{M}_S$, the category of S-modules developed in [2], together with the amusing consequence that $\mathcal{M}_S$ contains a copy of the (unstable!) category of topological spaces.

At this point we have several categories of spectra that are symmetric monoidal, with their smash products descending to the smash product in the stable category; let me mention in particular the S-modules of [2] and the symmetric spectra of [3]. These categories are much more nicely behaved than any of their predecessors, but their behavior is not absolutely ideal, because it can't be. This is a theorem of Gaunce Lewis's, whose paper [4] was published before any of the current batch of symmetric monoidal categories of spectra were developed. Suppose we have a candidate for a "good" category of spectra, which we ambiguously call $\mathcal{S}$. Lewis sets out the following pretty minimal list of properties for $\mathcal{S}$, all of which are devoutly to be desired:

(1) The category $\mathcal{S}$ has a symmetric monoidal product, which we call smash and write $\wedge$, as usual.
(2) Let $\mathcal{T}$ be the category of based topological spaces (in some convenient version such as compactly generated weak Hausdorff). Then there is a pair of functors $\Sigma^\infty : \mathcal{T} \to \mathcal{S}$ and $\Omega^\infty : \mathcal{S} \to \mathcal{T}$ with Σ^∞ being left adjoint to Ω^∞.
(3) The unit for the smash product in $\mathcal{S}$ is $\Sigma^\infty S^0$.
(4) Σ^∞ is a lax monoidal functor in the sense that there is a natural map

$$\Sigma^\infty(X \wedge Y) \to \Sigma^\infty X \wedge \Sigma^\infty Y,$$

subject to diagrams encoding commutation with the monoidal structure maps.

2000 *Mathematics Subject Classification.* 55P42.
Key words and phrases. ring spectra, smash product.

(5) There is a natural weak equivalence $\theta : \Omega^\infty \Sigma^\infty X \to QX$ (where QX is the usual stabilization construction) for which the diagram

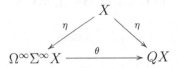

commutes, where η is used generically for the unit of an adjunction.

Theorem 1. (Lewis '89) *The above five properties are inconsistent.*

The proof is distressingly simple. Equivalent to property 4, there is a natural map

$$\Omega^\infty E_1 \wedge \Omega^\infty E_2 \to \Omega^\infty (E_1 \wedge E_2)$$

which also commutes with the monoidal structure maps. Suppose E is a commutative monoid in $\mathcal{S}$: what we would like to call a strictly commutative ring spectrum. Then the two maps

$$\Omega^\infty E \wedge \Omega^\infty E \longrightarrow \Omega^\infty (E \wedge E) \xrightarrow{\Omega^\infty \mu} \Omega^\infty E$$

and

$$S^0 \xrightarrow{\eta} \Omega^\infty \Sigma^\infty S^0 \longrightarrow \Omega^\infty E$$

make $\Omega^\infty E$ into a commutative monoid in $\mathcal{T}$, using the symmetric monoidal smash product of based spaces. Now the unit, in this case $\Sigma^\infty S^0$, is always a commutative monoid in a symmetric monoidal category, so in particular, $\Omega^\infty \Sigma^\infty S^0$ must be a commutative monoid in $\mathcal{T}$. From property 5, we now see that QS^0 is weakly equivalent to a commutative monoid. It follows from a theorem of Moore [5] that QS^0 is homotopic to a product of Eilenberg-Mac Lane spaces. Life would be a lot simpler if this were true...

As a consequence of this theorem, every "good" category of spectra has to edge around the fact that it can't satisfy all five of these properties simultaneously. Here's what $\mathcal{M}_S$, the category of S-modules does.

$\mathcal{M}_S$ is symmetric monoidal, so property 1 is satisfied, and there is an adjoint pair $(\Sigma^\infty, \Omega^\infty)$, so 2 is satisfied. We actually have an isomorphism

$$\Sigma^\infty (X \wedge Y) \cong \Sigma^\infty X \wedge_S \Sigma^\infty Y,$$

so property 3 is more than satisfied. And the unit for $\wedge_S$ is $\Sigma^\infty S^0$, so property 4 is satisfied, too. This leaves property 5 to fail, which it does in spectacular fashion: $\Omega^\infty \Sigma^\infty X$ is actually homeomorphic to X for all spaces X, no matter how badly behaved! Clearly Ω^∞ is not what we usually think, since Σ^∞ does look pretty much like what we think it should. In fact, what I've been calling $\Omega^\infty E$ is actually the space $\mathcal{M}_S(S, E)$ for an S-module E. The fact that $\mathcal{M}_S(S, \Sigma^\infty X) \cong X$ is a special case of the following theorem; see [1] for details:

Theorem 2. *The functor* $\Sigma^\infty : \mathcal{T} \to \mathcal{M}_S$ *induces a homeomorphism*

$$\mathcal{T}(X, Y) \to \mathcal{M}_S(\Sigma^\infty X, \Sigma^\infty Y)$$

for all spaces X *and* Y.

The proof reduces the general case to the specific one first mentioned, and then computes in an extremely explicit fashion.

As a consequence of this theorem, $\mathcal{M}_S$ has inside it a perfect copy of $\mathcal{T}$, the category of topological spaces, as the full subcategory of suspension spectra. This should seem bizarre, since the purpose of $\mathcal{M}_S$ is to model stable homotopy, and in $\mathcal{T}$, nothing has been stabilized. In fact, if we just use honest homotopy classes, which amounts to taking π_0 of the mapping spaces between spectra, we don't get stable homotopy, as we see from the presence of this copy of $\mathcal{T}$. The situation is saved by the requirement that we invert the weak equivalences, not just the ordinary homotopy equivalences, and doing so precisely stabilizes the maps between suspension spectra. See [1] for the details. As an added amusement, we find that if X is a CW complex, then in the model category structure on $\mathcal{M}_S$, the S-module $\Sigma^\infty X$ is homotopic to a cofibrant S-module precisely when X is contractible. From this point of view, it's cofibrant replacement that stabilizes the maps between suspension spectra.

In conclusion, I should mention that there is another candidate for the $(\Sigma^\infty, \Omega^\infty)$ adjunction between $\mathcal{T}$ and $\mathcal{M}_S$ which does satisfy $\Omega^\infty \Sigma^\infty X \simeq QX$. Obviously, this is the correct pair of functors to use when doing homotopy theory. However, if we use this pair, we find that $\Sigma^\infty S^0$ is not the unit of the smash product of S-modules, this being crucial to the proof of Lewis's theorem. Once again, we have to compromise when setting up the formal properties of a "good" category of spectra.

REFERENCES

[1] A. D. Elmendorf, Stabilization as a CW approximation, J. Pure and Appl. Algebra, vol. 140, 1999, 23–32

[2] A. D. Elmendorf, I. Kriz, M. A. Mandell, and J. P. May, Rings, modules, and algebras in stable homotopy theory, with an appendix by M. Cole, *Mathematical Surveys and Monographs, vol. 47*, American Mathematical Society, Providence, RI, 1997

[3] M. Hovey, B. Shipley, and J. Smith, Symmetric spectra, J. Amer. Math. Soc. vol. 13, 2000, 149–208

[4] L. Gaunce Lewis, Jr., Is there a convenient category of spectra?, J. Pure and Appl. Algebra, vol. 73, 1991, 233–246

[5] J. C. Moore, Semi-simplicial complexes and Postnikov systems, *Symposium Internacional de Topologia Algebraica, La Universidad Nacional Autónoma de México y la UNESCO* (1958), 232–247

DEPARTMENT OF MATHEMATICS, PURDUE UNIVERSITY CALUMET, HAMMOND, IN 46323

email-address: aelmendo@math.purdue.edu

PERMUTATIVE CATEGORIES AS A MODEL OF CONNECTIVE STABLE HOMOTOPY

A. D. ELMENDORF AND M. A. MANDELL

ABSTRACT. We aim to provide a more efficient way of processing multiplicative structure in the passage from permutative categories to spectra. In particular, we develop a multiplicative endomorphism operad for any permutative category, and show that under our passage to spectra, this endomorphism operad maps to the endomorphism operad of the associated spectrum. Together with the result that a permutative category supports bipermutative structure if and only if there is a map from a canonical E_∞ operad into the multiplicative endomorphism operad, this gives a relatively simple proof that every bipermutative category gives rise to an E_∞ ring spectrum. The other major result of our work so far is the development of a symmetric monoidal product on a certain category of permutative categories which captures all the homotopy information inherent in the category of all permutative categories.

This note describes some aspects of an on-going project to relate various structures on categories to corresponding structures on spectra. This is a preliminary report only; an exposition with full details is still in preparation.

One source of motivation for our project comes from a number of questions that Gunnar Carlsson asked about the K-theory spectra of permutative categories which are naturally raised by the development of associative smash products in [2]. First, we already knew that the spectrum given by a bipermutative category is an E_∞ ring spectrum, or equivalently a commutative S-algebra. Carlsson's first question was to ask what structure on a permutative category would give its associated spectrum a module structure over such a commutative S-algebra. More importantly, he asked what the underlying permutative category would be for the smash product and function spectrum for two such module permutative categories. We don't have full answers to these questions, but we do have some interesting progress to report.

A second source of motivation comes from a claim that Bob Thomason made in one of his last papers [8] that he had found a symmetric monoidal product on permutative categories that modelled the smash product of connective spectra. He proved in [8] that one recovers all weak homotopy types of connective spectra by taking the K-theory of permutative categories. Thomason was scheduled to talk about this at an Oberwolfach conference

---

2000 *Mathematics Subject Classification.* 19D23.

Key words and phrases. ring spectra, smash product, K-theory, symmetric monoidal category.

where Peter May spoke about the results later written up in [2]. Unfortunately, Thomason withdrew his talk, and died shortly thereafter. We have been unable to find anyone who knows what his construction was. Chuck Weibel has not found it in Thomason's papers.

Let's begin with the definition of a permutative category. For our purposes, it will be a small category A together with a direct sum functor $\oplus : A \times A \to A$ which is strictly associative, and a natural commutativity isomorphism $\gamma : a \oplus b \cong b \oplus a$ for which $\gamma^2 = $ id and the following diagram commutes:

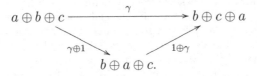

The experts will notice the absence of any unit condition in this definition. We find it convenient to omit this, since if one is needed, we can simply adjoin a disjoint unit. If there was one present already, the new one will be joined to it by the group completion process in passage to the K-theory spectrum, and we will lose no homotopy information by forgetting the original unit.

Next, we need a small result that characterizes permutative categories in terms an operad action. There is well-known operad "Σ" of sets whose nth term is the symmetric group Σ_n. Its algebras are precisely the monoids; with unit if you include $\Sigma_0 = *$ in your operad, without if you don't. Next, we apply a functor "E" from sets to categories. Given a set X, the objects of EX are the elements of the set X, and given two elements of X, say a and b, the hom-set $EX(a,b) = *$, a one point set. This completely determines the composition structure for EX. (We call this EX, because if G is a group, then the classifying space of EG is precisely the universal principal G-bundle usually called EG.) This functor E is right adjoint to the functor sending a (small) category to its set of objects, and therefore E preserves products as well as all other limits. Consequently, if we apply E to an operad of sets, such as Σ, we get an operad of categories, or Cat-operad.

Proposition 1. (Dunn) *Permutative categories are precisely the algebras over $E\Sigma$.*

This was apparently first noticed by Dunn [1], and the proof is simply to check definitions. We get for free the following corollary:

Corollary 2. *The free permutative category $\mathbb{P}B$ on a category B is given by the construction*

$$\mathbb{P}B = \coprod_{n \geq 1} E\Sigma_n \times_{\Sigma_n} B^n.$$

This defines a monad in **Cat**, *the category of small categories, whose algebras are precisely the permutative categories.*

(If you want your permutative categories to have units, you start your indexing at 0 instead of 1.)

This characterization of permutative categories gives us the first of the types of morphisms that we want to discuss.

Definition 3. *Let A and B be permutative categories. A* **strict** *morphism* $f : A \to B$ *is a map of* $\mathbb{P}$-*algebras.*

It is an easy exercise to see that this agrees with the usual definition of a strict morphism: a functor f for which $f(a_1 \oplus a_2) = f(a_1) \oplus f(a_2)$ and which converts the commutativity isomorphism in A into the one in B. We now have a category of permutative categories, which we call **Strict**, after the sort of morphisms we use. It is easy to enrich **Strict** over **Cat**. This category has some good properties and some weaknesses. One of the good properties is the following.

Corollary 4. Strict *is complete and cocomplete.*

Proof. Since the category of (small) categories is complete, and **Strict** is a category of algebras over a monad in **Cat**, **Strict** is automatically complete, with limits created in **Cat**. See [4], section VI.2, exercise 2. By essentially the same argument as in [2, II.7.2], the functor $\mathbb{P}$ preserves reflexive coequalizers, and it follows that **Strict** is cocomplete as well, with reflexive coequalizers created in **Cat**. See [2, II.7.4]. □

It's worth noting what colimits in **Strict** look like in general, at least on the level of objects. The objects of any permutative category form a monoid (without unit), and the forgetful functor from **Strict** to monoids has both a left and a right adjoint. (The right adjoint is just E again; the left adjoint is more complicated.) Consequently, the objects of a colimit of permutative categories are just the colimit of the objects as monoids. Note also that the objects of $\mathbb{P}B$ are the free monoid on the objects of B.

We find strict morphisms to be too restrictive. For one thing, if we try to turn the category $\mathbf{Strict}(A, B)$ into a permutative category in the obvious way, using $(f \oplus g)(a) := f(a) \oplus g(a)$, we find that $f \oplus g$ is no longer a strict morphism. In order to cure this defect, we consider **lax** morphisms.

Definition 5. *Let A and B be permutative categories. A* **lax** *morphism* $f : A \to B$ *is a functor together with a natural map*

$$\lambda : fa_1 \oplus fa_2 \to f(a_1 \oplus a_2)$$

for which the following two diagrams commute:

$$
\begin{array}{ccc}
fa_1 \oplus fa_2 \oplus fa_3 & \xrightarrow{\lambda \oplus 1} & f(a_1 \oplus a_2) \oplus fa_3 \\
{\scriptstyle 1 \oplus \lambda} \downarrow & & \downarrow {\scriptstyle \lambda} \\
fa_1 \oplus f(a_2 \oplus a_3) & \xrightarrow{\lambda} & f(a_1 \oplus a_2 \oplus a_3),
\end{array}
$$

$$fa_1 \oplus fa_2 \xrightarrow{\lambda} f(a_1 \oplus a_2)$$

$$\gamma \downarrow \qquad\qquad \downarrow f\gamma$$

$$fa_2 \oplus fa_1 \xrightarrow{\lambda} f(a_2 \oplus a_1).$$

Composition of lax maps is easy to define, as is an enrichment to **Cat**, and we get a category (actually, a 2-category) **Lax** of permutative categories and lax maps. While **Lax** is neither complete nor cocomplete, at least we can give the category **Lax**(A, B) the structure of a permutative category by means of the definition

$$(f \oplus g)(a) := fa \oplus ga;$$

the structure map is given by

$$(f \oplus g)(a_1) \oplus (f \oplus g)(a_2) = fa_1 \oplus ga_1 \oplus fa_2 \oplus ga_2 \xrightarrow{\cong} fa_1 \oplus fa_2 \oplus ga_1 \oplus ga_2$$

$$\xrightarrow{\lambda_f \oplus \lambda_g} f(a_1 \oplus a_2) \oplus g(a_1 \oplus ga_2) = (f \oplus g)(a_1 \oplus a_2).$$

(Notice that even if f and g are strict maps, the presence of the commutativity isomorphism in this structure map prevents $f \oplus g$ from being strict.) Attempting to provide this internal hom functor with a left adjoint leads us to bilinear maps.

Definition 6. *Let A, B, and C be permutative categories. A **bilinear** map $f : A \times B \to C$ is a functor which is lax in each variable separately, where the lax structure maps are related by the diagram below; in other words, it consists of a functor f and "distributivity" maps*

$$d_1 : f(a_1, b) \oplus f(a_2, b) \to f(a_1 \oplus a_2, b)$$
$$d_2 : f(a, b_1) \oplus f(a, b_2) \to f(a, b_1 \oplus b_2)$$

each subject to the two diagrams given before for a lax structure map and making the following diagram relating the two distributivity maps commute:

$$
\begin{array}{ccc}
& \xrightarrow{d_2 \oplus d_2} & f(a_1, b_1 \oplus b_2) \oplus f(a_2, b_1 \oplus b_2) \\
f(a_1, b_1) \oplus f(a_1, b_2) \oplus f(a_2, b_1) \oplus f(a_2, b_2) & & \downarrow d_1 \\
\uparrow 1 \oplus \gamma \oplus 1 & & f(a_1 \oplus b_1, a_2 \oplus b_2) \\
f(a_1, b_1) \oplus f(a_2, b_1) \oplus f(a_1, b_2) \oplus f(a_2, b_2) & & \uparrow d_2 \\
& \xrightarrow{d_1 \oplus d_1} & f(a_1 \oplus a_2, b_1) \oplus f(a_1 \oplus a_2, b_2).
\end{array}
$$

If we write **Bilin**$(A, B; C)$ for the bilinear maps from $A \times B$ to C, we find that

$$\mathbf{Lax}(A, \mathbf{Lax}(B, C)) \cong \mathbf{Bilin}(A, B; C).$$

Of course, this doesn't really provide the desired left adjoint, but it does provide the starting point for our treatment of ring objects within the category of permutative categories.

Our first set of results requires generalizing bilinear maps to n-linear maps. To be precise, an n-linear map has n "input" permutative categories $A_1, \ldots,$ A_n and an "output" permutative category B. An n-linear map is then a functor

$$f : A_1 \times \cdots \times A_n \to B$$

that is lax in each variable separately, with each pair of lax structure maps satisfying the same compatibility condition specified above in the pentagon diagram for a bilinear map. These should be thought of as being roughly analogous to n-linear maps of vector spaces, pretending we know nothing about tensor products. In particular, they compose in the same way: if we have n_j-linear maps $f_j : A_{j1} \times \cdots \times A_{jn_j} \to B_j$ for $1 \leq j \leq k$ and a k-linear map $g : B_1 \times \cdots \times B_k \to C$, then the composite $g \circ (f_1 \times \cdots \times f_k)$ is an $(n_1 + \cdots + n_k)$-linear map. The entire structure is an instance of the categorical notion of a "multicategory," but we restrict ourselves here to the special case where $A_1 = \cdots = A_n = B$, in which case we find ourselves with a sequence of categories

$$n\text{-}\mathbf{lin}(A, \ldots, A; A)$$

for each permutative category A. (The objects are the functors and structure maps just described, and the morphisms are structure-preserving natural transformations.) Writing this category as $\mathcal{M}_n(A)$, we find that it has an evident Σ_n-action, and there is a multiproduct

$$\gamma : \mathcal{M}_k(A) \times \mathcal{M}_{n_1}(A) \times \cdots \times \mathcal{M}_{n_k}(A) \to \mathcal{M}_{n_1 + \cdots + n_k}(A)$$

given by the formula above:

$$\gamma(g; f_1, \ldots, f_k) = g \circ (f_1 \times \cdots \times f_k).$$

It is now routine to verify that this sequence of categories forms an operad of categories, which we denote by $\mathcal{M}(A)$.

We think of $\mathcal{M}(A)$ as being the multiplicative endomorphism operad of A. This is justified by the our first main result, which again exploits our canonical operad $E\Sigma$, this time multiplicatively. First, we need the notion of a bipermutative category, which is a category equipped with two permutative structures $\oplus$ and $\otimes$, with distributivity maps

$$d_l : (a_1 \otimes b) \oplus (a_2 \otimes b) \to (a_1 \oplus a_2) \otimes b$$

and

$$d_r : (a \otimes b_1) \oplus (a \otimes b_2) \to a \otimes (b_1 \oplus b_2)$$

satisfying appropriate coherence conditions. (May [5] requires both maps to be isomorphisms and one of them to be the identity, but we find these restrictions unnecessary.) Now we have

Theorem 7. *Let A be a permutative category. Then bipermutative structures on A are in bijective correspondence with operad maps $E\Sigma \to \mathcal{M}(A)$.*

The idea of the proof is that the product in the multiplicative structure arises from the element $1 \in \Sigma_2$, which is an object of $E\Sigma_2$ and therefore gives a bilinear map $\otimes : A \times A \to A$. The distributivity maps are the bilinear structure maps, and the commutativity isomorphism is the image of the isomorphism between the two elements of Σ_2 in $E\Sigma_2$. One then checks that these data generate the rest of the map of operads.

Next, observe that if we take the nerve of the component categories of a Cat-operad, we get an operad of simplicial sets, since nerve is product preserving. Consequently, any time we have a bipermutative category A, we get a map of simplicial operads

$$NE\Sigma \to N\mathcal{M}(A).$$

This brings us to our next major theorem.

Theorem 8. *For any permutative category A, there is a construction of the K-theory spectrum KA as a symmetric spectrum, together with a map of simplicial operads*

$$N\mathcal{M}(A) \to \mathbf{End}^{\wedge}(KA),$$

where the target is the endomorphism operad of KA using the symmetric monoidal smash product of symmetric spectra.

In fact, we have much more: our construction of the K-theory spectrum preserves the entire multicategory structure, allowing us to treat rings, modules, and algebras from a unified perspective. We restrict our attention here to rings only; the key corollary in this case is as follows:

Corollary 9. *If A is a bipermutative category, then KA is equivalent to a strictly commutative ring symmetric spectrum.*

Proof. The composite map of operads

$$NE\Sigma \to N\mathcal{M}(A) \to \mathbf{End}^{\wedge}(KA)$$

makes KA an E_∞ ring in the category of symmetric spectra, and the homotopy category of E_∞ ring symmetric spectra is equivalent to the homotopy category of commutative ring symmetric spectra. $\square$

While this result is in the literature [6], the proof there is much more intricate and contains some errors, patches for which have never been published. Further, our treatment allows for a weaker notion of bipermutative category, and generalizes with little additional effort to modules and algebras, although we will not pursue this here.

The construction of the K-theory spectrum on which this all relies is a modification of May's description in [7], which is in turn based on ideas of Segal. The most convenient way for us to view May's description is by looking at $\mathcal{F}$, the category of finite based sets. May associates to each permutative category A a functor we'll call $JA : \mathcal{F} \to \mathbf{Cat}$. We can then use the iterated smash product functor $\wedge^n : \mathcal{F}^n \to \mathcal{F}$, a canonical functor $\Delta^{op} \to \mathcal{F}$ originally described by Segal and which presents S^1 as a simplicial set, the nerve functor from $\mathbf{Cat}$ to the category of simplicial sets, and finally diagonalization to obtain a sequence of simplicial sets which turn out to form a symmetric spectrum. May's construction of the central functor JA involves systems of objects of A indexed on subsets of the based sets forming the objects of $\mathcal{F}$. Composed with the smash product functor $\wedge^n : \mathcal{F}^n \to \mathcal{F}$, this requires indexing on subsets of n-fold products. Our modification, which allows us much greater multiplicative precision, is essentially a restriction of indexing sets to those subsets which are themselves n-fold products.

We turn now to the quest for a construction that actually represents the smash product of connective spectra. We have, from above, bilinear maps, and what is needed is an analogue of the tensor product from algebra: a universal bilinear target. We can construct one if we restrict universality to be with respect to strict maps.

Theorem 10. *Given permutative categories A and B, there is a permutative category written $\boxtimes_2(A, B)$, together with a bilinear map from $A \times B$ to $\boxtimes_2(A, B)$ inducing an isomorphism*

$$\mathbf{Bilin}(A \times B, C) \cong \mathbf{Strict}(\boxtimes_2(A, B), C).$$

The argument relies on the cocompleteness of $\mathbf{Strict}$. Note first that if we have categories X and Y, a permutative category C, and functors $F : X \to Y$ and $G : X \to C$, then we can form the pushout in $\mathbf{Strict}$

$$
\begin{array}{ccc}
\mathbb{P}X & \xrightarrow{\overline{G}} & C \\
{\scriptstyle \mathbb{P}F} \downarrow & & \downarrow \\
\mathbb{P}Y & \longrightarrow & F_*C,
\end{array}
$$

where $\overline{G}$ is the unique strict map extending G. Then it is easy to check that a strict map $\alpha : C \to C'$ factors through F_*C via a strict map if and only if $G \circ \alpha$ factors through Y via an ordinary functor. This allows us to glue morphisms and relations into a permutative category in a universal manner. To construct $\boxtimes_2(A, B)$, we start with the unit map $A \times B \to \mathbb{P}(A \times B)$, which is universal among all functors from $A \times B$ to a permutative category, but isn't bilinear. We write the image of the object (a, b) of $A \times B$ as $a \otimes b$. To build in a left distributivity map

$$d_l : (a_1 \otimes b) \oplus (a_2 \otimes b) \to (a_1 \oplus a_2) \otimes b,$$

we note that the source and target expressions each specify functors $A \times A \times B \to \mathbb{P}(A \times B)$, which we can package as a single functor

$$A \times A \times B \times \{\bullet \quad \bullet\} \to \mathbb{P}(A \times B),$$

which plays the role of G in our pushout diagram above. For F, we just use the functor induced by

$$\{\bullet \quad \bullet\} \to \{\bullet \to \bullet\},$$

and we get a pushout

$$\begin{array}{ccc}
\mathbb{P}(A \times A \times B \times \{\bullet \quad \bullet\}) & \longrightarrow & \mathbb{P}(A \times B) \\
\downarrow & & \downarrow \\
\mathbb{P}(A \times A \times B \times \{\bullet \to \bullet\}) & \longrightarrow & \mathbb{P}_1(A \times B)
\end{array}$$

in **Strict**. It follows that the composite

$$A \times B \to \mathbb{P}(A \times B) \to \mathbb{P}_1(A \times B)$$

is a universal map among those supporting a natural transformation d_l. Another pushout of this sort builds in d_r, and we get a map $A \times B \to \mathbb{P}_2(A \times B)$ universal among those supporting natural transformations d_l and d_r.

In order to force d_l and d_r to satisfy their necessary coherence conditions, we have to make some diagrams commute, and we use the same technique on all of them. As an example, we need an associativity diagram for d_r:

$$\begin{array}{ccc}
(a \otimes b_1) \oplus (a \otimes b_2) \oplus (a \otimes b_3) & \xrightarrow{d_r \oplus 1} & (a \otimes (b_1 \oplus b_2)) \oplus (a \otimes b_3) \\
{\scriptstyle 1 \oplus d_r} \downarrow & & \downarrow {\scriptstyle d_r} \\
(a \otimes b_1) \oplus (a \otimes (b_2 \oplus b_3)) & \xrightarrow{d_r} & a \otimes (b_1 \oplus b_2 \oplus b_3).
\end{array}$$

The upper left and lower right corner expressions each specify functors $A \times B^3 \to \mathbb{P}_2(A \times B)$, and the two ways around the square each specify a natural transformation, so we can package the square as a single functor

$$A \times B^3 \times \{\bullet \rightrightarrows \bullet\} \to \mathbb{P}_2(A \times B).$$

We want to force the square to commute, which means forcing the two natural transformations to coincide, so we push out along the functor induced by the collapse functor $\{\bullet \rightrightarrows \bullet\} \to \{\bullet \to \bullet\}$, and get a diagram

$$\begin{array}{ccc}
\mathbb{P}(A \times B^3 \times \{\bullet \rightrightarrows \bullet\}) & \longrightarrow & \mathbb{P}_2(A \times B) \\
\downarrow & & \downarrow \\
\mathbb{P}(A \times B^3 \times \{\bullet \to \bullet\}) & \longrightarrow & \mathbb{P}_3(A \times B).
\end{array}$$

The resulting functor $A \times B \to \mathbb{P}_3(A \times B)$ is universal among functors together with natural maps d_l and d_r for which d_r satisfies the associativity diagram. Further pushouts using exactly the same technique force the remaining required diagrams to commute, completing the construction of $\boxtimes_2(A \times B)$. It is interesting to note that the functor from **Strict** to the category of monoids that forgets morphisms and remembers only objects has a right adjoint, namely the translation category functor E, and therefore the "objects" functor preserves pushouts. Since all the pushouts in the construction of $\boxtimes_2(A, B)$ are along maps that are isomorphisms on objects, we see that $\boxtimes_2(A, B)$ has the same objects as $\mathbb{P}(A \times B)$, namely the free monoid on the objects of $A \times B$.

Similar constructions produce universal n-linear targets $\boxtimes_n(A_1, \ldots, A_n)$.

Combining the previous theorem with the earlier result about bilinear maps, we find that

$$\mathbf{Lax}(A, \mathbf{Lax}(B, C)) \cong \mathbf{Strict}(\boxtimes_2(A, B), C),$$

so $\boxtimes_2$ is not quite left adjoint to the hom functor in **Lax**. Further, $\boxtimes_2$ is not associative. In order to go further, we need to consider what might seem a trivial case of the $\boxtimes_n$ construction: the case $n = 1$. Examining the definition, we find that a 1-linear map is just a lax map, so the construction gives us a universal lax map

$$\eta : A \to \boxtimes_1 A$$

inducing an isomorphism

$$\mathbf{Lax}(A, B) \cong \mathbf{Strict}(\boxtimes_1 A, B).$$

As a result, $\boxtimes_1$ provides us with a left adjoint to the inclusion functor from **Strict** into **Lax**. Accordingly, we change notation and define S, the **strictification** functor, to be $\boxtimes_1$. We will also write S for the composite with the inclusion functor from **Strict** into **Lax**, so we get a functor

$$S : \mathbf{Strict} \to \mathbf{Strict}$$

which is a comonad, being the composite of a left and a right adjoint. *Our main category of interest is S-coalg, the category of coalgebras in Strict over the comonad S.*

The category S-**coalg** has more good properties than either **Strict** or **Lax**. The first one follows from:

Proposition 11. *S preserves equalizers in* **Strict**.

As an immediate consequence, we have

Corollary 12. *S-coalg is complete and cocomplete, with equalizers of S-coalgebras created in* **Strict** *(and therefore in* **Cat**.)

Proof. This is dual to the proof that **Strict** is complete and cocomplete. □

The proof of proposition 11 relies on an alternative description of S, as follows. Let $\mathcal{J}$ be the category with objects $\underline{n} = \{1, \ldots, n\}$ for $n \geq 1$, and morphisms the surjections. Then any permutative category A determines a lax functor $A^* : \mathcal{J} \to \mathbf{Cat}$, where $A^*(\underline{n}) = A^n$, and if $p : \underline{m} \to \underline{n}$ is a surjection, then $A^*(p) : A^m \to A^n$ is given on objects by the formula

$$A^*(p)(a_1, \ldots, a_m) = \times_{j=1}^{n} \bigoplus_{p(i)=j} a_i,$$

using the natural order on the set $p^{-1}(j)$, and similarly for morphisms. This is merely a lax functor because it preserves composition only up to natural isomorphism, not strictly. Now we can display SA as a construction that goes by various names in the literature, including wreath product, Grothendieck construction, and homotopy colimit, and which is usually written $\mathcal{J} \int A^*$. The objects consist of the free monoid (without unit) on the objects of A:

$$\coprod_{n \geq 1} |A|^n .$$

The morphisms from $(a_1, \ldots, a_m)$ to $(b_1, \ldots, b_n)$ consist of ordered pairs (f, p), where $p \in \mathcal{J}(\underline{m}, \underline{n})$, and f is a morphism in A^n with j^{th} component

$$f_j : \bigoplus_{p(i)=j} a_i \to b_j.$$

Given another morphism $(g, q) : (b_1, \ldots, b_n) \to (c_1, \ldots, c_s)$, we define $(g, q) \circ (f, p)$ to be $(h, q \circ p)$, where h has as its kth component the composite

$$\bigoplus_{qp(i)=k} a_i \cong \bigoplus_{q(j)=k} \bigoplus_{p(i)=j} a_i \xrightarrow{\oplus f_j} \bigoplus_{q(j)=k} b_j \xrightarrow{g_k} c_k.$$

The essential feature is the ability to use the permutative structure of A to rearrange the summands as indicated by the unlabelled isomorphism. We define the permutative structure on SA by concatenation on both objects and morphisms. It is now fairly straightforward to check that the inclusion of A as the full subcategory on the objects at filtration level 1 is a universal lax map, and therefore this is a legitimate description of SA. The proof of theorem 11 is now an explicit check using this description. One thing we would like to have, but do not, is a comparably explicit description of $\boxtimes_n(A_1, \ldots, A_n)$ for $n \geq 2$.

An indication that S-**coalg** is the "right" category of permutative categories to work with is the fact that **Lax** embeds in S-**coalg**. To see this, we just look at the composite of isomorphisms

$$\textbf{Lax}(A, B) \cong \textbf{Strict}(SA, B) \cong S\text{-}\textbf{coalg}(SA, SB)$$

and an easy check shows that this is simply induced by the functor S. An amusing addendum is that the natural forgetful map S-**coalg** $\to$ **Strict** is actually an embedding, so we have succeeded in displaying **Lax** as a full subcategory inside a complete and cocomplete subcategory of **Strict**. There are definitely other objects in S-**coalg** besides the strictifications SA, however; the free permutative categories $\mathbb{P}B$ are examples and so are all the universal n-linear targets $\boxtimes_n(A_1, \ldots, A_n)$ for $n > 1$. This is important because our construction of a tensor product, which we discuss next, does not preserve S-coalgebras of the form SA.

Another indication of the "rightness" of the category of S-coalgebras is that it models the connective stable category in the same way **Lax** and **Strict** do. Let $\mathrm{ho}\mathcal{S}^{\Sigma}$ denote the stable category, thought of as a localization of the category of symmetric spectra. We have inclusions of subcategories

$$S\text{-coalg} \to \textbf{Strict} \to \textbf{Lax},$$

as well as the K-theory functor $\textbf{Lax} \xrightarrow{K} \mathrm{ho}\mathcal{S}^{\Sigma}$, which is actually a composite

$$\textbf{Lax} \xrightarrow{K} \mathcal{S}^{\Sigma} \longrightarrow \mathrm{ho}\mathcal{S}^{\Sigma}.$$

Let $\mathcal{K}$ denote the class of morphisms in $\textbf{Lax}$ sent to isomorphisms by this composite K, as well as its restrictions to $\textbf{Strict}$ or $S\text{-coalg}$. Thomason showed in [8] that K induces an equivalence of categories from either $\textbf{Lax}[\mathcal{K}^{-1}]$ or $\textbf{Strict}[\mathcal{K}^{-1}]$ to $\mathrm{ho}\mathcal{S}_0^{\Sigma}$, the full subcategory of $\mathrm{ho}\mathcal{S}^{\Sigma}$ generated by the connective spectra. In fact, we may restrict further to $S\text{-coalg}[\mathcal{K}^{-1}]$ and still get such an equivalence. The reason is that the diagram

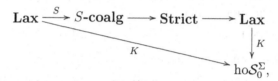

although it does not commute strictly, does commute up to natural isomorphism. This in turn follows from the following:

Proposition 13. *The universal lax map* $\eta : A \to SA$ *induces a homotopy equivalence of K-theory spectra.*

Proof. If we let $\varepsilon : SA \to A$ be the counit of the strictification adjunction, then $\varepsilon\eta = \mathrm{id}_A$. Further, although $\eta\varepsilon$ is not the identity (A and SA are certainly not isomorphic!), there is a natural transformation $\omega : \mathrm{id}_{SA} \to \eta\varepsilon$ of lax maps; in other words, ω is a morphism in $\textbf{Lax}(SA, SA)$. (What ω does is use the surjection $\{1, \ldots, n\} \to \{1\}$ to define a canonical morphism in SA from $(a_1, \ldots, a_n)$ to $\eta\varepsilon(a_1, \ldots, a_n) = (a_1 \oplus \cdots \oplus a_n)$.) Our construction of the K-theory spectrum is a continuous functor in the sense that it preserves the enrichment over simplicial sets derived from the nerve construction, and it follows that id_{KSA} and $K\eta \circ K\varepsilon$ are in the same component of $\mathcal{S}^{\Sigma}(KSA, KSA)$ and are consequently homotopic. Since $K\varepsilon \circ K\eta = K(\varepsilon\eta) = \mathrm{id}_{KA}$, it follows that $K\varepsilon$ and $K\eta$ are inverse homotopy equivalences. $\qquad\square$

The really nice thing about S-coalgebras is that we can shrink the universal bilinear target pairing $\boxtimes_2$ to a pairing that is associative. We call the construction the tensor product of S-coalgebras, and it is defined, annoyingly enough, as an equalizer. Given an S-coalgebra A, there are two canonical maps $A \to SA$: one is the universal lax map η that comes with the structure of S, and the other is the structure map of A as an S-coalgebra, which is a strict map we call χ. Given two S-coalgebras A and B, both $\boxtimes_2(\eta, \eta)$ and

$\boxtimes_2(\chi,\chi)$ give us maps of S-coalgebras from $\boxtimes_2(A,B)$ to $\boxtimes_2(SA,SB)$. We define $A \otimes B$ as the equalizer of these two maps:

$$A \otimes B \longrightarrow \boxtimes_2(A,B) \underset{\boxtimes_2(\eta,\eta)}{\overset{\boxtimes_2(\chi,\chi)}{\rightrightarrows}} \boxtimes_2(SA,SB).$$

Theorem 14. *Using this tensor product, S-**coalg** is a symmetric monoidal category, with $\mathbb{P}(*)$ as the unit.*

Since we want this tensor product to model the smash product of connective spectra, it is comforting to realize that the spectrum associated to $\mathbb{P}(*)$ is the sphere spectrum.

Given this theorem (whose proof will not appear here), we can now sketch some evidence that this tensor product actually models the smash product of connective spectra. First, it is a basic fact (and easy to show) that for any permutative categories A and B, we have $SA \otimes SB \cong \boxtimes_2(A,B)$; in fact, this is a special case of the more general

$$SA_1 \otimes \cdots \otimes SA_n \cong \boxtimes_n(A_1 \ldots A_n).$$

Consequently, we have an isomorphism (of categories)

$$n\text{-}\mathbf{lin}(A_1 \times \cdots \times A_n, B) \cong S\text{-}\mathbf{coalg}(SA_1 \otimes \cdots \otimes SA_n, SB).$$

Restricting to the special case in which $A_1 = \cdots = A_n = B$, we find that n-linear maps from n copies of A to A itself are just S-**coalg**$(SA^{\otimes n}, SA)$, so the endomorphism operad $\mathcal{M}(A)$ is just the endomorphism operad of SA using this symmetric monoidal tensor product of S-coalgebras. Since this is precisely the operad that gives rise to ring spectra on passage to K-theory, it seems likely that the tensor product of S-coalgebras represents the smash product of connective spectra.

In fact, this is just the tip of the iceberg, since in general, maps of S-coalgebras from $SA_1 \otimes \cdots \otimes SA_n$ to SB give us maps of symmetric spectra from $KA_1 \wedge \cdots \wedge KA_n$ to KB in a natural fashion. Using what are now fairly standard tricks with operads, we can now model modules, algebras, and the rest of the structure of algebra within the category of S-coalgebras in a way that passes cleanly to the multiplicative structure of symmetric spectra. Some of the details are still being written down, but a promising picture is emerging of a model in permutative categories for structured algebra in connective stable homotopy.

REFERENCES

[1] Gerald Dunn, E_n-monoidal categories and their group completions, J. Pure Appl. Algebra vol 95, 1994, 27–39
[2] A. D. Elmendorf, I. Kriz, M. A. Mandell, and J. P. May, Rings, modules, and algebras in stable homotopy theory, with an appendix by M. Cole, Mathematical Surveys and Monographs, vol. 47, American Mathematical Society, Providence, RI, 1997

[3] M. Hovey, B. Shipley, and J. Smith, Symmetric spectra, J. Amer. Math. Soc. vol 13, 2000, 149–208

[4] Saunders Mac Lane, Categories for the working mathematician, Springer-Verlag, New York, Berlin, 1971

[5] J. P. May, E_∞ ring spaces and E_∞ ring spectra, with contributions by Frank Quinn, Nigel Ray, and Jørgen Tornehave, Lecture Notes on Mathematics 577, Springer Verlag, Berlin, Heidelberg, New York, 1977

[6] J. P. May, Multiplicative infinite loop space theory, J. Pure Appl. Algebra, vol 26, 1982, 1–69

[7] J. P. May, The spectra associated to permutative categories, Topology vol 17, 1978, 225–228

[8] R. W. Thomason, Symmetric monoidal categories model all connective spectra, Theory Appl. Categ. vol 1, 1995, 78–118 (electronic)

DEPARTMENT OF MATHEMATICS, PURDUE UNIVERSITY CALUMET, HAMMOND, IN 46323, USA

email-address: aelmendo@math.purdue.edu

DEPARTMENT OF MATHEMATICS, UNIVERSITY OF CHICAGO, CHICAGO, IL 60637, USA

email-address: mandell@math.uchicago.edu

MORITA THEORY IN ABELIAN, DERIVED AND STABLE MODEL CATEGORIES

STEFAN SCHWEDE

CONTENTS

1. INTRODUCTION

The paper [Mo58] by Kiiti Morita seems to be the first systematic study of equivalences between module categories. Morita treats both contravariant equivalences (which he calls *dualities* of module categories) and covariant equivalences (which he calls *isomorphisms* of module categories) and shows that they always arise from suitable bimodules, either via contravariant hom functors (for 'dualities') or via covariant hom functors and tensor products (for 'isomorphisms'). The term 'Morita theory' is now used for results concerning equivalences of various kinds of module categories. The authors of the obituary article [AGH] consider Morita's theorem "probably one of the most frequently used single results in modern algebra".

In this survey article, we focus on the covariant form of Morita theory, so our basic question is:

> When do two 'rings' have 'equivalent' module categories ?

We discuss this question in different contexts:

- (Classical) When are the module categories of two rings equivalent as categories ?

- (Derived) When are the derived categories of two rings equivalent as triangulated categories ?
- (Homotopical) When are the module categories of two ring spectra Quillen equivalent as model categories ?

There is always a related question, which is in a sense more general:
What characterizes the category of modules over a 'ring' ?

The answer is, mutatis mutandis, always the same: modules over a 'ring' are characterized by the existence of a 'small generator', which plays the role of the free module of rank one. The precise meaning of 'small generator' depends on the context, be it an abelian category, a derived category or a stable model category. We restrict our attention to categories which have a single small generator; this keeps things simple, while showing the main ideas. Almost everything can be generalized to categories (abelian, derived or stable model categories) with a *set of small generators*. One would have to talk about *ringoids* (also called *rings with many objects*) and their differential graded and spectral analogues.

Background: for a historical perspective on Morita's work we suggest a look at the obituary article [AGH] by Arhangel'skii, Goodearl, and Huisgen-Zimmermann. The history of Morita theory for derived categories and 'tilting theory' is summarized in Section 3.1 of the book by König and Zimmermann [KZ]. Both sources contain lots of further references.

For general background material on derived and triangulated categories, see [SGA 4½] (Appendix by Verdier), [GM], [Ver96], or [Wei94]. We freely use the language of model categories, alongside with the concepts of Quillen adjoint pair and Quillen equivalence. For general background on model categories see Quillen's original article [Qui67], a modern introduction [DwSp95], or [Hov99] for a more complete overview.

Acknowledgments: The Morita theory in stable model categories which I describe in Section 4 is based on joint work with Brooke Shipley spread over many years and several papers; I would like to take this opportunity to thank her for the pleasant and fruitful collaboration. I would also like to thank Andy Baker and Birgit Richter for organizing the wonderful workshop *Structured ring spectra and their applications* in Glasgow.

2. Morita theory in abelian categories

To start, we review the covariant Morita theory for modules; this is essentially the content of Section 3 of [Mo58]. This and related material is treated in more detail in [Ba, II §3], [AF92, §22] or [Lam, §18].

Definition 2.1. Let $\mathcal{A}$ be an abelian category with infinite sums. An object M of $\mathcal{A}$ is *small* if the hom functor $\mathcal{A}(M, -)$ preserves sums; M is a *generator* if every object of $\mathcal{A}$ is an epimorphic image of a sum of (possibly infinitely many) copies of M.

The emphasis in the smallness definition is on *infinite* sums; finite sums are isomorphic to finite products, so they are automatically preserved by the hom functor. For modules over a ring, smallness is closely related to finite generation: every finitely generated module is small and for projective modules, 'small' and 'finitely generated' are equivalent concepts. Rentschler [Ren69] gives an example of a small module which is not finitely generated.

A generator can equivalently be defined by the property that the functor $\mathcal{A}(M, -)$ is faithful, compare [Ba, II Prop. 1.1]. A small projective generator is called a *progenerator*. The main example is when $\mathcal{A} = \mathsf{Mod}\text{-}R$ is the category of right modules over a ring R. Then the free module of rank one is a small projective generator. In this case, a general R-module M is a generator for $\mathsf{Mod}\text{-}R$ if and only if the free R-module of rank one is an epimorphic image of a sum of copies of M.

Here is one formulation of the classical Morita theorem for rings:

Theorem 2.2. *For two rings R and S, the following conditions are equivalent.*

(1) *The categories of right R-modules and right S-modules are equivalent.*
(2) *The category of right S-modules has a small projective generator whose endomorphism ring is isomorphic to R.*
(3) *There exists an R-S-bimodule M such that the functor*

$$- \otimes_R M \; : \; \mathsf{Mod}\text{-}R \; \longrightarrow \; \mathsf{Mod}\text{-}S$$

is an equivalence of categories.

If these conditions hold, then R and S are said to be Morita equivalent.

Here are some elementary remarks on Morita equivalence. Condition (1) above is symmetric in R and S. So if an R-S-bimodule M realizes an equivalence of module categories, then the inverse equivalence is also realized by an S-R-bimodule N. Since the equivalences are inverse to each other, $M \otimes_S N$ is then isomorphic to R as an R-bimodule and $N \otimes_R M$ is isomorphic to S as an S-bimodule. Moreover, M and N are then projective as right modules, and N is isomorphic to $\mathrm{Hom}_S(M, S)$ as a bimodule.

If R is Morita equivalent to S, then the opposite ring R^{op} is Morita equivalent to S^{op}. Indeed, suppose the R-S-bimodule M and the S-R-bimodule N satisfy

$$M \otimes_S N \; \cong \; R \quad \text{and} \quad N \otimes_R M \; \cong \; S$$

as bimodules. Since the category of right R^{op}-modules is isomorphic to the category of left R-modules, we can view M as an S^{op}-R^{op}-bimodule and N as an R^{op}-S^{op}-bimodule, and then they provide the equivalence of categories between $\mathsf{Mod}\text{-}R^{op}$ and $\mathsf{Mod}\text{-}S^{op}$. Similarly, if R is Morita equivalent to S and R' is Morita equivalent to S', then $R \otimes S$ is Morita equivalent to $R' \otimes S'$. Here, and in the rest of the paper, undecoratd tensor products are taken over $\mathbb{Z}$.

Invariants which are preserved under Morita equivalence include all concepts which can be defined from the category of modules without reference to the ring. Examples are the number of isomorphism classes of projective modules, of simple modules or of indecomposable modules, or the algebraic K-theory of the ring. The center $Z(R) = \{r \in R \,|\, rs = sr \text{ for all } s \in R\}$ is also Morita invariant, since the center of R is isomorphic to the endomorphism ring of the identity functor of Mod-R. A ring isomorphism

$$Z(R) \longrightarrow \operatorname{End}(\operatorname{Id}_{\mathsf{Mod}\text{-}R})$$

is obtained as follows: if $r \in Z(R)$ is a central element, then for every R-module M, multiplication by r is R-linear. So the collection of R-homomorphisms $\{\times r : M \longrightarrow M\}_{M \in \mathsf{Mod}\text{-}R}$ is a natural transformation from the identity functor to itself. For more details, see [Ba, II Prop 2.1] or [Lam, Remark 18.43]. In particular, if two *commutative* rings are Morita equivalent, then they are already isomorphic.

There is a variation of the Morita theorem 2.2 relative to a commutative ring k, with essentially the same proof. In this version R and S are k-algebras, condition (1) refers to a k-linear equivalence of module categories, condition (2) requires an isomorphism of k-algebras and in part (3), M has to be a k-*symmetric* bimodule, i.e., the scalars from the ground ring k act in the same way from the left (through R) and from the right (through S).

We sketch the **proof of Theorem 2.2** because it serves as the blueprint for analogous results in the contexts of differential graded rings and ring spectra. Suppose (1) holds and let

$$F : \mathsf{Mod}\text{-}R \longrightarrow \mathsf{Mod}\text{-}S$$

be an equivalence of categories. The free R-module of rank one is a small projective generator of the category of R-modules. Being projective, small or a generator are categorical conditions, so they are preserved by an equivalence of categories. So the S-module FR is a small projective generator of the category of S-modules. Since F is an equivalence of categories, it is in particular an additive fully faithful functor. So F restricts to an isomorphism of rings

$$F : R \cong \operatorname{End}_R(R) \overset{\cong}{\longrightarrow} \operatorname{End}_S(FR) .$$

Now assume condition (2) and let P be a small projective S-module which generates the category Mod-S. After choosing an isomorphism $f : R \cong \operatorname{End}_S(P)$, we can view P as an R-S-bimodule by setting $r \cdot x = f(r)(x)$ for $r \in R$ and $x \in P$. We show that P satisfies the conditions of (3) by showing that the adjoint functors $- \otimes_R P$ and $\operatorname{Hom}_S(P, -)$ are actually inverse equivalences.

The adjunction unit is the R-linear map

$$X \longrightarrow \operatorname{Hom}_S(P, X \otimes_R P) , \qquad x \longmapsto (y \longmapsto x \otimes y) .$$

For $X = R$, the map adjunction unit coincides with the isomorphism f, so it is bijective. Since P is small, source and target commute with sums, finite or infinite, so the unit is bijective for every free R-module. Since P is projective over S, both sides of the adjunction unit are right exact as functors of X. Every R-module is the cokernel of a morphism between free R-modules, so the adjunction unit is bijective in general.

The adjunction counit is the S-linear evaluation map

$$\mathrm{Hom}_S(P, Y) \otimes_R P \longrightarrow Y , \quad \phi \otimes x \longmapsto \phi(x) .$$

For $Y = P$, the counit is an isomorphism since the right action of R on $\mathrm{Hom}_S(P, P)$ arises from the isomorphism $R \cong \mathrm{Hom}_S(P, P)$. Now the argument proceeds as for the adjunction unit: both sides are right exact and preserves sums, finite or infinite, in the variable Y. Since P is a generator, every S-module is the cokernel of a morphism between direct sums of copies of P, so the counit is bijective in general.

Condition (1) is a special case of (3), so this finishes the proof of the Morita theorem.

Example 2.3. The easiest example of a Morita equivalence involves matrix algebras. Any free R-module of finite rank $n \geq 1$ is a small projective generator for the category of right R-modules. The endomorphism ring

$$\mathrm{End}_R(R^n) \cong M_n(R)$$

is the ring of $n \times n$ matrices with entries in R. So R and the matrix ring $M_n(R)$ are Morita equivalent. The bimodules which induce the equivalences of module categories can both be taken to be R^n, but viewed as 'row vectors' (or $1 \times n$ matrices) and 'column vectors' (or $n \times 1$ matrices) respectively.

Matrix rings do not provide the most general kind of Morita equivalences, as the example below shows. However, every ring Morita equivalent to R is isomorphic to a ring of the form $e M_n(R) e$ where $e \in M_n(R)$ is a *full idempotent* in the $n \times n$ matrix ring, i.e., we have $e^2 = e$ and $M_n(R) e M_n(R) = M_n(R)$. Indeed, if P is a small projective generator for a ring R, then P is a summand of a free module of finite rank n, say. Thus P is isomorphic to the image of an idempotent $n \times n$ matrix e, and then $\mathrm{End}_R(P) \cong e M_n(R) e$ as rings.

Example 2.4. The following example of a Morita equivalence which is not of matrix algebra type was pointed out to me by M. Künzer and N. Strickland. Consider a *commutative* ring R and an *invertible* R-module Q. In other words, there exists another R-module Q' and an isomorphism of R-modules $Q \otimes_R Q' \cong R$. Then tensor product with Q over R is a self-equivalence of the category of right R-modules (with quasi-inverse the tensor product with Q'). This self-equivalence is not isomorphic to the identity functor unless Q is free of rank one.

Because tensor product with an invertible module Q is an equivalence of categories, it follows that Q is a progenerator, with endomorphism ring isomorphic to R. Moreover, the 'inverse' module Q' is isomorphic to the R-linear dual $Q^* = \mathrm{Hom}_R(Q, R)$. Now we consider the direct sum $P = R \oplus Q$, which is another small projective generator for Mod-R. Then R is Morita equivalent to the endomorphism ring of P,

$$\mathrm{End}_R(P) \;=\; \mathrm{Hom}_R(R \oplus Q, R \oplus Q) \;.$$

As an R-module, $\mathrm{End}_R(P)$ is thus isomorphic to $R \oplus Q \oplus Q^* \oplus R$. So if Q is not free, then $\mathrm{End}_R(P)$ is not free over its center, hence not a matrix algebra.

For a specific example we consider the ring

$$R \;=\; \mathbb{Z}[u]/(u^2 - 5u) \;.$$

We set $Q = (2, u) \lhd R$, the ideal generated by 2 and u. Then Q is not free as an R-module, but it is invertible because the evaluation map

$$\mathrm{Hom}_R(Q, R) \otimes_R Q \;\longrightarrow\; R \;, \quad \phi \otimes x \longmapsto \phi(x)$$

is an isomorphism. Note that the inclusion $Q \longrightarrow R$ becomes an isomorphism after inverting 2; so after inverting 2 the module $P = R \oplus Q$ is free of rank 2 and hence the ring $\mathrm{End}_R(P)[\frac{1}{2}]$ is isomorphic to the ring of 2×2 matrices over $R[\frac{1}{2}]$.

The implication $(2) \Longrightarrow (1)$ in the Morita theorem 2.2 can be stated in a more general form, and then it gives a characterization of module categories as the cocomplete abelian categories with a small projective generator.

Theorem 2.5. *Let $\mathcal{A}$ be an abelian category with infinite sums and a small projective generator P. Then the functor*

$$\mathcal{A}(P, -) \;:\; \mathcal{A} \;\longrightarrow\; \text{Mod-}\mathrm{End}_{\mathcal{A}}(P)$$

is an equivalence of categories.

Proof. We give the same proof as in Bass' book [Ba, II Thm. 1.3]. Let us say that an object X of $\mathcal{A}$ is *good* if the map

$$\mathcal{A}(P, -) \;:\; \mathcal{A}(X, Y) \;\longrightarrow\; \mathrm{Hom}_{\mathrm{End}_{\mathcal{A}}(P)}(\mathcal{A}(P, X), \mathcal{A}(P, Y)) \qquad (2.6)$$

is bijective for every object Y of $\mathcal{A}$. We note that:

- The generator P is good since $\mathcal{A}(P, P)$ is the free $\mathrm{End}_{\mathcal{A}}(P)$-module of rank one.
- The class of good objects is closed under sums, finite or infinite: since P is small, $\mathcal{A}(P, -)$ preserves sums and both sides of the map (2.6) take direct sums in X to direct products.
- If $f : X \longrightarrow X'$ is a morphism between good objects in $\mathcal{A}$, then the cokernel of f is also good. This uses that P is projective, so $\mathcal{A}(P, -)$ is an exact functor and both sides of the map (2.6) are right exact in X.

Since P is a generator, every object can be written as the cokernel of a morphism between sums of copies of P. So every object of $\mathcal{A}$ is good, which precisely means that the hom functor $\mathcal{A}(P, -)$ is full and faithful.

It remains to check that every $\operatorname{End}_{\mathcal{A}}(P)$-module is isomorphic to a module in the image of the functor $\mathcal{A}(P, -)$. The free $\operatorname{End}_{\mathcal{A}}(P)$-module of rank one is the image of P. Since $\mathcal{A}(P, -)$ commutes with sums, every free module is in the image, up to isomorphism. Finally, every $\operatorname{End}_{\mathcal{A}}(P)$-module X has a presentation, so it occurs in an exact sequence of $\operatorname{End}_{\mathcal{A}}(P)$-modules

$$\bigoplus_I \operatorname{End}_{\mathcal{A}}(P) \xrightarrow{g} \bigoplus_J \operatorname{End}_{\mathcal{A}}(P) \longrightarrow X \longrightarrow 0 .$$

Since $\mathcal{A}(P, -)$ is full, the homomorphism g is isomorphic to $\mathcal{A}(P, f)$ for some morphism $f : \bigoplus_I P \longrightarrow \bigoplus_J P$ in $\mathcal{A}$. Since the functor $\mathcal{A}(P, -)$ is exact, X is the image of the cokernel of f. Thus $\mathcal{A}(P, -)$ is an equivalence of categories. □

Theorem 2.5 can be applied to the abelian category of right modules over a ring S; then we conclude that for every small projective generator P of Mod-S the functor

$$\operatorname{Hom}_S(P, -) : \operatorname{Mod-}S \longrightarrow \operatorname{Mod-End}_S(P)$$

is an equivalence of categories. This shows again that condition (2) in the Morita theorem 2.2 implies condition (1).

3. Morita theory in derived categories

Morita theory for derived categories is about the question:

> When are the derived categories $\mathcal{D}(R)$ and $\mathcal{D}(S)$
> of two rings R and S equivalent?

Here the derived category $\mathcal{D}(R)$ is defined from ($\mathbb{Z}$-graded and unbounded) chain complexes of right R-modules by formally inverting the quasi-isomorphisms, i.e., the chain maps which induce isomorphisms of homology groups. Of course, if R and S are Morita equivalent, then they are also derived equivalent. But it turns out that derived equivalences happen under more general circumstances.

Rickard [Ric89a, Ric91] developed a Morita theory for derived categories based on the notion of a tilting complex. Rickard's theorem did not come out of the blue, and he had built on previous work of several other people on *tilting modules*. Section 3.1 of the book by König and Zimmermann [KZ] gives a summary of the history in this area; this book also contains many more details, examples and references on the use of derived categories in representation theory.

We follow Keller's approach from [Kel94a], based on the (differential graded) endomorphism ring of a tilting complex. A similar approach to and more applications of Morita theory in derived categories can be found in the paper [DG02] by Dwyer and Greenlees.

3.1. The derived category. In this section, R is any ring. All chain complexes are $\mathbb{Z}$-graded and homological, i.e., the differential decreases the degree by 1.

Definition 3.1. A chain complex C of R-modules is *cofibrant* if there exists an exhaustive increasing filtration by subcomplexes

$$0 = C^0 \subseteq C^1 \subseteq \cdots \subseteq C^n \subseteq \cdots$$

such that each subquotient C^n/C^{n-1} consists of projective modules and has trivial differential. The *(unbounded) derived category* $\mathcal{D}(R)$ of the ring R has as objects the cofibrant complexes of R-modules and as morphisms the chain homotopy classes of chain maps.

Our definition of the derived category is different from the usual one. The more traditional way is to start with the homotopy category of all complexes, not necessarily cofibrant; then one uses a calculus of fractions to formally invert the class of quasi-isomorphisms. These two ways of constructing $\mathcal{D}(R)$ lead to equivalent categories.

The *shift functor* in $\mathcal{D}(R)$ is given by shifting a complex, i.e.,

$$(A[1])_n = A_{n-1}$$

with differential $d : (A[1])_n = A_{n-1} \longrightarrow A_{n-2} = (A[1])_{n-1}$ the *negative* of the differential of the original complex A. The *mapping cone* $C\varphi$ of a chain map $\varphi : A \longrightarrow B$ is defined by

$$(C\varphi)_n = B_n \oplus A_{n-1}, \quad d(x,y) = (dx + \varphi(y), -dy). \tag{3.2}$$

The mapping cone comes with an inclusion $i : B \longrightarrow C\varphi$ and a projection $p : C\varphi \longrightarrow A[1]$ which induce an isomorphism $(C\varphi)/B \cong A[1]$; if A and B are cofibrant, then so are the shift $A[1]$ and the mapping cone.

Remark 3.3. The following remarks are meant to give a better feeling for the notion of 'cofibrant complex' and the unbounded derived category.

(i) The concept of a 'cofibrant complex' is closely related to, but stronger than, a complex of projective modules. Indeed, if C is a cofibrant complex, then in every dimension $k \in \mathbb{Z}$, each subquotient C_k^n/C_k^{n-1} is projective. So C_k^n splits as the sum of the subquotients,

$$C_k^n \cong \bigoplus_{i=1}^{n} C_k^i/C_k^{i-1}.$$

Since C_k is the union of the submodules C_k^n, the module C_k also splits as the sum of the countably many quotients C_k^i/C_k^{i-1}. In particular, C_k is a sum of projective modules. Hence every cofibrant complex is dimensionwise projective. If C is a complex of projective modules which is *bounded below*, then it is also cofibrant. For example, if C is trivial in negative dimensions, then as the filtration we can simply take the (stupid) truncations of C, i.e.,

$$C_n^i = \begin{cases} C_n & \text{for } n < i \\ 0 & \text{for } n \geq i. \end{cases}$$

So for bounded below complexes, 'cofibrant' is equivalent to 'dimensionwise projective'.

On the other hand, not every complex which is dimensionwise projective is also cofibrant. The standard example is the complex C in which C_k is the free $\mathbb{Z}/4$-module of rank one for all $k \in \mathbb{Z}$, and where every differential $d : C_k \longrightarrow C_{k-1}$ is multiplication by 2.

(ii) Every quasi-isomorphism between cofibrant complexes is a chain homotopy equivalence. For every complex of R-modules X, there is a cofibrant complex X^c and a quasi-isomorphism $X^c \longrightarrow X$; together these two facts essentially prove that the derived category $\mathcal{D}(R)$ enjoys the universal property of the localization of the category of chain complexes of R-modules with the class of quasi-isomorphisms inverted. These properties are very analogous to the properties that CW-complexes have among all topological spaces: every weak equivalences between CW-complexes is a homotopy equivalence and every space admits a CW-approximation. This analogy is made precise in [KM, Part III].

(iii) The concept of a cofibrant chain complex is closely related to that of a *K-projective* complex as defined by Spaltenstein [Spa88, Sec. 1.1] (who attributes this notion to J. Bernstein; Keller [KZ, 8.1.1] calls this *homotopically projective*). A chain complex is K-projective if every chain map into an acyclic complex (i.e., a complex with trivial homology) is chain null-homotopic.

Every cofibrant complex is K-projective. Conversely, every K-projective complex X is chain homotopy equivalent to a cofibrant complex. Indeed, we can choose a cofibrant replacement, i.e., a cofibrant complex X^c and a quasi-isomorphism $q : X^c \longrightarrow X$; the mapping cone Cq is then acyclic. We have a short exact sequence of chain homotopy classes of chain maps in Cq,

$$[X^c[1], Cq] \longrightarrow [Cq, Cq] \longrightarrow [X, Cq] \ ;$$

both X and X^c are K-projective, so the left and right groups are trivial. Thus the identity map of Cq is null-homotopic and so the mapping cone of q is chain contractible. Thus the map $q : X^c \longrightarrow X$ is a chain homotopy equivalence.

(iv) We use the term 'cofibrant' complex because they are the cofibrant objects in the *projective* model category structure on chain complexes of R-modules [Hov99, 2.3.11]. In this model structure, the weak equivalences are the quasi-isomorphisms, the fibrations are the surjections and the cofibrations are the injections whose cokernel is cofibrant in the sense of Definition 3.1. In particular, every chain complex is fibrant in the projective model structure.

Since every object is fibrant and because the fibrations (= surjections) and weak equivalences (= quasi-isomorphisms) already have well-established names, it is unnecessary, and overly complicated, to use the language of model categories in order to work with the derived category of a ring.

(v) There is a 'dual' approach to the derived category $\mathcal{D}(R)$ as the homotopy category of 'fibrant' complexes (attention: these are *not* the fibrant objects in the projective model structure – there every complex is fibrant). This uses the notion of a 'K-injective' [Spa88, Sec. 1.1] or 'homotopically injective' [KZ, 8.1.1] complex, or the *injective model structure* on the category of complexes of R-modules [Hov99, 2.3.13]. It is often useful to have both descriptions available. Given arbitrary chain complexes C and D, we choose a cofibrant/K-projective resolution $C^c \xrightarrow{\sim} C$ and a fibrant/K-injective resolution $D \xrightarrow{\sim} D^f$. Then the maps induce isomorphisms of chain homotopy classes of chain maps

$$[C^c, D] \xrightarrow{\cong} [C^c, D^f] \xleftarrow{\cong} [C, D^f] .$$

There is an additive functor

$$[0] \; : \; \mathsf{Mod}\text{-}R \longrightarrow \mathcal{D}(R)$$

which is a fully faithful embedding onto the full subcategory of the derived category consisting of the complexes whose homology is concentrated in dimension zero. So we can think of the R-modules as sitting inside the derived category $\mathcal{D}(R)$. Had we defined the derived category from the category of all complexes of R-modules by formally inverting the quasi-isomorphisms, then we could define the complex $M[0]$ by putting the R-module M in dimension 0, and taking trivial chain modules everywhere else. With our present definition of $\mathcal{D}(R)$ we let $M[0]$ be a choice of resolution $P_\bullet$ of M by projective R-modules. Such a resolution is unique up to chain homotopy equivalence and it is cofibrant when viewed as a chain complex (by Remark 3.3 (i) above). Moreover, every R-linear map $M \longrightarrow N$ is covered by a unique chain homotopy class between the chosen projective resolutions. In other words, we really get a functor from R-modules to the derived category $\mathcal{D}(R)$, together with a natural isomorphism $H_0(M[0]) \cong M$.

The usual definition of Ext-groups involves a choice of projective resolution $P_\bullet$ of the source module M, and then $\mathrm{Ext}_R^n(M, N)$ can be defined as the chain homotopy classes of chain maps from the resolution $P_\bullet$ to N, shifted up into dimension n. We get the same result if N is also replaced by a projective resolution; this says that Ext-groups can be obtained from the derived category via

$$\mathrm{Ext}_R^n(M, N) \cong \mathcal{D}(R)(M[0], N[n]) \ . \tag{3.4}$$

The derived category of a ring has more structure. The category $\mathcal{D}(R)$ is additive since the homotopy relation for chain maps is additive. But $\mathcal{D}(R)$ is no longer an abelian category such as the category of chain complexes. Indeed, notions such as 'monomorphism', 'epimorphisms, 'kernels' for chain maps do not interact well with the passage to chain homotopy classes. The *distinguished triangles* in $\mathcal{D}(R)$ are what is left of the abelian structure on the category of chain complexes, and $\mathcal{D}(R)$ is an example of a *triangulated category*.

The distinguished triangles are the diagrams which are isomorphic in $\mathcal{D}(R)$ to a mapping cone triangle. More precisely, a diagram in $\mathcal{D}(R)$ of the form

$$X \xrightarrow{f} Y \xrightarrow{g} Z \xrightarrow{h} X[1] \tag{3.5}$$

is called a *distinguished triangle* if and only if there exists a chain map $\varphi : A \longrightarrow B$ between cofibrant complexes and isomorphisms $\iota_1 : A \cong X$, $\iota_2 : B \cong Y$ and $\iota_3 : C\varphi \cong Z$ in $\mathcal{D}(R)$ such that the diagram

$$
\begin{array}{ccccccc}
A & \xrightarrow{\varphi} & B & \xrightarrow{i} & C\varphi & \xrightarrow{p} & A[1] \\
\downarrow{\iota_1} & & \downarrow{\iota_2} & & \downarrow{\iota_3} & & \downarrow{\iota_1[1]} \\
X & \xrightarrow{f} & Y & \xrightarrow{g} & Z & \xrightarrow{h} & X[1]
\end{array}
$$

commutes in $\mathcal{D}(R)$.

We do not want to reproduce the complete definition of a triangulated category here; the data of a triangulated category consists of

(i) an additive category $\mathcal{T}$,
(ii) a self-equivalence $[1] : \mathcal{T} \longrightarrow \mathcal{T}$ called the *shift* functor and
(iii) a class of *distinguished triangles*, i.e., a collection of diagrams in $\mathcal{T}$ of the form (3.5).

This data is subject to several axioms which can be found for example in [Ver96], [Wei94, Sec. 10.2] [Nee01] or [KZ, 2.3].

Distinguished triangles are the source of many long exact sequences that come up in nature. Indeed, the axioms which we have suppressed imply in particular that for every distinguished triangle of the form (3.5) and every object W of $\mathcal{T}$ the sequence of abelian morphism groups

$$\mathcal{T}(W, X) \xrightarrow{f_*} \mathcal{T}(W, Y) \xrightarrow{g_*} \mathcal{T}(W, Z) \xrightarrow{h_*} \mathcal{T}(W, X[1])$$

is exact. One of the axioms also says that one can 'rotate' triangles, i.e., a sequence (3.5) is a distinguished triangle if and only if the sequence

$$Y \xrightarrow{g} Z \xrightarrow{h} X[1] \xrightarrow{-f[1]} Y[1]$$

is a distinguished triangle. So if we keep rotating a distinguished triangle in both directions and take morphisms from a fixed object W, we end up with a long exact sequence of abelian groups

$$\cdots T(W,X) \xrightarrow{f_*} T(W,Y) \xrightarrow{g_*} T(W,Z) \xrightarrow{h_*} T(W,X[1]) \qquad (3.6)$$

$$\xrightarrow{-f[1]_*} T(W,Y[1]) \xrightarrow{-g[1]_*} T(W,Z[1]) \xrightarrow{-h[1]_*} T(W,X[2]) \cdots$$

The axioms of a triangulated category also guarantee a similar long exact sequence when taking morphism *from* a triangle (and its rotations) *into* a fixed object W.

In the derived category $\mathcal{D}(R)$, the long exact sequence (3.6) becomes something more familiar when we take $W = R[0]$, the free R-module of rank one, viewed as a complex concentrated in dimension zero. For every chain complex C, cofibrant or not, the chain homotopy classes of morphisms from R to C are naturally isomorphic to the homology module $H_0 C$; so the long exact sequence (3.6) specializes to the long exact sequence of homology modules

$$\cdots \longrightarrow H_0(A) \xrightarrow{H_0(\varphi)} H_0(B) \longrightarrow H_0(C\varphi) \xrightarrow{\delta} H_{-1}(A) \longrightarrow \cdots .$$

Small generators. We will often require infinite direct sums in a triangulated category. The unbounded derived category $\mathcal{D}(R)$ has direct sums, finite and infinite. Indeed, the direct sum of any number of cofibrant complexes is again cofibrant (take the direct sum of the filtrations which are required in Definition 3.1), and this also represents the direct sum in $\mathcal{D}(R)$. This is one point where it is important to allow unbounded complexes. There are variants of the derived category which start with complexes which are bounded or bounded below. One also gets triangulated categories in much the same way as for $\mathcal{D}(R)$, but for example the countable family $\{R[-n]\}_{n\geq 0}$ has no direct sum in the bounded or bounded below derived categories.

In the Morita equivalence questions, a suitably defined notion of 'small generator' pops up regularly. The following concepts for triangulated categories are analogous to the ones for abelian categories in Definition 2.1.

Definition 3.7. Let T be a triangulated category with infinite coproducts. An object M of T is *small* if the hom functor $T(M, -)$ preserves sums; M is a *generator* if there is no proper full triangulated subcategory of T (with shift and triangles induced from T) which contains M and is closed under infinite sums.

As in abelian categories, the hom functor $T(M, -)$ automatically preserves finite sums. What we call 'small' is sometimes called *compact* or *finite* in the literature on triangulated categories. A triangulated category with infinite coproducts and a set of small generators is often called *compactly generated*.

The class of small objects in any triangulated category is closed under shifting in either direction, taking finite sums and taking direct summands. Moreover, if two of the three objects in a distinguished triangle are small, then so is the third one (one has to exploit that the morphisms from a distinguished triangle into a fixed object give rise to a long exact sequence).

There is a convenient criterion for when a *small* object M generates a triangulated category $\mathcal{T}$ with infinite coproducts: M generates $\mathcal{T}$ in the sense of Definition 3.7 if and only if it 'detects objects', i.e., an object X of $\mathcal{T}$ is trivial if and only if there are no graded maps from M to X, i.e. $\mathcal{T}(M[n], X) = 0$ for all $n \in \mathbb{Z}$. For the equivalence of the two conditions, see for example [SS03, Lemma 2.2.1].

The complex $R[0]$ consisting of the free module of rank one concentrated in dimension 0 is a small generator for the derived category $\mathcal{D}(R)$. Indeed, morphisms in $\mathcal{D}(R)$ out of the complex $R[0]$ represent homology, i.e., there is a natural isomorphism

$$\mathcal{D}(R)(R[n], C) \cong H_n(C)$$

for every cofibrant complex C. Since homology commutes with infinite sums, the complex $R[0]$ is small in $\mathcal{D}(R)$. Moreover, if all the morphism groups $\mathcal{D}(R)(R[n], C)$ are trivial as n ranges over the integers, then the complex C is acyclic, hence contractible, and so it is trivial in the derived category $\mathcal{D}(R)$. In other words, mapping out of shifted copies of the complex $R[0]$ detects whether an object in $\mathcal{D}(R)$ is trivial or not, so $R[0]$ is also a generator, by the previous criterion.

There is a nice characterization of the small objects in the derived category of a ring. Every bounded complex of finitely generated projective modules is built from summands of the small object $R[0]$ by shifts and extensions in triangles. Since the class of small objects is closed under these operations, a bounded complex of finitely generated projective modules is small. Conversely, these are the only small objects, up to isomorphism in $\mathcal{D}(R)$:

Theorem 3.8. *Let R be a ring. A complex of R-modules is small in the derived category $\mathcal{D}(R)$ if and only if it is quasi-isomorphic to a bounded complex of finitely generated projective R-modules.*

The proof that every small object in $\mathcal{D}(R)$ is quasi-isomorphic to a bounded complex of finitely generated projective modules is more involved. It is a special case of a result about triangulated categories $\mathcal{T}$ with a set of small generators. Neeman [Nee92] showed that every small object in $\mathcal{T}$ is a direct summand of an iterated extension of finitely many shifted generators. The proof can also be found in [Kel94a, 5.3].

There are non-trivial triangulated categories in which only the zero objects are small, see for example [Kel94b] or [HS99, Cor. B.13]. If a triangulated category has a set of generators, then the coproduct of all of them is a single generator. However, an infinite coproduct of non-trivial small objects is not small. So the property of having a single small generator is something special. In fact we see in Theorem 4.16 below that this condition characterizes the module categories over ring spectra among the stable model categories. A triangulated category need not have a *set* of generators whatsoever (one could consider all objects, but in general these form a proper class). For example $K(\mathbb{Z})$, the homotopy category of chain complexes of abelian groups, is not generated by a set [Nee01, E.3].

Equivalences of triangulated categories. A functor between triangulated categories is called exact if it commutes with shift and preserves distinguished triangles. More precisely, $F: \mathcal{S} \longrightarrow \mathcal{T}$ is *exact* it is is equipped with a natural isomorphism $\iota_X : F(X[1]) \cong F(X)[1]$ such that for every distinguished triangle (3.5) the sequence

$$F(X) \xrightarrow{F(f)} F(Y) \xrightarrow{F(g)} F(Z) \xrightarrow{\iota_X \circ F(h)} F(X)[1]$$

is again a distinguished triangle. An exact functor is automatically additive. An *equivalence of triangulated categories* is an equivalence of categories which is exact and whose inverse functor is also exact.

Exact equivalences between derived categories preserve all concepts which can be defined from $\mathcal{D}(R)$ using only the triangulated structure. One such invariant is the Grothendieck group $K_0(R)$, defined as the free abelian group generated by the isomorphism classes of finitely generated projective R-modules, modulo the relation

$$[P] + [Q] = [P \oplus Q] .$$

For any compactly generated triangulated category $\mathcal{T}$, the Grothendieck group $K_0(\mathcal{T})$ is defined as the free abelian group generated by the isomorphism classes of small objects in $\mathcal{T}$, modulo the relation

$$[X] + [Z] = [Y]$$

for every distinguished triangle

$$X \longrightarrow Y \longrightarrow Z \longrightarrow X[1]$$

involving small objects X, Y and Z (the morphisms in the triangle do not affect the relation). The split triangle

$$X \xrightarrow{(1,0)} X \oplus Y \xrightarrow{\binom{0}{1}} Y \xrightarrow{0} X[1]$$

is always distinguished, so the relation $[X \oplus Y] = [X] + [Y]$ holds in $K_0(\mathcal{T})$. So for every ring R, the assignment

$$K_0(R) \longrightarrow K_0(\mathcal{D}(R)) , \quad [P] \longmapsto [P[0]] \qquad (3.9)$$

defines a group homomorphism. This is in fact an isomorphism, see [Gr77, Sec. 7]. The inverse takes the class in $K_0(\mathcal{D}(R))$ of a bounded complex C of finitely generated projective modules to its 'Euler characteristic',

$$\sum_{n \in \mathbb{Z}} (-1)^n [C_n] \in K_0(R) .$$

It is much less obvious that constructions such as the center of a ring, Hochschild and cyclic homology and the higher Quillen K-groups are also invariants of the derived category. In contrast to the Grothendieck group K_0, there is no construction which produces these groups from the triangulated structure of $\mathcal{D}(R)$ only. The proof that two derived equivalent rings share these invariants uses the 'tilting theory' which we outline in the next section. More precisely, if R and S are derived equivalent flat algebras over some commutative ground ring, then there exists a *two-sided tilting complex*, i.e., a chain complex C of R-S-bimodules such that the functor $- \otimes_R C$ induces a (possibly different) derived equivalence [Ric91]. Tensor product with the bimodule complex C then induces an equivalence of K-theory spaces by the work of Thomason-Trobaugh [TT, Thm. 1.9.8]. Without the flatness assumption, the Waldhausen categories of small, cofibrant chain complexes can be related through an intermediate category of differential graded modules, to still obtain an equivalence of K-theory spaces; for more details we refer to [DuSh]. Similarly, Hochschild homology and cohomology (see [Ric91, Prop. 2.5], or, including the Gerstenhaber bracket, see [Kel03]) and cyclic homology (see [Kel96, Kel98]) are isomorphic for derived equivalent rings which are flat algebras over some commutative ground ring. The invariance of the center under derived equivalence is established in [Ric89a, Prop. 9.2] or [KZ, Prop. 6.3.2].

There is a general argument which we will use several times to verify that certain triangulated functors are equivalences, so we state it as a separate proposition. This Proposition 3.10 is a version of 'Beilinson's Lemma' [Bei78] and is typically applied when F is the total derived functor of a suitable left adjoint. In the following proposition, it is crucial that the functor F be defined and exact on the entire triangulated category $\mathcal{S}$. It is easy to find non-equivalent triangulated categories $\mathcal{S}$ and $\mathcal{T}$ with infinite sums and small generators P and Q respectively such that

$$\mathcal{S}(P, P)_* \cong \mathcal{T}(Q, Q)_*$$

as graded rings. For example one can take a differential graded ring A with a non-trivial triple Massey product and consider derived categories $\mathcal{S} = \mathcal{D}(A)$ and $\mathcal{T} = \mathcal{D}(H^*A)$ (where the cohomology ring of A is given the trivial differential).

Proposition 3.10. *Let $F\colon \mathcal{S} \longrightarrow \mathcal{T}$ be an exact functor between triangulated categories with infinite sums. Suppose that F preserves infinite sums and $\mathcal{S}$ has a small generator P such that*

(i) *FP is a small generator of $\mathcal{T}$ and*
(ii) *for all integers n, the map*

$$F\colon \mathcal{S}(P[n], P) \longrightarrow \mathcal{T}(FP[n], FP)$$

is bijective

Then F is an equivalence of categories.

Proof. We consider the full subcategory of $\mathcal{S}$ consisting of those Y for which the map

$$F \ : \ \mathcal{S}(P[n], Y) \ \longrightarrow \ \mathcal{T}(FP[n], FY) \qquad\qquad (3.11)$$

is bijective for all $n \in \mathbb{Z}$. By assumption this subcategory contains P. Since F is exact, the subcategory is closed under extensions. Since P and FP are small and F preserves coproducts, this subcategory is also closed under coproducts. Since P generates $\mathcal{S}$, the map (3.11) is thus bijective for arbitrary Y.

Similarly for arbitrary but fixed Y the full subcategory of $\mathcal{S}$ consisting of those X for which the map $F : \mathcal{S}(X, Y) \longrightarrow \mathcal{T}(FX, FY)$ is bijective is closed under extensions and coproducts. By the first part, it also contains P, so this subcategory is all of $\mathcal{S}$. In other words, F is full and faithful.

Now we consider the full subcategory of $\mathcal{T}$ of objects which are isomorphic to an object in the image of F. This subcategory contains the generator FP and it is closed under shifts and coproducts since these are preserved by F. We claim that this subcategory is also closed under extensions. Since FP generates $\mathcal{T}$, this shows that F is essentially surjective and hence an equivalence.

To prove the last claim we consider a distinguished triangle

$$X \ \xrightarrow{\ f\ } \ Y \ \longrightarrow \ Z \ \longrightarrow \ X[1] \ .$$

Since the subcategory under consideration is closed under isomorphism and shift in either direction we can assume that $X = F(X')$ and $Y = F(Y')$ are objects in the image of F. Since F is full there exists a map $f' : X' \longrightarrow Y'$ satisfying $F(f') = f$. We can then choose a mapping cone for the map f' and a compatible map from Z to $F(\mathrm{Cone}(f'))$ which is necessarily an isomorphism. $\qquad\square$

3.2. Derived equivalences after Rickard and Keller. In this section we state and prove Rickard's "Morita theory for derived categories". Rickard shows in [Ric89a, Thm. 6.4] that the existence of a *tilting complex* is necessary and sufficient for an equivalence between the unbounded derived categories of two rings. A tilting complex is a special small generator of the derived category, see Definition (3.12) below. The idea to use differential graded algebras in the proof is due to Keller [Kel94a], and we closely follow his approach.

The notion of a tilting complex comes up naturally when we examine the properties of the preferred generator $R[0]$ of the derived category $\mathcal{D}(R)$. First of all, the free R-module of rank one, considered as a complex concentrated in dimension zero, is a small generator of the derived category $\mathcal{D}(R)$. Since R is a free module, it has no self-extensions. Because Ext groups can be identified with morphisms in the derived category (see (3.4)), this means that the graded self-maps of the complex $R[0]$ are concentrated in dimension zero:

$$\mathcal{D}(R)(R[n], R) \;=\; 0 \quad \text{for } n \neq 0.$$

A tilting complex is any complex which also has these properties. Hence the definition is made so that the image of $R[0]$ under an equivalence of triangulated categories is a tilting complex.

Definition 3.12. A *tilting complex* for a ring R is a bounded complex T of finitely generated projective R-modules which generates the derived category $\mathcal{D}(R)$ and whose graded ring of self maps $\mathcal{D}(R)(T,T)_*$ is concentrated in dimension zero.

Special kinds of tilting complexes are the *tilting modules*; we give examples of tilting modules and tilting complexes in Section 3.3. The following theorem is due to Rickard [Ric89a, Thm. 6.4].

Theorem 3.13. *For two rings R and S the following conditions are equivalent.*

(1) The unbounded derived categories of R and S are equivalent as triangulated categories.

(2) There is a tilting complex T in $\mathcal{D}(S)$ whose endomorphism ring $\mathcal{D}(S)(T,T)$ is isomorphic to R.

Moreover, conditions (1) and (2) are implied by the condition

(3) There exists a chain complex of R-S-bimodules M such that the derived tensor product functor

$$- \otimes_R^L M \;:\; \mathcal{D}(R) \;\longrightarrow\; \mathcal{D}(S)$$

is an equivalence of categories.

If R or S is flat as an abelian group, then all three conditions are equivalent.

Instead of using the unbounded derived category, one can replace condition (1) by an equivalence between the full subcategories of homologically bounded below or small objects inside the derived categories, see for example [Ric89a, Thm. 6.4]. There is a version relative to a commutative ring k. Then R and S are k-algebras, conditions (1) and (3) then refer to k-linear equivalences of derived categories, condition (2) requires an isomorphism of k-algebras and in the addendum, one of R or S has to be flat as a k-module.

Remark 3.14. A derived equivalence F from $\mathcal{D}(R)$ to $D(S)$ which is not already a Morita equivalence maps the R-modules inside $\mathcal{D}(R)$ (i.e., complexes with homology concentrated in dimension 0) "transversely" to the S-modules inside $\mathcal{D}(S)$; more precisely, for an R-module M, the complex $F(M[0])$ can have non-trivial homology in several, or even in infinitely many dimensions; we give an example in 3.26 below. However, the homology of $F(M[0])$ is always bounded below.

A related point is that we can *not* recover the module category Mod-R from $\mathcal{D}(R)$, viewed as an abstract triangulated category. This is because we cannot make sense of "complexes with homology concentrated in dimension 0" unless we specify a homology functor like H_0, or we single out the class of complexes with homology in non-negative dimensions. This sort of extra structure is called a *t-structure* [BBD, 1.3] on a triangulated category. Every t-structure has a *heart*, an abelian category which plays the role of complexes in $\mathcal{D}(R)$ with homology concentrated in dimension zero.

The most involved part of the tilting theorem is the implication (2)$\Longrightarrow$(1), i.e., showing that a tilting complex gives rise to a derived equivalence. The proof we give is due to Keller; in the original paper [Kel94a], his setup is more general (he works in differential graded categories in order to allow 'many generator' versions). In the special case of interest for us, the exposition simplifies somewhat [KZ, Ch. 8]. Given a tilting complex T in $\mathcal{D}(S)$, the comparison between the derived categories of R and S passes through the derived category of a certain *differential graded ring* (generalizing the derived category of a an ordinary ring), namely the *endomorphism DG ring* $\mathrm{End}_S(T)$ of the tilting complex T (generalizing the endomorphism ring). So we start by introducing these new characters.

Definition 3.15. A *differential graded ring* is a $\mathbb{Z}$-graded ring A together with a differential d of degree -1 which satisfies the Leibniz rule

$$d(a \cdot b) = d(a) \cdot b + (-1)^{|a|} a \cdot d(b) \qquad (3.16)$$

for all homogeneous elements $a, b \in A$. A *differential graded right module* (or *DG module* for short) over a differential graded ring A consists of a graded right A-module together with a differential d of degree -1 which satisfies the Leibniz rule (3.16), but where now a is a homogeneous element of the module and b is a homogeneous element of A. A *homomorphism* of DG modules is a homomorphism of graded A-modules which is also a chain map. A *chain homotopy* between homomorphisms of DG modules is a homomorphism of graded A-modules of degree 1 which is also a chain homotopy.

A differential graded A-module M is *cofibrant* if there exists an exhaustive increasing filtration by sub DG modules

$$0 = M^0 \subseteq M^1 \subseteq \cdots \subseteq M^n \subseteq \cdots$$

such that each subquotient M^n/M^{n-1} is a direct summand of a direct sum of shifted copies of A. The *derived category* $\mathcal{D}(A)$ of the differential graded ring A has as objects the cofibrant DG modules over A and as morphisms the chain homotopy classes of DG module homomorphisms.

Up to chain homotopy equivalence, the cofibrant DG modules are the ones which have Keller's 'property (P)' in [Kel94a, 3.1]. A cofibrant differential graded module is sometimes called 'semi-free' or a 'cell module' [KM, Part III] (up to direct summands).

Remark 3.17. We need some facts about differential graded rings and modules which are not very difficult to prove, but which we do not want to discuss in detail.

(i) Several of the remarks from 3.3 carry over from rings to DG rings. A cofibrant DG A-module is projective as a graded A-module, ignoring the differential. Every quasi-isomorphism between cofibrant DG modules is a chain homotopy equivalence, and every DG module can be approximated up to quasi-isomorphism by a cofibrant one. A DG module is called *homotopically projective* if every homomorphism into an acyclic DG module is null-homotopic. Then a DG module is homotopically projective if and only if it is chain homotopy equivalent, as a DG module, to a cofibrant DG module.

(ii) The derived category $\mathcal{D}(A)$ of a differential graded ring A is naturally a triangulated category. The shift functor is again given by reindexing a DG module, and distinguished triangles arise from mapping cones as for the derived category of a ring. The only thing to note is that for a homomorphism $f : M \longrightarrow N$ of DG modules over A, the mapping cone becomes a graded A-module as the direct sum $N \oplus M[1]$, and this A-action satisfies the Leibniz rule with respect to the mapping cone differential (3.2).

(iii) Suppose that $f : A \longrightarrow B$ is a homomorphism of differential graded rings, i.e., f is a multiplicative chain homomorphism. Then extension of scalars $M \mapsto M \otimes_A B$ is exact on cofibrant differential graded modules (since the underlying *graded* modules over the graded ring underlying A are projective), it takes cofibrant modules to cofibrant modules, and it preserves the chain homotopy relation. So extension of scalars induces an exact functor on the level of derived categories

$$\mathcal{D}(A) \underset{f^*}{\overset{-\otimes_A^L B}{\rightleftarrows}} \mathcal{D}(B) \ , \tag{3.18}$$

called the *left derived functor*. This derived functor has an exact right adjoint f^* induced by restriction of scalars along f. This is not completely obvious with our definition of the derived category, since a cofibrant differential graded B-module is usually *not* cofibrant when viewed as a DG module over A via f.

If $f : A \longrightarrow B$ is a *quasi-isomorphism* of differential graded rings, then the derived functors of restriction and extension of scalars (3.18) are inverse *equivalences* of triangulated categories.

(iv) Suppose A is a differential graded ring whose homology is concentrated in dimension zero. Then A is quasi-isomorphic, as a differential graded ring, to the zeroth homology ring $H_0 A$. Indeed, a chain of two quasi-isomorphisms is given by

$$A \xleftarrow{\text{inclusion}} A_+ \xrightarrow{\text{projection}} H_0(A) \ .$$

Here A_+ is the differential graded sub-ring of A given by

$$(A_+)_n = \begin{cases} A_n & \text{for } n > 0 \\ \operatorname{Ker}(d : A_0 \longrightarrow A_{-1}) & \text{for } n = 0, \text{ and} \\ 0 & \text{for } n < 0. \end{cases}$$

Since the homology of A is trivial in negative dimensions, the inclusion $A_+ \longrightarrow A$ is a quasi-isomorphism. Since A_+ is trivial in negative dimensions, the projection $A_+ \longrightarrow H_0(A_+)$ is a homomorphism of differential graded rings, where the target is concentrated in dimension zero. This projection is also a quasi-isomorphism since the homology of A, and hence that of A_+, is trivial in positive dimensions.

Homomorphism complexes. Let A be a DG ring and let M and N be DG modules over A, not necessarily cofibrant. We defined the *homomorphism complex* $\operatorname{Hom}_A(M, N)$ as follows. In dimension $n \in \mathbb{Z}$, the chain group $\operatorname{Hom}_A(M, N)_n$ is the group of graded A-module homomorphisms of degree n, i.e.,

$$\operatorname{Hom}_A(M, N)_n = \operatorname{Hom}_A(M[n], N) .$$

The differentials of M and N do not play any role in the definition of the chain groups, but they enter in the formula for the differential which makes $\operatorname{Hom}_A(M, N)$ into a chain complex. This differential $d \colon \operatorname{Hom}_A(M, N)_n \longrightarrow \operatorname{Hom}_A(M, N)_{n-1}$ is defined by

$$d(f) = d_N \circ f - (-1)^n f \circ d_M . \tag{3.19}$$

Here f is a graded A-module map of degree n and the composites $d_N \circ f$ and $f \circ d_M$ are then graded A-module maps of degree $n - 1$.

With this definition, the 0-cycles in $\operatorname{Hom}_A(M, N)$ are those graded A-module maps f which satisfy $d_N \circ f - f \circ d_M = 0$, so they are precisely the DG homomorphisms from M to N. Moreover, if $f, g : M \longrightarrow N$ are two DG A-module maps, then the difference $f - g$ is a coboundary in the complex $\operatorname{Hom}_A(M, N)$ if and only if f is chain homotopic to g. So we have established a natural isomorphism

$$H_0\left(\operatorname{Hom}_A(M, N)\right) \cong [M, N]$$

between the zeroth homology of the complex $\operatorname{Hom}_A(M, N)$ and the chain homotopy classes of DG A-homomorphisms from M to N.

Now suppose that we have a third DG module L. Then the composition of graded A-module maps gives a bilinear pairing between the homomorphism complexes

$$\circ : \operatorname{Hom}_A(N, L)_m \times \operatorname{Hom}_A(M, N)_n \longrightarrow \operatorname{Hom}_A(M, L)_{m+n} .$$

Moreover, composition and the differential (3.19) satisfy the Leibniz rule, i.e., for graded A-module maps $f : M \longrightarrow N$ of degree n and $g : N \longrightarrow L$ of degree m we have

$$d(g \circ f) = dg \circ f + (-1)^m g \circ df$$

as graded maps from M to L.

The following consequences are crucial for the remaining step in the tilting theorem:

• for every DG A-module M, the endomorphism complex $\mathrm{Hom}_A(M, M) = \mathrm{End}_A(M)$ is a differential graded ring under composition and M is a differential graded $\mathrm{End}_A(M)$-A-bimodule;

• for every DG A-module N, the homomorphism complex $\mathrm{Hom}_A(M, N)$ is a differential graded module over $\mathrm{End}_A(M)$ under composition. Moreover the functor $\mathrm{Hom}_A(M, -) : \mathsf{Mod}\text{-}A \longrightarrow \mathsf{Mod}\text{-}\mathrm{End}_A(M)$ is right adjoint to tensoring with the $\mathrm{End}_A(M)$-A-bimodule M;

• if M is cofibrant, then the functor $\mathrm{Hom}_A(M, -)$ is exact and takes quasi-isomorphisms of DG A-modules to quasi-isomorphisms. Moreover, its left adjoint $- \otimes_{\mathrm{End}_A(M)} M$ preserves cofibrant objects and chain homotopies. So there exists a derived functor on the level of derived categories

$$- \otimes^L_{\mathrm{End}_A(M)} M \; : \; \mathcal{D}(\mathrm{End}_A(M)) \longrightarrow \mathcal{D}(A) \, ,$$

an exact functor which preserves infinite sums.

The following theorem is a special case of Lemma 6.1 in [Kel94a].

Theorem 3.20. *Let A be a DG ring and M a cofibrant A-module which is a small generator for the derived category $\mathcal{D}(A)$. Then the derived functor*

$$- \otimes^L_{\mathrm{End}_A(M)} M \; : \; \mathcal{D}(\mathrm{End}_A(M)) \longrightarrow \mathcal{D}(A) \tag{3.21}$$

is an equivalence of triangulated categories.

Proof. The total left derived functor (3.21) is an exact functor between triangulated categories which preserves infinite sums. Moreover, it takes the free $\mathrm{End}_A(M)$-module of rank one — which is a small generator for the derived category of $\mathrm{End}_A(M)$ — to the small generator M for $\mathcal{D}(A)$. The induced map of graded endomorphism rings

$$- \otimes^L_{\mathrm{End}_A(M)} M \; : \; \mathcal{D}(\mathrm{End}_A(M))(\mathrm{End}_A(M), \mathrm{End}_A(M))_* \longrightarrow \mathcal{D}(A)(M, M)_*$$

is an isomorphism (both sides are isomorphic to the homology ring of $\mathrm{End}_A(M)$). So Proposition 3.10 shows that this derived functor is an equivalence of triangulated categories. $\square$

After all these preparations we can give the

Proof of the tilting theorem 3.13. Clearly, condition (3) implies condition (1). Now we assume condition (1) and we choose an exact equivalence F from the derived category $\mathcal{D}(R)$ to $\mathcal{D}(S)$. The defining properties of a tilting complex are preserved under exact equivalences of triangulated categories. Since $R[0]$, the free R-module of rank one, concentrated in dimension zero, is a tilting complex for the ring R, its image $T = F(R[0])$ is a tilting complex for S. Moreover, F restricts to a ring isomorphism

$$F : R \cong \mathcal{D}(R)(R[0], R[0]) \xrightarrow{\cong} \mathcal{D}(S)(T, T) .$$

Hence condition (2) holds.

For the implication (2)$\Longrightarrow$(1) we are given a tilting complex T in $\mathcal{D}(S)$ and an isomorphism of rings $\mathcal{D}(S)(T, T) \cong R$. The complex T is naturally a differential graded $\mathrm{End}_S(T)$-S-bimodule, and by Theorem 3.20, the derived functor

$$- \otimes^L_{\mathrm{End}_S(T)} T : \mathcal{D}(\mathrm{End}_S(T)) \longrightarrow \mathcal{D}(S)$$

is an equivalence of triangulated categories. The isomorphism of graded rings

$$H_*\left(\mathrm{End}_S(T)\right) \cong \mathcal{D}(S)(T, T)_*$$

and the defining property of a tilting complex show that the homology of $\mathrm{End}_S(T)$ is concentrated in dimension zero. So there is a chain of two quasi-isomorphisms between $\mathrm{End}_S(T)$ and the ring $H_0\left(\mathrm{End}_S(T)\right) \cong \mathcal{D}(S)(T, T) \cong R$. Restriction and extension of scalars along these quasi-isomorphisms gives a chain of equivalences between the derived categories of the differential graded ring $\mathrm{End}_S(T)$ and the derived category of the ordinary ring $\mathcal{D}(S)(T, T)$. Putting all of this together we end up with a chain of three equivalences of triangulated categories:

$$\mathcal{D}(R) \cong \mathcal{D}(\mathrm{End}_S(T)_+) \cong \mathcal{D}(\mathrm{End}_S(T)) \cong \mathcal{D}(S) .$$

It remains to prove the implication (2)$\Longrightarrow$(3), assuming that R or S is flat. Let T be a tilting complex in $\mathcal{D}(S)$ and $f : \mathcal{D}(S)(T, T) \longrightarrow R$ an isomorphism of rings. The homology of $\mathrm{End}_S(T)$ is isomorphic to the graded self maps of T in $\mathcal{D}(S)$, so it is concentrated in dimension 0. So the inclusion of the DG sub-ring $\mathrm{End}_S(T)_+$ into the endomorphism DG ring $\mathrm{End}_S(T)$ induces an isomorphism on homology, compare Remark 3.17 (iv). Since $\mathrm{End}_S(T)_+$ is trivial in negative dimensions, there is a unique morphisms of DG rings $\mathrm{End}_S(T)_+ \longrightarrow R$ which realizes the isomorphism f on H_0. We choose a *flat resolution* of $\mathrm{End}_S(T)_+$, i.e., a DG ring E and a quasi-isomorphism of DG rings $E \xrightarrow{\simeq} \mathrm{End}_S(T)_+$, such that the functor $E \otimes -$ preserves quasi-isomorphisms between chain complexes of abelian groups (see for example [Kel99, 3.2 Lemma (a)]). We end up with a chain of two quasi-isomorphisms of DG rings

$$R \xleftarrow{\simeq} E \xrightarrow{\simeq} \mathrm{End}_S(T) .$$

The complex T is naturally a DG $\mathrm{End}_S(T)$-S-bimodule, and we restrict the left action to E and view T as a DG E-S-bimodule. We choose a cofibrant replacement $T^c \xrightarrow{\sim} T$ as a DG E-S-bimodule. Then we obtain the desired complex of R-S-bimodules by

$$M = R \otimes_E T^c .$$

Tensoring with M over R has a total left derived functor

$$- \otimes_R^L M \ : \ \mathcal{D}(R) \ \longrightarrow \ \mathcal{D}(S) \tag{3.22}$$

(although this is not obvious with our definition since M need not be cofibrant as a complex of right S-modules, and then $- \otimes_R M$ does not takes values in cofibrant complexes). In order to show that this derived functor is an exact equivalence we use that the diagram of triangulated categories

$$\begin{CD}
\mathcal{D}(E) @>{-\otimes_E^L \mathrm{End}_S(T)}>> \mathcal{D}(\mathrm{End}_S(T)) \\
@V{-\otimes_E^L R}VV @VV{-\otimes_{\mathrm{End}_S(T)}^L T}V \\
\mathcal{D}(R) @>>{-\otimes_R^L M}> \mathcal{D}(S)
\end{CD} \tag{3.23}$$

commutes up to natural isomorphism. Indeed, two ways around the square are given by derived tensor product with the E-S-bimodules M respectively T, so it suffices to find a chain of quasi-isomorphisms of DG bimodules between M and T.

Since E is cofibrant as a complex of abelian groups and the composite map $E \longrightarrow \mathrm{End}_S(T)_+ \longrightarrow R$ is a quasi-isomorphism, $E \otimes S^{op}$ models the derived tensor product of R and S. If one of R or S are flat, then $R \otimes S^{op}$ also models the derived tensor product, so that the map

$$E \otimes S^{op} \ \longrightarrow \ R \otimes S^{op}$$

is a quasi-isomorphism of DG rings. Since T^c is cofibrant as an $E \otimes S^{op}$-module, the induced map

$$T^c \ = \ (E \otimes S^{op})_{E\otimes S^{op}} T^c \ \longrightarrow \ (R \otimes S^{op})_{E\otimes S^{op}} T^c \ \cong \ R \otimes_E T^c \ = \ M$$

is a quasi-isomorphism. So we have a chain of two quasi-isomorphisms of E-S-bimodules

$$T \ \xleftarrow{\simeq} \ T^c \ \xrightarrow{\simeq} \ M \ .$$

The left and upper functors in the commutative square (3.23) are derived from extensions of scalars along quasi-isomorphisms of DG rings; thus they are exact equivalence of triangulated categories. The right vertical derived functor is an exact equivalence by Theorem 3.20. So we conclude that the lower horizontal functor (3.22) in the square (3.23) is also an exact equivalence of triangulated categories. This establishes condition (3). $\square$

3.3. Examples. Historically, tilting modules seem to have been the first examples of derived equivalences which are not Morita equivalences. I will not try to give an account of the history of tilting modules and rather refer to [KZ, Sec. 3.1] or [AGH].

Definition 3.24. Let R be a finite dimensional algebra over a field. A *tilting module* is a finitely generated R-module T with the following properties.

(i) T has projective dimension 0 or 1,

(ii) T has no self-extensions, i.e., $\mathrm{Ext}_R^1(T,T) = 0$,

(iii) there is an exact sequence of right R-modules

$$0 \longrightarrow R \longrightarrow T_1 \longrightarrow T_2 \longrightarrow 0$$

such that T_1 and T_2 are direct summands of a finite sum of copies of T.

Note that if the tilting module T is actually projective, then condition (ii) is automatic and the exact sequence required in (iii) splits. So then the free R-module of rank one is a summand of a finite sum of copies of T, and hence T is a finitely generated projective generator for Mod-R. So R is then Morita equivalent to the endomorphism ring of the tilting module T, by the Morita theorem 2.2. For a self-injective algebra, for example a group algebra over a field, the converse also holds; indeed, every module of finite projective dimension over a self-injective algebra is already projective. So for these algebras, tilting is the same as Morita equivalence.

If the projective dimension of the tilting module T is 1, then we do not get an equivalence between the modules over R and $S = \mathrm{End}_R(T)$, but we get a derived equivalence. Indeed, since R is noetherian, condition (i) implies that T has a 2-step resolution $P_1 \longrightarrow P_0$ by two finitely generated projective R-modules; this resolution is a small object in the derived category $\mathcal{D}(R)$, and its graded self-maps in $\mathcal{D}(R)$ are concentrated in dimension 0 by condition (ii). Condition (iii) implies that the complex $R[0]$ is contained in the triangulated subcategory generated by the resolution, and the resolution is thus a generator for $\mathcal{D}(R)$, hence a tilting complex.

Example 3.25. For an example of a non-projective tilting module we fix a field k and we let A be the algebra of upper triangular 3×3 matrices over k,

$$A = \left\{ \begin{pmatrix} x_{11} & x_{12} & x_{13} \\ 0 & x_{22} & x_{23} \\ 0 & 0 & x_{33} \end{pmatrix} \mid x_{ij} \in k \right\} .$$

Up to isomorphism, there are three indecomposable projective right A-modules, namely the row vectors

$$P^1 = \{(y_1, y_2, y_3) \mid y_1, y_2, y_3 \in k\}$$

and its A-submodules

$$P^2 = \{(0, y_2, y_3) \mid y_2, y_3 \in k\} \quad \text{and} \quad P^3 = \{(0, 0, y_3) \mid y_3 \in k\} .$$

These projectives are the covers of three corresponding simple modules, namely

$$S^1 = P^1/P^2, \quad S^2 = P^2/P^3, \quad \text{and} \quad S^3 = P^3 .$$

In particular, S^3 is projective and S^1 and S^2 have projective dimension 1.

We define the tilting module T as the direct sum

$$T = P^1 \oplus P^2 \oplus S^2 .$$

The projective resolution

$$0 \longrightarrow P^1 \oplus P^2 \oplus P^3 \xrightarrow{\text{inclusion}} P^1 \oplus P^2 \oplus P^2 \longrightarrow S^2 \longrightarrow 0$$

can be used to calculate $\mathrm{Ext}_A^1(T,T) = \mathrm{Ext}_A^1(S^2,T) = 0$. Since $P^1 \oplus P^2 \oplus P^3$ is a free A-module of rank one, this short exact sequence verifies tilting condition (iii) in Definition 3.24 for the A-module T.

Altogether this shows that T is a tilting module for A of projective dimension one. So A 'tilts' to the endomorphism algebra of T; this endomorphism algebra can be calculated directly, and it comes out to be another subalgebra of the 3×3 matrices over k, namely

$$\mathrm{End}_A(T) \cong \left\{ \begin{pmatrix} x_{11} & x_{12} & x_{13} \\ 0 & x_{22} & 0 \\ 0 & 0 & x_{33} \end{pmatrix} \mid x_{ij} \in k \right\}.$$

(hint: the modules P^1 and S^2 do not map to each other nor to P^2, and the remaining relevant morphism spaces are 1-dimensional over k.) The algebras A and $\mathrm{End}_A(T)$ are *not* Morita equivalent. Indeed, both have exactly three isomorphism classes of indecomposable projective modules, but in one case these modules are 'directed' (i.e., linearly ordered under the existence of non-trivial homomorphisms), whereas in the other case two of these indecomposable projectives do not map to each other non-trivially.

The preceding example, and many other ones, are often described using representations of quivers. Indeed, the upper triangular matrices A and the tilted algebra $\mathrm{End}_A(T)$ are isomorphic to the path algebras of the A_3-quivers

$$\bullet \longrightarrow \bullet \longrightarrow \bullet \qquad \text{respectively} \qquad \bullet \longleftarrow \bullet \longrightarrow \bullet \,.$$

Example 3.26. We obtain a tilting complex whose homology is concentrated in more then one dimension by 'spreading out' the free module of rank one. Let $R = R_1 \times R_2$ be the product of two rings. Let $P_1 = R_1 \times 0$ and $P_2 = 0 \times R_2$ be the two "blocks", i.e., the projective R-bimodules corresponding to the central idempotents $(1,0)$ and $(0,1)$ in R. Then $R = P_1 \oplus P_2$ as an R-bimodule, and there are no non-trivial R-homomorphisms between P_1 and P_2. Now take $T = P_1[0] \oplus P_2[n]$ for some number $n \neq 0$. This is a complex of R-modules with trivial differential whose homology is concentrated in two dimensions. Moreover, the complex T is a small generator for the derived category $\mathcal{D}(R)$. But the only non-trivial self-maps of T are of degree 0 since P_1 and P_2 don't map to each other. Hence T is a tilting complex which is not (quasi-isomorphic to) a tilting module. The endomorphisms of T are again the ring R, so it is a non-trivial self-tilting complex of R. Under the equivalence $\mathcal{D}(R) \cong D(R_1) \times D(R_2)$, the self-equivalence induced by T is the identity on the first factor and the n-fold shift on the second factor.

More examples of tilting complexes can be found in Sections 4 and 5 of [Ric89b] or Chapter 5 of [KZ].

4. Morita theory in stable model categories

Now we carry the Morita philosophy one step further: we sketch Morita theory for ring spectra and for stable model categories. As a summary one can say that essentially everything which we have said for rings and differential graded rings works, suitably interpreted, for ring spectra as well.

First a few words about what we mean by a ring spectrum. The stable homotopy category of algebraic topology has a symmetric monoidal smash product; the monoids are homotopy-associative ring spectra, and they represent multiplicative cohomology theories. While the notion of a homotopy-associative ring spectrum is useful for many things, it does not have a good enough module theory for our present purpose. One can certainly consider spectra with a homotopy-associative action of a homotopy-associative ring spectrum; but the mapping cone of a homomorphism between such modules does not inherit a *natural* action of the ring spectrum, and the category of such modules does not form a triangulated category.

So in order to carry out the Morita-theory program we need a highly structured model for the category of spectra which admits a symmetric monoidal and homotopically well behaved smash product — before passing to the homotopy category ! The first examples of such categories were the *S*-modules [EKMM] and the symmetric spectra [HSS]; by now several more such categories have been constructed [Lyd98, MMSS]. All these spectra categories are *model categories* in the sense of Quillen [Qui67] and the appropriate notion of equivalence is that of a 'Quillen equivalence' [Hov99, Def. 1.3.12] since these equivalences preserve the 'homotopy theory', not just the homotopy category; all known model categories of spectra are Quillen equivalent in a monoidal fashion.

For definiteness, we work in one specific category of spectra with nice smash product, namely the *symmetric spectra* based on simplicial sets, as introduced by Hovey, Shipley and Smith [HSS]. The monoids are called *symmetric ring spectra*, and I personally think that they are the simplest kind of ring spectra; as far as their homotopy category is concerned, symmetric ring spectra are equivalent to the older notion of A_∞-*ring spectrum*, and *commutative* symmetric ring spectra are equivalent to E_∞-*ring spectra*. The good thing is that operads are not needed anymore.

However, using symmetric spectra is not essential and the results described in this section could also be developed in more or less the same way in any other of the known model categories of spectra with compatible smash product. Alternatively, we could have taken an axiomatic approach and use the term 'spectra' for any stable, monoidal model category in which the unit object 'looks and feels' like the sphere spectrum. Indeed, we are essentially only using the following properties of the category of symmetric spectra:

(i) there is a symmetric monoidal *smash product*, which makes symmetric spectra into a *monoidal model category* ([Hov99, 4.2.6], [SS00]);

(ii) the model structure is *stable* (Definition 4.1);

(iii) the unit $\mathbb{S}$ of the smash product is a small generator (Definition 3.7) of the homotopy category of spectra;

(iv) the (derived) space of self maps of the unit object $\mathbb{S}$ is weakly equivalent to $QS^0 = \text{hocolim}_n \, \Omega^n S^n$ and in the homotopy category, there are no maps of negative degree from $\mathbb{S}$ to itself.

A large part of the material in this section is taken from a joint paper with Shipley [SS03]. Two other papers devoted to Morita theory in the context of ring spectra are [DGI] by Dwyer, Greenlees and Iyengar and [BL] by Baker and Lazarev.

4.1. **Stable model categories.** Recall from [Qui67, I.2] or [Hov99, 6.1] that the homotopy category of a pointed model category supports a suspension and a loop functor. In short, for any object X the map to the zero object can be factored

$$X \longrightarrow C \xrightarrow{\;\simeq\;} *$$

as a cofibration followed by a weak equivalence. The suspension of X is then defined as the quotient of the cofibration, $\Sigma X = C/X$. Dually, the loop object ΩX is the fiber of a fibration from a weakly contractible object to X. On the level of homotopy categories, the suspension and loop constructions become functorial, and Σ is left adjoint to Ω.

Definition 4.1. A *stable model category* is a pointed model category for which the functors Ω and Σ on the homotopy category are inverse equivalences.

The homotopy category of a stable model category has a large amount of extra structure, some of which is relevant for us. First of all, it is naturally a triangulated category, see [Hov99, 7.1.6] for a detailed proof. The rough outline is as follows: by definition of 'stable' the suspension functor is a self-equivalence of the homotopy category and it defines the shift functor. Since every object is a two-fold suspension, hence an abelian co-group object, the homotopy category of a stable model category is additive. Furthermore, by [Hov99, 7.1.11] the cofiber sequences and fiber sequences of [Qui67, 1.3] coincide up to sign in the stable case, and they define the distinguished triangles. The model categories which we consider have all limits and colimits, so the homotopy categories have infinite sums and products. Objects of a stable model category are called 'generators' or 'small' if they have this property as objects of the triangulated homotopy category, compare Definition 3.7.

A Quillen adjoint functor pair between stable model categories gives rise to total derived functors which are exact functors with respect to the triangulated structure; in other words both total derived functors commute with suspension and preserve distinguished triangles.

Examples 4.2.

(1) **Chain complexes.** In the previous section, we have already seen an

important class of examples from algebra, namely the category of chain complexes over a ring R. This category actually has several different stable model structures: the *projective* model structure (see Remark 3.3 (iv)) and the *injective* model structure (see Remark 3.3 (v)) have as weak equivalences the quasi-isomorphisms. There is a clash of terminology here: the homotopy category in the sense of homotopical algebra is obtained by formally inverting the weak equivalences; so for the projective and injective model structures, this gives the unbounded derived category $\mathcal{D}(R)$. But the category of unbounded chain complexes admits another model structure in which the weak equivalences are the *chain homotopy equivalences*, see e.g. [CH, Ex. 3.4]. Thus for this model structure, the homotopy category is what is commonly called the homotopy category, often denoted by $K(R)$. The derived category $\mathcal{D}(R)$ is a quotient of the homotopy category $K(R)$; the derived category $\mathcal{D}(R)$ has a single small generator, but for example the homotopy category of chain complexes of abelian groups $K(\mathbb{Z})$ does not have a set of generators whatsoever, compare [Nee01, E.3.2]. The three stable model structures on chain complexes of modules have been generalized in various directions to chain complexes in abelian categories or to other differential graded objects, see [CH], [Bek00] and [Hov01a].

(2) **The stable module category of a Frobenius ring.** A different kind of algebraic example — not involving chain complexes — is formed by the stable module categories of Frobenius rings. A Frobenius ring A is defined by the property that the classes of projective and injective A-modules coincide. Important examples are finite dimensional self-injective algebras over a field, in particular finite dimensional Hopf-algebras, such as group algebras of finite groups. The *stable module category* has as objects the A-modules (*not* chain complexes of modules). Morphisms in the stable category or represented by module homomorphisms, but two homomorphisms are identified if their difference factors through a projective (= injective) A-module.

Fortunately the two different meanings of 'stable' fit together nicely: the stable *module* category is the homotopy category associated to a stable *model* category structure on the category of A-modules, see [Hov99, Sec. 2]. The cofibrations are the monomorphisms, the fibrations are the epimorphisms, and the weak equivalences are the maps which become isomorphisms in the stable category. Every finitely generated module is small when considered as an object of the stable module category. As in the case of chain complexes of modules, there is usually no point in making the model structure explicit since the cofibration, fibrations and weak equivalences coincide with certain well-known concepts.

Quillen equivalences between stable module categories arise under the name of *stable equivalences of Morita type* ([Bro94, Sec. 5], [KZ, Ch. 11]). For simplicity, suppose that A and B are two finite-dimensional self-injective algebras over a field k; then A and B are in particular Frobenius rings. Consider an

A-B-bimodule M (by which we mean a k-symmetric bimodule, also known as a right module over $A^{op} \otimes_k B$), which is projective as left A-module and as a right B-module separately. Then the adjoint functor pair

$$\text{Mod-}A \underset{\text{Hom}_B(M,-)}{\overset{-\otimes_A M}{\rightleftarrows}} \text{Mod-}B \tag{4.3}$$

is Quillen adjoint pair with respect to the 'stable' model structures.

A stable equivalence of Morita type consists of an A-B-bimodule M and a B-A-bimodule N such that both M and N are projective as left and right modules separately, and such that there are direct sum decompositions

$$N \otimes_A M \cong B \oplus X \quad \text{and} \quad M \otimes_B N \cong A \oplus Y$$

as bimodules, where Y is a projective A-A-bimodule and B is a projective B-B-bimodule. In this situation, the functors $- \otimes_A M$ and $- \otimes_B N$ induce inverse equivalences of the stable module categories. Moreover, the Quillen adjoint pair (4.3) is a Quillen equivalence.

Rickard observed [Ric89b] that a derived equivalence between self-injective, finite-dimensional algebras also gives rise to a stable equivalence of Morita type.

In the above algebraic examples, there is no real need for the language of model categories; moreover, 'Morita theory' is covered by Keller's paper [Kel94a], which uses differential graded categories. A whole new world of stable model categories comes from homotopy theory, see the following list. The associated homotopy categories yield triangulated categories which are not immediately visible to the eyes of an algebraist, since they do not arise from abelian categories.

(3) **Spectra.** The prototypical example of a stable model category (which is not 'algebraic'), is 'the' category of spectra. We review one model, the *symmetric spectra* of Hovey, Shipley and Smith [HSS] in more detail in Section 4.2. Many other model categories of spectra have been constructed, see for example [BF78, Rob87a, Jar97, EKMM, Lyd98, MMSS]. All known model categories of spectra Quillen equivalent (see e.g. [HSS, Thm. 4.2.5], [Sch01a] or [MMSS]), and their common homotopy category is referred to as *the* stable homotopy category. The sphere spectrum is a small generator for stable homotopy category.

(4) **Modules over ring spectra.** Modules over an S-algebra [EKMM, VII.1], over a symmetric ring spectrum [HSS, 5.4.2], or over an orthogonal ring spectrum [MMSS] form stable model categories. We recall symmetric ring spectra and their module spectra in Section 4.2. In each case a module is small if and only if it is weakly equivalent to a retract of a finite cell module. The free module of rank one is a small generator. More generally there are stable model categories of modules over 'symmetric ring spectra with several objects', or *spectral categories*, see [SS03, A.1].

(5) **Equivariant stable homotopy theory.** If G is a compact Lie group, there is a category of G-equivariant coordinate free spectra [LMS86] which is a stable model category. Modern versions of this model category are the G-equivariant orthogonal spectra of [MM02] and G-equivariant S-modules of [EKMM]. In this case the equivariant suspension spectra of the coset spaces G/H_+ for all closed subgroups $H \subseteq G$ form a set of small generators.

(6) **Presheaves of spectra.** For every Grothendieck site Jardine [Jar87] constructs a stable model category of presheaves of Bousfield-Friedlander type spectra; the weak equivalences are the maps which induce isomorphisms of the associated sheaves of stable homotopy groups. For a general site these stable model categories do not seem to have a set of small generators. A similar model structure for presheaves of symmetric spectra is developed in [Jar00a].

(7) **The stabilization of a model category.** Modulo technicalities, every pointed model category gives rise to an associated stable model category by 'inverting' the suspension functor, i.e., by passage to internal spectra. This has been carried out, under different hypotheses, in [Sch97] and [Hov01b].

(8) **Bousfield localization.** Following Bousfield [Bou75], localized model structures for modules over an S-algebra are constructed in [EKMM, VIII 1.1]. Hirschhorn [Hir03] shows that under quite general hypotheses the localization of a model category is again a model category. The localization of a stable model category is stable and localization preserves generators. Smallness need not be preserved.

(9) **Motivic stable homotopy.** In [MV, Voe98] Morel and Voevodsky introduced the $\mathbb{A}^1$-local model category structure for schemes over a base. An associated stable homotopy category of $\mathbb{A}^1$-local T-spectra (where $T = \mathbb{A}^1/(\mathbb{A}^1 - 0)$ is the 'Tate-sphere') is an important tool in Voevodsky's proof of the Milnor conjecture [Voe]. There are several stable model categories underlying this motivic stable homotopy category, see for example [Jar00b], [Hov01b], [Hu03] or [DØR].

4.2. Symmetric ring and module spectra. In this section we give a quick introduction to symmetric spectra and symmetric ring and module spectra. I recommend reading the original, self-contained paper by Hovey, Shipley and Smith [HSS]. At several points, our exposition differs from theirs; for example, we let the spheres act from the right.

Definition 4.4. [HSS] A symmetric spectrum consists of the following data:

- a sequence of pointed simplicial sets X_n for $n \geq 0$
- for each $n \geq 0$ a base-point preserving action of the symmetric group Σ_n on X_n
- pointed maps $\alpha_{p,q} : X_p \wedge S^q \longrightarrow X_{p+q}$ for $p, q \geq 0$ which are $\Sigma_p \times \Sigma_q$-equivariant; here $S^1 = \Delta^1/\partial\Delta^1$, $S^q = (S^1)^{\wedge q}$ and Σ_q permutes the factors.

This data is subject to the following conditions:

- under the identification $X_n \cong X_n \wedge S^0$, the map $\alpha_{n,0} : X_n \wedge S^0 \longrightarrow X_n$ is the identity,
- for $p, q, r \geq 0$, the following square commutes

$$
\begin{array}{ccc}
X_p \wedge S^q \wedge S^r & \xrightarrow{\alpha_{p,q} \wedge \mathrm{Id}} & X_{p+q} \wedge S^r \\
{\scriptstyle \cong} \downarrow & & \downarrow {\scriptstyle \alpha_{p+q,r}} \\
X_p \wedge S^{q+r} & \xrightarrow[\alpha_{p,q+r}]{} & X_{p+q+r} \ .
\end{array}
\tag{4.5}
$$

A *morphism* $f : X \longrightarrow Y$ of symmetric spectra consists of Σ_n-equivariant pointed maps $f_n : X_n \longrightarrow Y_n$ for $n \geq 0$, which are compatible with the structure maps in the sense that $f_{p+q} \circ \alpha_{p,q} = \alpha_{p,q} \circ (f_q \wedge \mathrm{Id}_{S^q})$ for all $p, q \geq 0$. The category of symmetric spectra is denoted by Sp^Σ.

The definition we have just given is somewhat redundant, and Hovey, Shipley and Smith use a more economical definition in [HSS, Def. 1.2.2]. Indeed, the commuting square (4.5), shows that all action maps $\alpha_{p,q}$ are given by composites of the maps $\alpha_{p,1} : X_p \wedge S^1 \longrightarrow X_{p+1}$ for varying p.

A first example is the symmetric *sphere spectrum* $\mathbb{S}$ given by $\mathbb{S}_n = S^n$, where the symmetric group permutes the factors and $\alpha_{p.q} : S^p \wedge S^q \longrightarrow S^{p+q}$ is the canonical isomorphism. More generally, every pointed simplicial set K gives rise to a *suspension spectrum* $\Sigma^\infty K$ via

$$
(\Sigma^\infty K)_n = K \wedge S^n \ ;
$$

then we have $\mathbb{S} \cong \Sigma^\infty S^0$.

A symmetric spectrum is *cofibrant* if it has the left lifting property for levelwise acyclic fibrations. More precisely, A is cofibrant if the following holds: for every morphism $f : X \longrightarrow Y$ of symmetric spectra such that $f_n : X_n \longrightarrow Y_n$ is a weak equivalence and Kan fibration for all n, and for every morphism $\iota : A \longrightarrow X$, there exists a morphism $\bar{\iota} : A \longrightarrow Y$ such that $\iota = f\bar{\iota}$. Suspension spectra are examples of cofibrant symmetric spectra. An equivalent definition uses the *latching space* $L_n A$, a simplicial set which roughly is the 'stuff coming from dimensions below n'; see [HSS, 5.2.1] for the precise definition. A symmetric spectrum A is cofibrant if and only if for all n, the map $L_n A \longrightarrow A_n$ is injective and symmetric group Σ_n is freely on the complement of the image, see [HSS, Prop. 5.2.2]. An Ω-*spectrum* is defined by the properties that each simplicial set X_n is a Kan complex and all the maps $X_n \longrightarrow \Omega(X_{n+1})$ adjoint to $\alpha_{n,1}$ are weak homotopy equivalences. The *stable homotopy category* has as objects the cofibrant symmetric Ω-spectra and as morphisms the homotopy classes of morphisms of symmetric spectra.

Although we just gave a perfectly good definition of the stable homotopy category, in order to work with it one needs an ambient model category structure. One such model structure is the *stable model structure* of [HSS, Thm. 3.4.4]. A morphism of symmetric spectra is a *stable equivalence* if it induces isomorphisms on all *cohomology theories* represented by (injective)

Ω-spectra, see [HSS, Def, 3.1.3] for the precise statement. There is a notion of *cofibration* such that a symmetric spectrum X is cofibrant in the above sense if and only if the map from the trivial symmetric spectrum to X is a cofibration. The Ω-spectra then coincide with the stably fibrant symmetric spectra. There are other model structure for symmetric spectra with the same class of weak (=stable) equivalences, hence with the same homotopy category, for example the *S-model structure* which is hinted at in [HSS, 5.3.6].

Stable equivalences versus π_*-isomorphisms. One of the tricky points with symmetric spectra is the relationship between stable equivalences and π_*-isomorphisms. The stable equivalences are defined as the morphisms which induce isomorphisms on all *cohomology theories*; there is the strictly smaller class of morphisms which induce isomorphisms on stable homotopy groups. The *k-th stable homotopy group* of a symmetric spectrum X is defined as the colimit

$$\pi_k X \;=\; \operatorname{colim}_n \pi_{n+k}|X_n| \,,$$

where $|X_n|$ denotes the geometric realization of the simplicial set X_n. The colimit is taken over the maps

$$\pi_{n+k}|X_n| \xrightarrow{\;-\wedge S^1\;} \pi_{n+k+1}\left(|X_n| \wedge S^1\right) \xrightarrow{\;(\alpha_{n,1})_*\;} \pi_{n+k+1}|X_{n+1}| \,. \qquad (4.6)$$

While every π_*-isomorphism of symmetric spectra is a stable equivalence [HSS, Thm. 3.1.11], the converse is not true. The standard example is the following: consider the symmetric spectrum $F_1 S^1$ freely generated by the circle S^1 in dimension 1. Explicitly, $F_1 S^1$ is given by

$$(F_1 S^1)_n \;=\; \Sigma_n^+ \wedge_{\Sigma_{n-1}} S^{n-1} \wedge S^1 \,.$$

So $(F_1 S^1)_n$ is a wedge of n copies of S^n and in the stable range, i.e., up to roughly dimensions $2n$, the homotopy groups of $(F_1 S^1)_n$ are a direct sum of n copies of the homotopy groups of S^n. Moreover, in the stable range, the map in the colimit system (4.6) is a direct summand inclusion into $(n+1)$ copies of the homotopy groups of S^n. Thus in the colimit, the stable homotopy groups of the symmetric spectrum $F_1 S^1$ are a countably infinite direct sum of copies of the stable homotopy groups of spheres. Since $F_1 S^1$ is freely generated by the circle S^1 in dimension 1, it ought to be a desuspension of the suspension spectrum of the circle. However, the necessary symmetric group actions 'blow up' such free objects with the effect that the stable homotopy groups are larger than they should be. This example indicates that inverting only the π_*-isomorphisms would leave too many stable homotopy types, and the resulting category could not be equivalent to the usual stable homotopy category.

Smash product. One of the main features which distinguishes *symmetric* spectra from the more classical spectra is the internal smash product. The smash product of symmetric spectra can be described via its universal property, analogous to universal property of the tensor product over a commutative ring. Indeed, if R is a commutative ring and M and N are right R-modules, then a bilinear map to another left R-module W is a map $b : M \times N \longrightarrow W$ such that for each $m \in M$ the map $b(m, -) : N \longrightarrow W$ and each $n \in N$ the map $b(-, n) : M \longrightarrow W$ are R-linear. The tensor product $M \otimes_R N$ is the universal example of a right R-module together with a bilinear map from $M \times N$.

Let us define a *bilinear morphism* $b : (X, Y) \longrightarrow Z$ from two symmetric spectra X and Y to a symmetric spectrum Z to consist of a collection of $\Sigma_p \times \Sigma_q$-equivariant maps of pointed simplicial sets

$$b_{p,q} : X_p \wedge Y_q \longrightarrow Z_{p+q}$$

for $p, q \geq 0$, such that for all $p, q, r \geq 0$, the following diagram commutes

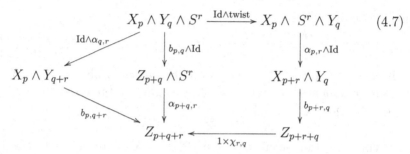

$$(4.7)$$

The automorphism $1 \times \chi_{r,q}$ of Z_{p+q+r} may look surprising at first sight. Here $1 \times \chi_{r,q} \in \Sigma_{p+r+q}$ denotes the block permutation which fixes the first p elements, and which moves the next q elements past the last r elements. This can be viewed as a topological version of the sign rule which says that when two symbols of degree q and r are permuted past each other, the sign $(-1)^{qr}$ should appear as well. The block permutation $\chi_{r,q}$ has sign $(-1)^{qr}$ and it compensates the upper vertical interchange of Y_q and S^r. A good way to keep track of such permutations is to carefully distinguish between indices such as $r + q$ and $q + r$. Of course these two numbers are equal, but the fact that one arises naturally instead of the other reminds us that a block permutation should be inserted.

The smash product $X \wedge Y$ is the universal example of a symmetric spectrum with a bilinear morphism from X and Y. In other words, it comes with a bilinear morphism $\iota : (X, Y) \longrightarrow X \wedge Y$ such that for every symmetric spectrum Z the map

$$\mathcal{S}p^{\Sigma}(X \wedge Y, Z) \longrightarrow \text{Bi-}\mathcal{S}p^{\Sigma}((X, Y), Z) \tag{4.8}$$

is bijective. If we suppose that such a universal object exist, this property characterizes the smash product and the maps $\iota_{p,q} : X_p \wedge Y_q \longrightarrow (X \wedge Y)_{p+q}$ up to canonical isomorphism. An actual construction as a certain coequalizer is given in [HSS, Def. 2.2.3]; in this article, we will only use the universal property of the smash product.

We use the universal property to derive that the smash product is functorial and symmetric monoidal. For example, let $f : X \longrightarrow Y$ and $f' : X' \longrightarrow Y'$

be morphisms of symmetric spectra. Then the collection of maps of pointed simplicial sets

$$\left\{ X_p \wedge X'_q \xrightarrow{f_q \wedge f'_q} Y_p \wedge Y'_q \xrightarrow{\iota_{p,q}} (Y \wedge Y')_{p+q} \right\}_{p,q \geq 0}$$

form a bilinear morphism $(X, X') \longrightarrow Y \wedge Y'$, so it corresponds to a unique morphism of symmetric spectra $f \wedge f' : X \wedge X' \longrightarrow Y \wedge Y'$. The universal property implies functoriality in both arguments. For the proof of the associativity of the smash product we notice that the family

$$\left\{ X_p \wedge Y_q \wedge Z_r \xrightarrow{\iota_{p,q} \wedge \mathrm{Id}} (X \wedge Y)_{p+q} \wedge Z_r \xrightarrow{\iota_{p+q,r}} ((X \wedge Y) \wedge Z)_{p+q+r} \right\}_{p,q,r \geq 0}$$

and the family

$$\left\{ X_p \wedge Y_q \wedge Z_r \xrightarrow{\mathrm{Id} \wedge \iota_{q,r}} X_p \wedge (Y \wedge Z)_{q+r} \xrightarrow{\iota_{p,q+r}} (X \wedge (Y \wedge Z))_{p+q+r} \right\}_{p,q,r \geq 0}$$

both have the universal property of a *tri-linear* morphism out of X, Y and Z. The uniqueness of universal objects gives a preferred isomorphism of symmetric spectra

$$(X \wedge Y) \wedge Z \cong X \wedge (Y \wedge Z) .$$

The symmetry isomorphism $X \wedge Y \cong Y \wedge X$ corresponds to the bilinear morphism

$$\left\{ X_p \wedge Y_q \xrightarrow{\mathrm{twist}} Y_q \wedge X_p \xrightarrow{\iota_{q,p}} (Y \wedge X)_{q+p} \xrightarrow{\chi_{q,p}} (Y \wedge X)_{p+q} \right\}_{p,q \geq 0} . \quad (4.9)$$

The block permutation $\chi_{q,p}$ is crucial here: without it we would not get a bilinear morphism is the sense of diagram (4.7). In much the same spirit, the universal properties can be used to provide unit isomorphisms $\mathbb{S} \wedge X \cong X \cong X \wedge \mathbb{S}$, to verify the coherence conditions of a symmetric monoidal structure, and to establish an isomorphism of suspension spectra

$$(\Sigma^\infty K) \wedge (\Sigma^\infty L) \cong \Sigma^\infty (K \wedge L) .$$

The symmetric monoidal structure given by the smash product of symmetric spectra is *closed* in the sense that internal function objects exist as well. For each pair of symmetric spectra X and Y there is a symmetric function spectrum $\mathrm{Hom}(X, Y)$ [HSS, 2.2.9], and there are natural composition morphisms

$$\circ \; : \; \mathrm{Hom}(Y, Z) \wedge \mathrm{Hom}(X, Y) \longrightarrow \mathrm{Hom}(X, Z)$$

which are associative and unital with respect to a unit map $\mathbb{S} \longrightarrow \mathrm{Hom}(X, X)$. Moreover, the usual adjunction isomorphism

$$Sp^\Sigma(X \wedge Y, Z) \cong Sp^\Sigma(X, \mathrm{Hom}(Y, Z))$$

relates the smash product and function spectra.

Ring and module spectra. The smash product of symmetric spectra leads to the concomitant concepts *symmetric ring spectra, module spectra* and *algebra spectra.*

Definition 4.10. A *symmetric ring spectrum* is a symmetric spectrum R together with morphisms of symmetric spectra

$$\eta : \mathbb{S} \longrightarrow R \quad \text{and} \quad \mu : R \wedge R \longrightarrow R,$$

called the unit and multiplication map, which satisfy certain associativity and unit conditions (see [McL, VII.3]). A ring spectrum R is *commutative* if the multiplication map is unchanged when composed with the twist, or the symmetry isomorphism (4.9), of $R \wedge R$. A morphism of ring spectra is a morphism of spectra commuting with the multiplication and unit maps. If R is a symmetric ring spectrum, a *right R-module* is a spectrum N together with an action map $N \wedge R \longrightarrow N$ satisfying associativity and unit conditions (see again [McL, VII.4]). A morphism of right R-modules is a morphism of spectra commuting with the action of R. We denote the category of right R-modules by Mod-R.

With the universal property of the smash product we can make the structure of a symmetric ring spectrum more explicit. The multiplication map $\mu : R \wedge R \longrightarrow R$ corresponds to a family of pointed, $\Sigma_p \times \Sigma_q$-equivariant maps

$$\mu_{p,q} : R_p \wedge R_q \longrightarrow R_{p+q}$$

for $p, q \geq 0$, which are bilinear in the sense of diagram (4.7). The maps are supposed to be associative and unital with respect to the maps $\eta_p : S^p \longrightarrow R_p$ which constitute the unit map $\eta : \mathbb{S} \longrightarrow R$.

The commutativity isomorphism of the smash product involves the block permutation $\chi_{q,p}$, see (4.9). So the multiplication of a symmetric ring spectrum is commutative if and only if the following diagrams commute for all $p, q \geq 0$

$$
\begin{array}{ccc}
R_p \wedge R_q & \xrightarrow{\ \mu_{p,q}\ } & R_{p+q} \\
{\scriptstyle\text{twist}}\downarrow & & \downarrow{\scriptstyle\chi_{p,q}} \\
R_q \wedge R_p & \xrightarrow{\ \mu_{q,p}\ } & R_{q+p}
\end{array}
$$

The block permutation $\chi_{p,q}$ has sign $(-1)^{pq}$, so this diagram is reminiscent of the sign rule in a graded ring which is commutative in the graded sense.

The unit $\mathbb{S}$ of the smash product is a ring spectrum in a unique way, and $\mathbb{S}$-modules are the same as symmetric spectra. The smash product of two ring spectra is naturally a ring spectrum. For a ring spectrum R the opposite ring spectrum R^{op} is defined by composing the multiplication with the twist map $R \wedge R \longrightarrow R \wedge R$ (so in terms of the bilinear maps $\mu_{p,q} : R_p \wedge R_q \longrightarrow R_{p+q}$, a block permutation appears). The definitions of left modules and bimodules is hopefully clear; left R-modules and R-T-bimodule can also be defined as right modules over the opposite ring spectrum R^{op}, respectively right modules over the ring spectrum $R^{\mathrm{op}} \wedge T$.

A formal consequence of having a closed symmetric monoidal smash product is that the category of R-modules inherits a smash product and function objects. The smash product $M \wedge_R N$ of a right R-module M and a left R-module N can be defined as the coequalizer, in the category of symmetric spectra, of the two maps

$$M \wedge R \wedge N \Longrightarrow M \wedge N$$

given by the action of R on M and N respectively. Alternatively, one can characterize $M \wedge_R N$ as the universal example of a symmetric spectrum which receives a bilinear map from M and N which is R-*balanced*, i.e., all the diagrams

$$
\begin{array}{ccc}
M_p \wedge R_q \wedge N_r & \xrightarrow{\mathrm{Id} \wedge \alpha_{q,r}} & M_p \wedge N_{q+r} \\
{\scriptstyle \alpha_{p,q} \wedge \mathrm{Id}} \downarrow & & \downarrow {\scriptstyle \iota_{p,q+r}} \\
M_{p+q} \wedge N_r & \xrightarrow{\iota_{p+q,r}} & (M \wedge N)_{p+q+r}
\end{array}
\tag{4.11}
$$

commute. If M happens to be a T-R-bimodule and N an R-S-bimodule, then $M \wedge_R N$ is naturally a T-S-bimodule. If R is a commutative ring spectrum, the notions of left and right module coincide and agree with the notion of a symmetric bimodule. In this case $\wedge_R$ is an internal symmetric monoidal smash product for R-modules. There are also internal function spectra and function modules, enjoying the 'usual' adjointness properties with respect to the various smash products.

The modules over a symmetric ring spectrum R inherit a model category structure from symmetric spectra, see [HSS, Cor. 5.4.2] and [SS00, Thm. 4.1 (1)]. More precisely, a morphism of R-modules is called a weak equivalence (resp. fibration) if the underlying morphism of symmetric spectra is a stable equivalence (resp. stable fibration). The cofibrations are then determined by the left lifting property with respect to all acyclic fibrations in Mod-R. This model structure is stable, so the homotopy category of modules over a ring spectrum is a triangulated category. The free module of rank one is a small generator.

For a map $R \longrightarrow S$ of ring spectra, there is a Quillen adjoint functor pair analogous to restriction and extension of scalars: any S-module becomes an R-module if we let R act through the map. This functor has a left adjoint taking an R-module M to the S-module $M \wedge_R S$. If $R \longrightarrow S$ is a stable equivalence, then the functors of restriction and extension of scalars are a Quillen equivalence between the categories of R-modules and S-modules, see [HSS, Thm. 5.4.5] and [SS00, Thm. 4.3].

Example 4.12 (Monoid ring spectra). If M is a simplicial monoid, and R is a symmetric ring spectrum, we define a symmetric spectrum $R[M]$ by

$$R[M]_n = R_n \wedge M^+ \, ,$$

where M^+ denotes the underlying simplicial set of M, with disjoint basepoint added. The unit map is the composite of the unit map of R and the wedge summand inclusion indexed by the unit of M; the multiplication map $R[M] \wedge R[M] \longrightarrow R[M]$ is induced from the bilinear morphism

$$(R_p \wedge M^+) \wedge (R_q \wedge M^+) \cong (R_p \wedge R_q) \wedge (M \times M)^+ \xrightarrow{\mu_{p,q} \wedge \text{mult.}} R_{p+q} \wedge M^+ .$$

The construction of the monoid ring over $\mathbb{S}$ is left adjoint to the functor which takes a symmetric ring spectrum R to the simplicial monoid R_0.

Example 4.13 (Matrix ring spectra). Let R be a symmetric ring spectrum and consider the wedge (coproduct)

$$R \times n = \underbrace{R \vee \cdots \vee R}_{n}$$

of n copies of the free R-module of rank 1. In the usual stable model structure, the free module of rank 1 is cofibrant, hence so is $R \times n$. We choose a fibrant replacement $R \times n \xrightarrow{\sim} (R \times n)^{\mathrm{f}}$. The ring spectrum of $n \times n$ *matrices* over R is defined as the endomorphism ring spectrum of $(R \times n)^{\mathrm{f}}$,

$$M_n(R) = \text{End}_R((R \times n)^{\mathrm{f}}) .$$

The stable equivalence type of the matrix ring spectrum $M_n(R)$ is independent of the choice of fibrant replacement, see Corollary A.2.4 of [SS03]. Moreover, the underlying spectrum of $M_n(R)$ is isomorphic, in the stable homotopy category, to a sum of n^2 copies of R.

Example 4.14 (Eilenberg-Mac Lane spectra). For an abelian group A, the *Eilenberg-Mac Lane spectrum HA* is defined by

$$(HA)_n = A \otimes \mathbb{Z}[S^n] ,$$

i.e., the underlying simplicial set of the dimensionwise tensor product of A with the reduced free simplicial abelian generated by the simplicial n-sphere. The symmetric groups acts by permuting the smash factors of S^n. The homotopy groups of the symmetric spectrum HA are concentrated in dimension zero, where we have a natural isomorphism $\pi_0 HA \cong A$.

For two abelian groups A and B, a natural morphism of symmetric spectra

$$HA \wedge HB \longrightarrow H(A \otimes B)$$

is obtained, by the universal property (4.8), from the bilinear morphism

$$(HA)_n \wedge (HB)_m = (A \otimes \mathbb{Z}[S^n]) \wedge (B \otimes \mathbb{Z}[S^m])$$
$$\longrightarrow (A \otimes B) \otimes \mathbb{Z}[S^{n+m}] = (H(A \otimes B))_{n+m}$$

given by

$$\left(\sum_i a_i \cdot x_i \right) \wedge \left(\sum_j b_j \cdot x'_j \right) \longmapsto \sum_{i,j} (a_i \otimes b_j) \cdot (x_i \wedge x'_j) .$$

A unit map $\mathbb{S} \longrightarrow H\mathbb{Z}$ is given by the inclusion of generators. With respect to these maps, H becomes a lax symmetric monoidal functor from the category of abelian groups to the category of symmetric spectra. As a formal consequence, H turns a ring R into a symmetric ring spectrum with multiplication map

$$HR \wedge HR \longrightarrow H(R \otimes R) \longrightarrow HR .$$

Similarly, an R-module structure on A gives rise to an HR-module structure on HA.

The definition of the symmetric spectrum HA makes just as much sense when A is a *simplicial* abelian group; thus the Eilenberg-Mac Lane functor makes simplicial rings into symmetric ring spectra, respecting possible commutativity of the multiplications. With a little bit of extra care, the Eilenberg-Mac Lane construction can also be extended to a differential graded context, compare [SS03, App. B] and [Ri03].

For a fixed ring B, the modules over the Eilenberg-Mac Lane ring spectrum HB of a ring B have the same homotopy theory as complexes of B-modules. The first results of this kind were obtained by Robinson for A_∞-ring spectra [Rob87b], and later for S-algebras in [EKMM, IV Thm. 2.4]; in both cases, equivalences of triangulated homotopy categories are constructed. But more is true: for any ring B, Theorem 5.1.6 of [SS03] provides a chain of two Quillen equivalences between the categories of unbounded chain complexes of B-modules and the HB-module spectra.

Example 4.15 (Cobordism spectra). We define a commutative symmetric ring spectrum MO whose stable homotopy groups are isomorphic to the ring of cobordism classes of smooth closed manifolds. We set

$$(MO)_n = EO(n)^+ \wedge_{O(n)} S^n .$$

Here $O(n)$ is the n-th orthogonal group consisting of Euclidean automorphisms of $\mathbb{R}^n$. The space $EO(n)$ is the geometric realization of the simplicial object of topological groups which in dimension k is the k-fold product of copies of $O(n)$, and where are face maps are projections. Thus $EO(n)$ is a topological group with a homomorphism $O(n) \longrightarrow EO(n)$ coming from the inclusion of 0-simplices. The underlying space of $EO(n)$ is contractible and has two commuting actions of $O(n)$ from the left and the right. The right $O(n)$-action is used to form the orbit space $(MO)_n$, where we think of S^n as the one-point compactification of $\mathbb{R}^n$ with its natural left $O(n)$-action. Thus the space $(MO)_n$ still has a left $O(n)$-action, which we restrict to an action of the symmetric group Σ_n, sitting inside $O(n)$ as the coordinate permutations. Topologically, $(MO)_n$ is nothing but the Thom space of the tautological bundle over the classifying space $BO(n)$.

The unit of the ring spectrum MO is given by the maps

$$S^n \cong O(n)^+ \wedge_{O(n)} S^n \longrightarrow EO(n)^+ \wedge_{O(n)} S^n = (MO)_n$$

using the 'vertex map' $O(n) \longrightarrow EO(n)$. There are multiplication maps

$$(MO)_p \wedge (MO)_q \longrightarrow (MO)_{p+q}$$

which are induced from the identification $S^p \wedge S^q \cong S^{p+q}$ which is equivariant with respect to the group $O(p) \times O(q)$, viewed as a subgroup of $O(p+q)$. The fact that these multiplication maps are associative and commutative uses that

- for topological groups G and H, the simplicial model of EG comes with a natural, associative and commutative isomorphism $E(G \times H) \cong EG \times EH$;
- the group monomorphisms $O(p) \times O(q) \longrightarrow O(p+q)$ are strictly associative, and the following diagram commutes

$$
\begin{array}{ccc}
O(p) \times O(q) & \longrightarrow & O(p+q) \\
\text{twist} \downarrow & & \downarrow \text{conj. by } \chi_{p,q} \\
O(q) \times O(p) & \longrightarrow & O(q+p)
\end{array}
$$

where the right vertical map is conjugation by the permutation matrix of the block permutation $\chi_{p,q}$.

In very much the same way we obtain commutative symmetric ring spectra model for the oriented cobordism spectrum MSO and the spin cobordism spectrum $MSpin$. The complex cobordism ring spectrum MU does not fit in here directly; one has to vary the notion of a symmetric spectrum slightly, and consider only symmetric spectra which are defined 'in even dimensions'.

4.3. Characterizing module categories over ring spectra.

Several of the examples of stable model categories in Section 4.1 already come as categories of modules over suitable rings or ring spectra. This is no coincidence. In fact, every stable model category with a single small generator has the same homotopy theory as the modules over a ring spectrum. This is an analog of Theorem 2.5, which characterizes module categories over a ring as the cocomplete abelian category with a small projective generator.

To an object P in a sufficiently nice stable model category $\mathcal{C}$ we can associate a symmetric *endomorphism ring spectrum* $\text{End}_{\mathcal{C}}(P)$; among other things, this ring spectrum comes with an isomorphism of graded rings

$$\pi_* \text{End}_{\mathcal{C}}(P) \cong \text{Ho}(\mathcal{C})(P, P)_*$$

between the homotopy groups of $\text{End}_{\mathcal{C}}(P)$ and the morphism of P in the triangulated homotopy category of $\mathcal{C}$. The precise characterization is as follows:

Theorem 4.16. *Let $\mathcal{C}$ be a stable model category which is simplicial, cofibrantly generated and proper. If $\mathcal{C}$ has a small generator P, then there exists a chain of Quillen equivalences between $\mathcal{C}$ and the model category of $\text{End}_{\mathcal{C}}(P)$-modules,*

$$\mathcal{C} \simeq_Q \text{Mod-End}_{\mathcal{C}}(P) .$$

Unfortunately, the theorem is currently only known under the above technical hypothesis: the stable model category in question should be simplicial (see [Qui67, II.2], [Hov99, 4.2.18]), *cofibrantly generated* (see [Hov99, Sec. 2.1] or [SS00, Sec. 2]) and proper (see [BF78, Def. 1.2] or [HSS, Def. 5.5.2]). The conditions enter in the construction of 'good' endomorphism ring spectra. I suspect however, that these hypothesis are not essential and can be eliminated with a clever use of framing techniques. For example, in [SS02, Sec. 6], we use framings to construct function spectra in a arbitrary stable model category; that construction does however not yield *symmetric* spectra, and there is no good composition pairing. The condition of being a *simplicial* model category can be removed by appealing to [RSS] or [Dug] where suitable model categories are replaced by Quillen equivalent simplicial model categories.

This theorem is a special case of the more general result which applies to stable model categories with a set of small generators (as opposed to a single small generator), see [SS03, Thm. 3.3.3].

Spectral model categories. In the algebraic situations which we considered in Sections 2 and 3, the key point is to have a good notion of endomorphism ring or endomorphism DG ring together with a 'tautological' functor

$$\text{Hom}_{\mathcal{A}}(P, -) \; : \; \mathcal{A} \; \longrightarrow \; \text{Mod-End}_{\mathcal{A}}(P) \; . \qquad (4.17)$$

Then it is a matter of checking that when P is a small generator, the functor $\text{Hom}(P, -)$ is either an equivalence of categories (in the context of abelian categories) or induces an equivalence of derived categories (in the context of DG categories). For abelian categories the situation is straightforward, and the ordinary endomorphism ring does the job. In the differential graded context already a little complication comes in because the categorical hom functor $\text{Hom}_{A}(P, -)$ need not preserve quasi-isomorphisms in general.

For stable model categories, the key construction is again to have an *endomorphism ring spectrum* $\text{End}_{\mathcal{C}}(P)$ together with a homotopically well-behaved homomorphism functor (4.17) to modules over the endomorphism ring spectrum. This is easy for the following class of *spectral model categories* where composable function spectra are part of the data. A spectral model category is analogous to a simplicial model category [Qui67, II.2], but with the category of simplicial sets replaced by symmetric spectra. Roughly speaking, a spectral model category is a pointed model category which is compatibly enriched over the stable model category of spectra. In particular there are 'tensors' $K \wedge X$ and 'cotensors' X^{K} of an object X of $\mathcal{C}$ and a symmetric spectrum K, and function symmetric spectra $\text{Hom}_{\mathcal{C}}(A, Y)$ between two objects of $\mathcal{C}$. The compatibility is expressed by the following axiom which takes the place of [Qui67, II.2 SM7]; there are two equivalent 'adjoint' forms of this axiom, compare [Hov99, Lemma 4.2.2] or [SS03, 3.5].

(Pushout product axiom) For every cofibration $A \longrightarrow B$ in $\mathcal{C}$ and every cofibration $K \longrightarrow L$ of symmetric spectra, the *pushout product map*

$$L \wedge A \cup_{K \wedge A} K \wedge B \longrightarrow L \wedge B$$

is a cofibration; the pushout product map is a weak equivalence if in addition $A \longrightarrow B$ is a weak equivalence in $\mathcal{C}$ or $K \longrightarrow L$ is a stable equivalence of symmetric spectra.

A *spectral Quillen pair* is a Quillen adjoint functor pair $L : \mathcal{C} \longrightarrow \mathcal{D}$ and $R : \mathcal{D} \longrightarrow \mathcal{C}$ between spectral model categories together with a natural isomorphism of symmetric homomorphism spectra

$$\mathrm{Hom}_{\mathcal{C}}(A, RX) \cong \mathrm{Hom}_{\mathcal{D}}(LA, X)$$

which on the vertices of the 0-th level reduces to the adjunction isomorphism. A spectral Quillen pair is a *spectral Quillen equivalence* if the underlying Quillen functor pair is an ordinary Quillen equivalence.

A spectral model category is the same as a 'Sp^{Σ}-model category' in the sense of [Hov99, Def. 4.2.18]; Hovey's condition 2 of [Hov99, 4.2.18] is automatic since the unit $\mathbb{S}$ for the smash product of symmetric spectra is cofibrant. Similarly, a spectral Quillen pair is a 'Sp^{Σ}-Quillen functor' in Hovey's terminology. Examples of spectral model categories are module categories over a ring spectrum, and the category of symmetric spectra over a suitable simplicial model category [SS03, Thm. 3.8.2].

A spectral model category is in particular a *simplicial* and *stable* model category. Moreover, for X a cofibrant and Y a fibrant object of a spectral model category $\mathcal{C}$ there is a natural isomorphism of graded abelian groups $\pi_* \mathrm{Hom}_{\mathcal{C}}(X, Y) \cong \mathrm{Ho}(\mathcal{C})(X, Y)_*$. These facts are discussed in Lemma 3.5.2 of [SS03].

For an object P in a spectral model category, the function spectrum $\mathrm{End}_{\mathcal{C}}(P) = \mathrm{Hom}_{\mathcal{C}}(P, P)$ is naturally a ring spectrum; the multiplication is a special case of the composition product

$$\circ : \mathrm{Hom}_{\mathcal{C}}(Y, Z) \wedge \mathrm{Hom}_{\mathcal{C}}(X, Y) \longrightarrow \mathrm{Hom}_{\mathcal{C}}(X, Z) .$$

Via the composition pairing, the function symmetric spectrum $\mathrm{Hom}_{\mathcal{C}}(P, X)$ becomes a right module over the symmetric ring spectrum $\mathrm{End}_{\mathcal{C}}(P)$ for any object X. In order for the endomorphism ring spectrum $\mathrm{End}_{\mathcal{C}}(P)$ to have the correct homotopy type, the object P should be both cofibrant and fibrant. In that case, the ring of homotopy groups $\pi_* \mathrm{End}_{\mathcal{C}}(P)$ is isomorphic to $\mathrm{Ho}(\mathcal{C})(P, P)_*$, the ring of graded self maps of P in the homotopy category of $\mathcal{C}$. Moreover, the homotopy type of the endomorphism ring spectrum then depends only on the weak equivalence type of the object (see [SS03, Cor. A.2.4]). Note that this is not completely obvious since taking endomorphisms is *not* a functor.

If P is a cofibrant object of a spectral model category, then the functor

$$\mathrm{Hom}_{\mathcal{C}}(P, -) : \mathcal{C} \longrightarrow \mathsf{Mod}\text{-}\mathrm{End}_{\mathcal{C}}(P)$$

is the right adjoint of a Quillen adjoint functor pair, see [SS03, 3.9.3 (i)]. The left adjoint is denoted

$$- \wedge_{\mathrm{End}_{\mathcal{C}}(P)} P \; : \; \mathsf{Mod\text{-}End}_{\mathcal{C}}(P) \; \longrightarrow \; \mathcal{C} \; . \tag{4.18}$$

For spectral model categories, the proof of Theorem 4.16 is now straightforward, and very analogous to the proofs of Theorem 2.5 and Theorem 3.20; indeed, to obtain the following theorem, one applies Proposition 3.10 to the total left derived functor of the left Quillen functor (4.18).

Theorem 4.19. *Let $\mathcal{C}$ be a spectral model category and P a cofibrant-fibrant object. If P is a small generator for $\mathcal{C}$, then the adjoint pair $\mathrm{Hom}_{\mathcal{C}}(P, -)$ and $- \wedge_{\mathrm{End}_{\mathcal{C}}(P)} P$ forms a spectral Quillen equivalence.*

The remaining step is worked out in Theorem 3.8.2 of [SS03], which proves that every simplicial, cofibrantly generated, proper stable model category is Quillen equivalent to a spectral model category, namely the category $\mathcal{S}p(\mathcal{C})$ of *symmetric spectra over $\mathcal{C}$*. The proof is technical and we will not go into details here. Theorem 4.16 follows by combining [SS03, Theorem 3.8.2] and Theorem 4.19 to obtain a diagram of model categories and Quillen equivalences (the left adjoints are on top)

$$\mathcal{C} \; \underset{\mathrm{Ev}_0}{\overset{\Sigma^{\infty}}{\rightleftarrows}} \; \mathcal{S}p(\mathcal{C}) \; \underset{\mathrm{Hom}_{\mathcal{C}}(P,-)}{\overset{-\wedge_{\mathrm{End}_{\mathcal{C}}(P)}P}{\rightleftarrows}} \; \mathsf{Mod\text{-}End}_{\mathcal{C}}(P) \; .$$

4.4. Morita context for ring spectra. Now we come to 'Morita theory for ring spectra', by which we mean the question when two symmetric spectra have Quillen equivalent module categories. For ring spectra, there is a significant difference between a Quillen equivalence of the module categories and an equivalence of the homotopy categories. The former implies the latter, but not conversely. The same kind of difference already exists for differential graded rings, but it is not visible for ordinary rings (see Example 4.5 (5)).

We call a symmetric spectrum X *flat* if the functor $X \wedge -$ preserves stable equivalences of symmetric spectra. If X is cofibrant, or more generally S-cofibrant in the sense of [HSS, 5.3.6], then X is flat, see [HSS, 5.3.10]. Every symmetric ring spectrum has a 'flat resolution': we may take a cofibrant approximation in the stable model structure of symmetric ring spectra [HSS, 5.4.3]; the underlying symmetric spectrum of the approximation is cofibrant, thus flat.

Theorem 4.20. (Morita context) *The following are equivalent for two symmetric ring spectra R and S.*

(1) There exists a chain of spectral Quillen equivalences between the categories of R-modules and S-modules.

(2) There is a small, cofibrant and fibrant generator of the model category of S-modules whose endomorphism ring spectrum is stably equivalent to R.

Both conditions are implied by the following condition.

(3) *There exists an R-S-bimodule M such that the derived smash product functor*

$$- \wedge_R^L M \; : \; \mathrm{Ho}(\mathsf{Mod}\text{-}R) \longrightarrow \mathrm{Ho}(\mathsf{Mod}\text{-}S)$$

is an equivalence of categories.

If moreover R or S is flat as a symmetric spectrum, then all three conditions are equivalent.

Again there is a version of the Morita context 4.20 relative to a commutative symmetric ring spectrum k. In that case, R and S are k-algebras, condition (1) refers to k-linear spectral Quillen equivalences, condition (2) requires a stable equivalence of k-algebras, the bimodule M in (3) has to be k-symmetric and in the addendum, one of R or S has to be flat as a k-module.

Proof of Theorem 4.20. (2)$\Longrightarrow$ (1): Modules over a symmetric ring spectrum form a spectral model category; so this implication is a special case of Theorem 4.19, combined with the fact that stably equivalent ring spectra have Quillen equivalent module categories.

(1) $\Longrightarrow$ (2): To simplify things we suppose that there exists a single spectral Quillen equivalence

$$\mathsf{Mod}\text{-}R \; \underset{\Phi}{\overset{\Lambda}{\rightleftarrows}} \; \mathsf{Mod}\text{-}S$$

with Λ the left adjoint. The general case of a chain of such Quillen equivalences is treated in [SS03, Thm. 4.1.2]. We choose a trivial cofibration $\iota : \Lambda(R) \longrightarrow \Lambda(R)^{\mathrm{f}}$ of S-modules such that $M := \Lambda(R)^{\mathrm{f}}$ is fibrant; since M is isomorphic in the homotopy category of S-modules to the image of the free R-module of rank one under the equivalence of homotopy categories, M is a small generator for the homotopy category of S-modules. It remains to show that the endomorphism ring spectrum of M is stably equivalent to R.

We define $\mathrm{End}_S(\iota)$, the endomorphism ring spectrum of the S-module map $\iota : \Lambda(R) \longrightarrow M$, as the pullback in the diagram of symmetric spectra

$$\begin{array}{ccc}
\mathrm{End}_S(\iota) & \longrightarrow & \mathrm{End}_S(M) \\
\downarrow & & \downarrow{\scriptstyle \iota^*} \\
R & \underset{\iota_*}{\longrightarrow} & \mathrm{Hom}_S(\Lambda(R), M)
\end{array} \qquad (4.21)$$

The right vertical map ι^* is obtained by applying $\mathrm{Hom}_S(-, M)$ to the acyclic cofibration ι; since M is stably fibrant, ι^* is acyclic fibration of symmetric spectra. Since Λ and Φ form a Quillen equivalence, the adjoint $\hat{\iota} : R \longrightarrow \Phi(M)$ of ι is a stable equivalence of R-modules. The lower horizontal map ι_*

is the composite stable equivalence

$$R \cong \mathrm{Hom}_R(R, R) \xrightarrow{\mathrm{Hom}_R(R, \hat{\iota})} \mathrm{Hom}_R(R, \Phi(M)) \cong \mathrm{Hom}_S(\Lambda(R), M) \ .$$

All maps in the pullback square (4.21) are thus stable equivalences and the morphism connecting $\mathrm{End}_S(\iota)$ to R and $\mathrm{End}_S(M)$ are homomorphisms of symmetric ring spectra (whereas the lower right corner $\mathrm{Hom}_S(\Lambda(R), M)$ has no multiplication). So R is indeed stably equivalent, as a symmetric ring spectrum, to the endomorphisms of M.

(3)$\Longrightarrow$(1): If M happens to be cofibrant as a right S-module, then smashing with M over R is a left Quillen equivalence from R-modules to S-modules. Since we did not assume that M is cofibrant over S, we have to be content with a chain of two Quillen equivalences, which we get as follows.

Let M be an R-S-bimodule as in condition (3). We choose a cofibrant approximation $\iota : R^c \longrightarrow R$ in the stable model structure of symmetric ring spectra and we view M as an R^c-S-bimodule by restriction of scalars. Then we choose a cofibrant approximation $M^c \longrightarrow M$ as R^c-S-bimodules. Since the underlying symmetric spectrum of R^c is cofibrant [SS00, 4.1 (3)], $R^c \wedge S$ is cofibrant as a right S-module, and thus every cofibrant R^c-S-bimodule is cofibrant as a right S-module. In particular, this holds for M^c, and so we have a chain of two spectral Quillen pairs

$$\text{Mod-}R \underset{\iota^*}{\overset{-\wedge_{R^c} R}{\rightleftarrows}} \text{Mod-}R^c \underset{\mathrm{Hom}_S(M^c, -)}{\overset{-\wedge_{R^c} M^c}{\rightleftarrows}} \text{Mod-}S \ .$$

The left pair is a Quillen equivalence since the approximation map $\iota : R^c \longrightarrow R$ is a stable equivalence. For every cofibrant R^c-module X, the map

$$X \wedge_{R^c} M^c \longrightarrow X \wedge_{R^c} M \cong (X \wedge_{R^c} R) \wedge_R M$$

is a stable equivalence. This means that the diagram of homotopy categories and derived functors

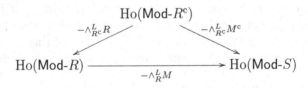

commutes up to natural isomorphism. Thus the right Quillen pair above induces an equivalence of homotopy categories, so it is a Quillen equivalence.

(2)$\Longrightarrow$(3), assuming that R or S is flat: Let T be a cofibrant and fibrant small generator of $\mathrm{Ho}(\text{Mod-}S)$ such that R is stably equivalent to the endomorphism ring spectrum of T. We choose a cofibrant approximation $R^c \xrightarrow{\simeq} R$ in the model category of symmetric ring spectra. Since T is cofibrant and fibrant, its endomorphism ring spectrum is fibrant. So any isomorphism between R and $\mathrm{End}_S(T)$ in the homotopy category of symmetric ring spectra can be represented by a chain of two stable equivalences

$$R \xleftarrow{\simeq} R^c \xrightarrow{\simeq} \mathrm{End}_S(T) \ .$$

The module T is naturally an $\mathrm{End}_S(T)$-S-bimodule, and we restrict the left action to R^c and view T as an R^c-S-bimodule. We choose a cofibrant replacement $T^c \xrightarrow{\sim} T$ as an R^c-S-bimodule. Then we set

$$M \ = \ R \wedge_{R^c} T^c \ ,$$

an R-S-bimodule. We have no reason to suppose that M is cofibrant as a right S-module, so we cannot assume that the functor $- \wedge_R M : \mathsf{Mod}\text{-}R \longrightarrow \mathsf{Mod}\text{-}S$ is a left Quillen functor. Nevertheless, smashing with M over R takes stable equivalences between cofibrant R-modules to stable equivalences, so it has a total left derived functor

$$- \wedge_R^L M \ : \ \mathrm{Ho}(\mathsf{Mod}\text{-}R) \ \longrightarrow \ \mathrm{Ho}(\mathsf{Mod}\text{-}S) \ ;$$

we claim that this functor is an equivalence.

Since R^c is cofibrant as a symmetric ring spectrum, it is also cofibrant as a symmetric spectrum [SS00, 4.1 (3)], so $R^c \wedge S^{op}$ models the derived smash product of R and S. If one of R or S is flat, then $R \wedge S^{op}$ also models the derived smash product, so that the map

$$R^c \wedge S^{op} \ \longrightarrow \ R \wedge S^{op}$$

is a stable equivalence of symmetric ring spectra. Since T^c is cofibrant as an $R^c \wedge S^{op}$-module, the induced map

$$T^c = (R^c \wedge S^{op})_{R^c \wedge S^{op}} T^c \to (R \wedge S^{op})_{R^c \wedge S^{op}} T^c \cong R \wedge_{R^c} T^c = M \qquad (4.22)$$

is a stable equivalence. We smash the stable equivalence (4.22) from the left with an R^c-module X to get a natural map of S-modules

$$X \wedge_{R^c} T^c \ \to \ X \wedge_{R^c} M \cong (X \wedge_{R^c} R) \wedge_R M \ . \qquad (4.23)$$

If X is cofibrant as an R^c-module, then $X \wedge_{R^c} -$ takes stable equivalences of left R^c-modules to stable equivalences, so in this case, the map (4.23) is a stable equivalence. Thus the diagram of triangulated categories and derived functors

$$\mathrm{Ho}(\mathsf{Mod}\text{-}R^c) \qquad (4.24)$$

with maps $-\wedge_{R^c}^L R$ and $-\wedge_{R^c}^L T^c$ to $\mathrm{Ho}(\mathsf{Mod}\text{-}R)$ and $\mathrm{Ho}(\mathsf{Mod}\text{-}S)$, and bottom map $-\wedge_R^L M$

commutes up to natural isomorphism.

The left diagonal functor in the diagram (4.24) is derived from extensions of scalars along stable equivalences of ring spectra; such extension of scalars is a left Quillen equivalences, so the derived functor $- \wedge_{R^c}^L R$ is an exact equivalence of triangulated categories. We argued in the previous implication that any cofibrant $R^c \wedge S^{op}$-module such as T^c has an underlying cofibrant right S-module. So smashing with T^c over R^c is a left Quillen functor. Since T^c is isomorphic to T in the homotopy category of S-modules, T^c is a small generator of Ho(Mod-S). So the right diagonal derived functor in (4.24) is an exact equivalence by Proposition 3.10, applied to the free R^c-module of rank 1. So we conclude that the lower horizontal functor in the diagram (4.24) is also an exact equivalence of triangulated categories. This establishes condition (3). $\qquad\square$

4.5. Examples. (1) **Matrix ring spectra.** As for classical rings (compare Example 2.3), matrix ring spectra give rise to the simplest kind of Morita equivalence. Indeed over any a ring spectrum R, the 'free module of rank n', i.e., the wedge of n copies of R, is a small generator for the homotopy category of R-modules. The endomorphism ring spectrum of a (stably fibrant replacement) of $R \times n$ is the $n \times n$ matrix ring spectrum as we defined it in Example 4.13. So R and

$$M_n(R) = \operatorname{End}_R((R \times n)^f)$$

are Morita equivalent as ring spectra.

(2) **Upper triangular matrices.** In Example 3.25 we saw that the upper triangular 3×3 matrices over a field are derived equivalent, but not Morita equivalent, to its sub-algebra of matrices of the form

$$\left\{ \begin{pmatrix} x_{11} & x_{12} & x_{13} \\ 0 & x_{22} & 0 \\ 0 & 0 & x_{33} \end{pmatrix} \mid x_{ij} \in k \right\} .$$

We cannot directly define the algebra of 3×3 matrices over a ring spectrum; the problem is that the usual basis of elementary matrices is closed under multiplication, but the unit matrix is a *sum* basis elements, not a single basis element.

The algebras of Example 3.25 can also be obtained as the path algebras of the two quivers

$$Q = \left\{ 1 \xrightarrow{\ \alpha\ } 2 \xrightarrow{\ \beta\ } 3 \right\} \qquad \text{respectively} \quad \left\{ 1 \longleftarrow 2 \longrightarrow 3 \right\} .$$

With this in mind we can now run Example 3.25 with the ground field replaced by a commutative symmetric ring spectrum R.

We construct the 'path algebra' indirectly via 'representations of the quiver Q over R'. A representation of Q over R is a collection

$$M = \left\{ M_1 \xrightarrow{\ \alpha\ } M_2 \xrightarrow{\ \beta\ } M_3 \right\}$$

of three R-modules and two homomorphisms. A morphism $f : M \longrightarrow N$ of representations consists of R-homomorphisms $f_i : M_i \longrightarrow N_i$ for $i = 1, 2, 3$ satisfying $\alpha f_1 = f_2 \alpha$ and $\beta f_2 = f_3 \beta$. There is a stable model structure on the category of such representations in which a morphism $f : M \longrightarrow N$ is a stable equivalence or fibration if and only if each R-homomorphism $f_i : M_i \longrightarrow N_i$ is a stable equivalence or fibration for all $i = 1, 2, 3$. Moreover, f is a cofibration if and only if the morphisms

$$f_1 : M_1 \longrightarrow N_1 \, , \quad f_2 \cup \alpha : M_2 \cup_{M_1} N_1 \longrightarrow N_2 \quad \text{and} \quad f_3 \cup \beta : M_3 \cup_{M_2} N_2 \longrightarrow N_3$$

are cofibrations of R-modules.

We consider the 'free' or 'projective' representations P^i given by $P^1 = \{R \overset{=}{\longrightarrow} R \overset{=}{\longrightarrow} R\}$, $P^2 = \{* \to R \overset{=}{\longrightarrow} R\}$ respectively $P^3 = \{* \to * \to R\}$. Then P^i represents the evaluation functor, i.e., we have a natural isomorphism

$$\operatorname{Hom}_{Q\text{-rep}}(P^i, M) \; \cong \; M_i \, .$$

This implies that the wedge of these three representations is a small generator of the homotopy category of Q-representations. The three projective representations are cofibrant; we let $M_3^\triangle(R)$ denote the endomorphism ring spectrum of a stably fibrant approximation of their wedge,

$$M_3^\triangle(R) \;=\; \operatorname{End}_{Q\text{-rep}}((P^1 \vee P^2 \vee P^3)^{\mathrm{f}}) \, .$$

This symmetric ring spectrum deserves to be called the 'upper triangular 3×3 matrices' over R.

Now we find a different small generator for the model category of Q-representations which is the analog of the tilting module in Example 3.25. We note that the inclusion $P^3 \longrightarrow P^2$ is not a cofibration, and the quotient $P^2/P^3 = \{* \longrightarrow R \longrightarrow *\}$ is not cofibrant. A cofibrant approximation of this quotient is given by the representation $S^2 = \{* \longrightarrow R \longrightarrow CR\}$ in which the second map is the inclusion of R into its cone. We form the representation

$$T \;=\; P^1 \vee P^2 \vee S^2 \, .$$

Then S^2 represents the functor which sends a Q-representation M to the homotopy fiber of the morphism $\beta : M_2 \longrightarrow M_3$, and thus

$$\operatorname{Hom}_{Q\text{-rep}}(T, M) \;\cong\; M_1 \times M_2 \times \operatorname{hofibre}(\beta : M_2 \longrightarrow M_3) \, .$$

This implies that T is also a small generator for the homotopy category of Q-representation over R.

We conclude that the upper triangular matrix algebra $M_3^\triangle(R)$ is Quillen equivalent to the endomorphism ring spectrum of a stably fibrant approximation of the representation T. Just as the endomorphisms of the generator $P^1 \vee P^2 \vee P^3$ should be thought of as upper triangular matrices, the endomorphisms of the generator $P^1 \vee P^2 \vee S^2$ are analogous to a certain algebra of 3×3 matrices over R, namely the ones of the form

$$\begin{pmatrix} R & R & R \\ * & R & * \\ * & * & R \end{pmatrix} \, .$$

Another example is obtained as follows. We consider the representation $S^1 = \{R \longrightarrow CR \overset{=}{\longrightarrow} CR\}$ which is a cofibrant replacement of the representation

P^1/P^2 and which represents the homotopy fiber of the morphism $\alpha : M_1 \longrightarrow M_2$. Then

$$T' = S^1 \vee S^2 \vee P^3$$

is another small generator for the homotopy category of Q-representation over R. So $M_3^{\triangle}(R)$ is also Quillen equivalent to the derived endomorphism ring spectrum of T', which is an algebra of 3×3 matrices of the form

$$\begin{pmatrix} R & \Omega R & * \\ * & R & \Omega R \\ * & * & R \end{pmatrix} ,$$

where each symbol '$*$' designates an entry in a stably contractible spectrum.

(3) **Uniqueness results for stable homotopy theory.** Theorem 4.16 characterizes module categories over ring spectra among stable model categories. This also yields a characterization of the model category of spectra: a stable model category is Quillen equivalent to the category of symmetric spectra if and only if it has a small generator P for which the unit map of ring spectra $\mathbb{S} \longrightarrow \operatorname{End}(P)$ is a stable equivalence. The technical conditions of being simplicial, cofibrantly generated and proper can be eliminated as in [SS02] with the use of *framings* [Hov99, Chpt. 5]. The paper [SS02] also gives other necessary and sufficient conditions for when a stable model category is Quillen equivalent to spectra – some of them in terms of the homotopy category and the natural action of the stable homotopy groups of spheres. In [Sch01b], this result is extended to a uniqueness theorem showing that the 2-local stable homotopy category has only one underlying model category up to Quillen equivalence; the odd-primary version is work in progress. In another direction, the uniqueness result is extended to include the monoidal structure in [Sh2].

(4) **Chain complexes and Eilenberg-Mac Lane spectra.** For a ring R, the category of chain complexes of R-modules (under quasi-isomorphisms) is Quillen equivalent to the category of modules over the Eilenberg-Mac Lane ring spectrum HR. An explicit chain of two Quillen equivalences can be found in Theorem B.1.11 of [SS03]; I don't know if it is possible to compare the two categories by a single Quillen equivalence.

This result can also be viewed as an instance of Theorem 4.16: the free R-module of rank one, considered as a complex concentrated in dimension zero, is a small generator for the unbounded derived category of R. Since the homotopy groups of its endomorphism ring spectrum (as an object of the model category of chain complexes) are concentrated in dimension zero, the endomorphism ring spectrum is stably equivalent to the Eilenberg-Mac Lane ring spectrum for R (Proposition B.2.1 of [SS03]).

(5) **A generalized tilting theorem.** We interpret and generalize the tilting theory from the perspective of stable model categories. A *tilting object* in a stable model category $\mathcal{C}$ as a small generator T such that the graded homomorphism group $\operatorname{Ho}(\mathcal{C})(T, T)_*$ in the homotopy category is concentrated

in dimension zero. The following 'generalized tilting theorem' of [SS03, Thm. 5.1.1] then shows that the existence of a tilting object is necessary and sufficient for a stable model category to be Quillen equivalent or derived equivalent to the category of unbounded chain complexes over a ring.

Generalized tilting theorem. *Let $\mathcal{C}$ be a stable model category which is simplicial, cofibrantly generated and proper, and let R be a ring. Then the following conditions are equivalent:*

(1) There is a chain of Quillen equivalences between $\mathcal{C}$ and the model category of chain complexes of R-modules.

(2) The homotopy category of $\mathcal{C}$ is triangulated equivalent to the derived category $\mathcal{D}(R)$.

(3) The model category $\mathcal{C}$ has a tilting object whose endomorphism ring in $\mathrm{Ho}(\mathcal{C})$ is isomorphic to R.

In the derived category of a ring, a tilting object is the same as a tilting complex, and the result reduces to Rickard's tilting theorem 3.13.

The generalized tilting situation enjoys one very special feature. In general, the notion of Quillen equivalence is considerably stronger than triangulated equivalence of homotopy categories; two examples are given in [Sch01b, 2.1 and 2.2]. Hence it is somewhat remarkable that for complexes of modules over rings, the two notions are in fact equivalent. In general the homotopy category determines the homotopy groups of the endomorphism ring spectrum, but not its homotopy type. The real reason behind the equivalences of conditions (1) and (2) above is the fact that in contrast to arbitrary ring spectra, Eilenberg-Mac Lane spectra are determined up to stable equivalence by their homotopy groups. We explain this in more detail in Section 5 of [SS03].

(6) **Frobenius rings.** As in Example 4.2 (2) we consider a Frobenius ring and assume that the stable module category has a small generator. Then we are in the situation of Theorem 4.16; however this example is completely algebraic, and there is no need to consider ring spectra to identify the stable module category as the derived category of a suitable 'ring'. In fact Keller shows [Kel94a, 4.3] that in such a situation there exists a differential graded algebra and an equivalence between the stable module category and the unbounded derived category of the differential graded algebra.

(7) **Smashing Bousfield localizations.** Let E be a spectrum and consider the E-local model category structure on some model category of spectra (see e.g. [EKMM, VIII 1.1]). This is another stable model category in which the localization of the sphere spectrum $L_E\mathbb{S}$ is a generator. This localized sphere is small if the localization is *smashing*, i.e., if a certain natural map $X \wedge L_E\mathbb{S} \longrightarrow L_E X$ is a stable equivalence for all X. So for a smashing localization the E-local model category of spectra is Quillen equivalent to modules over the ring spectrum $L_E\mathbb{S}$ (which is the endomorphism ring spectrum of the localized sphere in the localized model structure).

(8) **Finite localization.** Suppose P is a small object of a triangulated category $\mathcal{T}$ with infinite coproducts. Then there always exists an idempotent localization functor L_P on $\mathcal{T}$ whose acyclics are precisely the objects of the localizing subcategory generated by P (compare [Mil92] or the proofs of [SS03, Lemma 2.2.1] or [HPS97, Prop. 2.3.17]). These localizations are often referred to as *finite Bousfield localizations* away from P.

This type of localization has a refinement to the model category level. Suppose $\mathcal{C}$ is a stable model category and P a small object, and let L_P denote the associated localization functor on the homotopy category of $\mathcal{C}$. By 4.19, or rather the refined version [SS03, Thm. 3.9.3 (ii)], the acyclics for L_P are equivalent to the homotopy category of $\mathrm{End}_{\mathcal{C}}(P)$-modules, the equivalence arising from a Quillen adjoint functor pair. Furthermore the counit of the derived adjunction

$$\mathrm{Hom}_{\mathcal{C}}(P, X) \wedge^L_{\mathrm{End}_{\mathcal{C}}(P)} P \longrightarrow X$$

is the acyclicization map and its cofiber is a model for the localization $L_P X$.

(9) $K(n)$-**local spectra.** Even if a Bousfield localization is not smashing, Theorem 4.16 might be applicable. As an example we consider Bousfield localization with respect to the n-th Morava K-theory $K(n)$ at a fixed prime. The localization of the sphere is still a generator, but for $n > 0$ it is not small in the local category, see [HPS97, 3.5.2]. However the localization of any finite type n spectrum F is a small generator for the $K(n)$-local category [HS99, 7.3]. Hence the $K(n)$-local model category is Quillen equivalent to modules over the endomorphism ring spectrum of $L_{K(n)}F$.

I would like to conclude with a few words about **invariants** of ring spectra which are preserved under Quillen equivalence (but *not* in general under equivalences of triangulated homotopy categories). Such invariants include the algebraic K-theory [DuSh], topological Hochschild homology and topological cyclic homology.

In the classical framework, the center of a ring is invariant under Morita and derived equivalence. As a general philosophy for spectral algebra, definitions which use elements are not well-suited for generalization to ring spectra. So how do we define the 'center' of a ring spectrum, such that it only depends, up to stable equivalence, on the Quillen equivalence class of the module category? The center of an ordinary ring R is isomorphic to the endomorphism ring of R, considered as a bimodule over itself, via

$$\mathrm{End}_{R \otimes R^{\mathrm{op}}}(R) \longrightarrow Z(R), \quad f \longmapsto f(1).$$

So we define the *center* of a ring spectrum R as the endomorphism ring spectrum of a cofibrant-fibrant replacement of R, considered as a bimodule over itself,

$$Z(R) = \mathrm{End}_{R \wedge R^{\mathrm{op}}}(R^{\mathrm{cf}}, R^{\mathrm{cf}}). \tag{4.25}$$

In this definition, R should be flat as a symmetric spectrum, in order for the smash product $R \wedge R^{\mathrm{op}}$ to have the 'correct homotopy type'.

This kind of center of a ring spectrum is homotopy commutative; slightly more is true: the center (4.25) is often called the *topological Hochschild cohomology spectrum* of the ring spectrum R, and its multiplication extends to an action of an operad weakly equivalent to the operad of little discs, see [McCS, Thm. 3]. However, the above center is usually *not* stably equivalent to a *commutative* symmetric ring spectrum (or what is the same, E_∞-homotopy commutative); the Gerstenhaber operations on the homotopy of $Z(R)$ are obstruction to higher order commutativity. So is it just a coincidence that the classical center of an ordinary ring is commutative ? Or is there some 'higher', E_∞-commutative, center of a ring spectrum, yet to be discovered ?

REFERENCES

[AF92]   F. W. Anderson and K. R. Fuller, *Rings and categories of modules (second edition)*, Graduate Texts in Mathematics, **13**, Springer-Verlag, New York, 1992, viii+376 pp.

[AGH]   A. V. Arhangel'skii, K. R. Goodearl, and B. Huisgen-Zimmermann, *Kiiti Morita (1915–1995)*, Notices Amer. Math. Soc. 44 (1997), no. 6, 680–684.

[Ba]   H. Bass, *Algebraic K-theory*. W. A. Benjamin, Inc., New York-Amsterdam 1968 xx+762 pp.

[BL]   A. Baker, A. Lazarev, *Topological Hochschild cohomology and generalized Morita equivalence*, Preprint (2002). http://arXiv.org/abs/math.AT/0209003

[BBD]   A. A. Beilinson, J. Bernstein, P. Deligne, *Faisceaux pervers*. Analysis and topology on singular spaces, I (Luminy, 1981), 5–171, Astérisque **100**, Soc. Math. France, Paris, 1982.

[Bei78]   A. A. Beilinson, *Coherent sheaves on P^n and problems in linear algebra*, Functional Anal. Appl. 12 (1978), no. 3, 214–216 (1979).

[Bek00]   T. Beke, *Sheafifiable homotopy model categories*. Math. Proc. Cambridge Philos. Soc. **129** (2000), 447–475.

[Bou75]   A. K. Bousfield, *The localization of spaces with respect to homology*, Topology **14** (1975), 133–150.

[BF78]   A. K. Bousfield and E. M. Friedlander, *Homotopy theory of Γ-spaces, spectra, and bisimplicial sets*, Geometric applications of homotopy theory (Proc. Conf., Evanston, Ill., 1977), II (M. G. Barratt and M. E. Mahowald, eds.), Lecture Notes in Math., **658**, Springer, Berlin, 1978, pp. 80–130.

[Bro94]   M. Broué, *Equivalences of blocks of group algebras*. Finite-dimensional algebras and related topics (Ottawa, ON, 1992), 1–26, NATO Adv. Sci. Inst. Ser. C Math. Phys. Sci., 424, Kluwer Acad. Publ., Dordrecht, 1994.

[CH]   D. Christensen, M. Hovey, *Quillen model structures for relative homological algebra*, Math. Proc. Cambridge Philos. Soc. **133** (2002), 261–293.

[SGA 4½]   P. Deligne, *Cohomologie étale*. Séminaire de Géométrie Algébrique du Bois-Marie SGA 4½. Avec la collaboration de J. F. Boutot, A. Grothendieck, L. Illusie et J. L. Verdier. Lecture Notes in Mathematics, Vol. 569. Springer-Verlag, Berlin-New York, 1977. iv+312pp.

[Dug]   D. Dugger, *Replacing model categories with simplicial ones*, Trans. Amer. Math. Soc. **353** (2001), 5003–5027.

[DuSh]   D. Dugger and B. Shipley, *K-theory and derived equivalences*, Preprint (2002). http://arXiv.org/abs/math.KT/0209084

[DØR]   B. Dundas, P. A. Østvær and O. Röndigs, *Enriched functors and motivic stable homotopy theory*. Preprint (2002). http://www.math.ntnu.no/~dundas/

[DwSp95] W. G. Dwyer and J. Spalinski, *Homotopy theories and model categories*, Handbook of algebraic topology (Amsterdam), North-Holland, Amsterdam, 1995, pp. 73–126.

[DG02] W. G. Dwyer, J. Greenlees, *Complete modules and torsion modules*, Amer. J. Math. 124 (2002), 199–220.

[DGI] W. G. Dwyer, J. Greenlees, S. Iyengar. *Duality in algebra and topology*, Preprint (2002). http://www.shef.ac.uk/~pm1jg/

[EKMM] A. D. Elmendorf, I. Kriz, M. A. Mandell, and J. P. May, *Rings, modules, and algebras in stable homotopy theory. With an appendix by M. Cole*, Mathematical Surveys and Monographs, **47**, American Mathematical Society, Providence, RI, 1997, xii+249 pp.

[GM] S. I. Gelfand, Y. I. Manin, *Methods of homological algebra.* Translated from the 1988 Russian original. Springer-Verlag, Berlin, 1996. xviii+372 pp.

[Gr77] A. Grothendieck, *Groupes des classes des catégories abéliennes et triangulées, complexes parfaits*. Exposé VIII, Séminaire de Géométrie Algébrique du Bois-Marie SGA 5, Lecture Notes in Math., **589**, Springer, Berlin, 1977, pp. 351–371.

[Hir03] P. S. Hirschhorn, *Model Categories and Their Localizations*, Mathematical Surveys and Monographs **99**, American Mathematical Society, 2003, 457 pp.

[Hov99] M. Hovey, *Model categories*, Mathematical Surveys and Monographs, **63**, American Mathematical Society, Providence, RI, 1999, xii+209 pp.

[Hov01a] M. Hovey, *Model category structures on chain complexes of sheaves*, Trans. Amer. Math. Soc. **353** (2001), 2441–2457.

[Hov01b] M. Hovey, *Spectra and symmetric spectra in general model categories*, J. Pure Appl. Algebra **165** (2001), 63–127.

[HPS97] M. Hovey, J. H. Palmieri, and N. P. Strickland, *Axiomatic stable homotopy theory*, Mem. Amer. Math. Soc. **128** (1997), no. 610.

[HSS] M. Hovey, B. Shipley, and J. Smith, *Symmetric spectra*, J. Amer. Math. Soc. **13** (2000), 149–208.

[HS99] M. Hovey and N. P. Strickland, *Morava K-theories and localisation*, Mem. Amer. Math. Soc. **139** (1999)

[Hu03] P. Hu, *S-modules in the category of schemes.* Mem. Amer. Math. Soc. **161** (2003), no. 767, viii+125 pp.

[Jar87] J. F. Jardine, *Stable homotopy theory of simplicial presheaves*, Canad. J. Math. **39** (1987), 733–747.

[Jar97] J. F. Jardine, *Generalized étale cohomology theories*, Progress in Mathematics **146**, Birkhäuser Verlag, Basel, 1997, x+317 pp.

[Jar00a] J. F. Jardine, *Presheaves of symmetric spectra*, J. Pure Appl. Algebra **150** (2000), 137–154.

[Jar00b] J. F. Jardine, *Motivic symmetric spectra*, Doc. Math. **5** (2000), 445–553.

[Kel94a] B. Keller, *Deriving DG categories*, Ann. Sci. École Norm. Sup. (4) **27** (1994), 63–102.

[Kel94b] B. Keller, *A remark on the generalized smashing conjecture*, Manuscripta Math. **84** (1994), 193–198.

[Kel96] B. Keller, *Invariance of cyclic homology under derived equivalence.* Representation theory of algebras (Cocoyoc, 1994), 353–361, CMS Conf. Proc., 18, Amer. Math. Soc., Providence, RI, 1996.

[Kel98] B. Keller, *Invariance and localization for cyclic homology of DG algebras.* J. Pure Appl. Algebra **123** (1998), 223–273.

[Kel99] B. Keller, *On the cyclic homology of exact categories*, J. Pure Appl. Algebra **136** (1999), 1–56.

[Kel03]   B. Keller, *Hochschild cohomology and derived Picard groups*, Preprint (2003).
           http://www.math.jussieu.fr/~keller/

[KZ]   S. König, A. Zimmermann, *Derived equivalences for group rings*. With contribu-
           tions by B. Keller, M. Linckelmann, J. Rickard and R. Rouquier. Lecture Notes
           in Mathematics, 1685. Springer-Verlag, Berlin, 1998. x+246 pp.

[KM]   I. Kriz and J. P. May, *Operads, algebras, modules and motives*. Astérisque No.
           233, (1995), iv+145pp.

[Lam]   T. Y. Lam, *Lectures on modules and rings*. Graduate Texts in Mathematics, 189.
           Springer-Verlag, New York, 1999. xxiv+557 pp.

[LMS86]   L. G. Lewis, Jr., J. P. May, and M. Steinberger, *Equivariant stable homotopy
           theory*, Lecture Notes in Mathematics, **1213**, Springer-Verlag, 1986.

[Lyd98]   M. Lydakis, *Simplicial functors and stable homotopy theory*, Preprint (1998).
           http://hopf.math.purdue.edu/

[McCS]   J. McClure, J. Smith, *A solution of Deligne's Hochschild cohomology conjecture*,
           Recent progress in homotopy theory (Baltimore, MD, 2000), 153–193, Contemp.
           Math. **293**, Amer. Math. Soc., Providence, RI, 2002.

[McL]   S. Mac Lane, *Categories for the working mathematician*. Graduate Texts in
           Mathematics, Vol. 5. Springer-Verlag, New York-Berlin, 1971. ix+262 pp.

[MM02]   M. A. Mandell and J. P. May, *Equivariant orthogonal spectra and S-modules*,
           Mem. Amer. Math. Soc. **159** (2002), x+108 pp.

[MMSS]   M. A. Mandell, J. P. May, S. Schwede and B. Shipley, *Model categories of diagram
           spectra*, Proc. London Math. Soc. **82** (2001), 441-512.

[Mil92]   H. Miller, *Finite localizations*, Bol. Soc. Mat. Mexicana (2) **37** (1992), (Papers
           in honor of José Adem), 383–389.

[MV]   F. Morel and V. Voevodsky, $\mathbb{A}^1$-*homotopy theory of schemes*, Inst. Hautes Études
           Sci. Publ. Math. **90** (2001), 45–143.

[Mo58]   K. Morita, *Duality for modules and its applications to the theory of rings with
           minimum condition*. Sci. Rep. Tokyo Kyoiku Daigaku Sect. A 6 (1958), 83–142.

[Nee92]   A. Neeman, *The connection between the K-theory localization theorem of Thoma-
           son, Trobaugh and Yao and the smashing subcategories of Bousfield and Ravenel*.
           Ann. Sci. École Norm. Sup. (4) **25** (1992), 547–566.

[Nee01]   A. Neeman, *Triangulated categories*, Annals of Mathematics Studies, 148. Prince-
           ton University Press, Princeton, NJ, 2001. viii+449 pp.

[Qui67]   D. G. Quillen, *Homotopical algebra*, Lecture Notes in Mathematics, **43**, Springer-
           Verlag, 1967.

[Ren69]   R. Rentschler, *Sur les modules M tels que* Hom(M, −) *commute avec les sommes
           directes*, C. R. Acad. Sci. Paris Sér. A-B **268** (1969), A930–A933.

[RSS]   C. Rezk, S. Schwede and B. Shipley, *Simplicial structures on model categories
           and functors*, Amer. J. Math. **123** (2001), 551–575.

[Ri03]   B. Richter, *Symmetry properties of the Dold-Kan correspondence*, Math. Proc.
           Cambridge Philos. Soc. **134** (2003), 95–102.

[Ric89a]   J. Rickard, *Morita theory for derived categories*, J. London Math. Soc. (2) **39**
           (1989), 436–456.

[Ric89b]   J. Rickard, *Derived categories and stable equivalence*. J. Pure Appl. Algebra **61**
           (1989), 303–317.

[Ric91]   J. Rickard, *Derived equivalences as derived functors*, J. London Math. Soc. (2)
           43 (1991), 37–48.

[Rob87a]   A. Robinson, *Spectral sheaves: a model category for stable homotopy theory*, J.
           Pure Appl. Algebra **45** (1987), 171–200.

[Rob87b]   A. Robinson, *The extraordinary derived category*, Math. Z. **196** (1987), no.2,
           231–238.

[Sch97]   S. Schwede, *Spectra in model categories and applications to the algebraic cotangent complex*, J. Pure Appl. Algebra **120** (1997), 77–104.

[Sch01a]   S. Schwede, *S-modules and symmetric spectra*, Math. Ann. **319** (2001), 517–532.

[Sch01b]   S. Schwede, *The stable homotopy category has a unique model at the prime 2*, Adv. Math. **164** (2001), 24-40.

[SS00]   S. Schwede and B. Shipley, *Algebras and modules in monoidal model categories*, Proc. London Math. Soc. **80** (2000), 491-511.

[SS02]   S. Schwede and B. Shipley, *A uniqueness theorem for stable homotopy theory*, Math. Z. **239** (2002), 803-828.

[SS03]   S. Schwede and B. Shipley, *Stable model categories are categories of modules*, Topology **42** (2003), 103-153.

[Sh2]   B. Shipley, *Monoidal uniqueness of stable homotopy theory*, Adv. in Math. **160** (2001), 217-240.

[Spa88]   N. Spaltenstein, *Resolutions of unbounded complexes*, Compositio Math. **65** (1988), 121–154.

[TT]   R. W. Thomason, T. Trobaugh, *Higher algebraic K-theory of schemes and of derived categories*, The Grothendieck Festschrift, Vol. III, 247–435, Progr. Math., 88, Birkhäuser Boston, Boston, MA, 1990.

[Ver96]   J.-L. Verdier, *Des catégories dérivées des catégories abéliennes*, Astérisque **239** (1997). With a preface by Luc Illusie, Edited and with a note by Georges Maltsiniotis, xii+253 pp.

[Voe]   V. Voevodsky, *The Milnor Conjecture*, Preprint (1996)

[Voe98]   V. Voevodsky, $\mathbb{A}^1$-*homotopy theory*, Doc. Math. ICM **I**, 1998, 417-442.

[Wei94]   C. A. Weibel, *An introduction to homological algebra*, Cambridge Studies in Advanced Mathematics **38**. Cambridge University Press, Cambridge, 1994. xiv+450 pp.

SFB 478 GEOMETRISCHE STRUKTUREN IN DER MATHEMATIK, WESTFÄLISCHE WILHELMS-UNIVERSITÄT MÜNSTER, GERMANY

E-mail address: `sschwede@math.uni-muenster.de`

HIGHER COHERENCES FOR EQUIVARIANT K-THEORY

MICHAEL JOACHIM

ABSTRACT. Let G be a compact Lie group. We show that concepts of operator theory can be used to define an E_∞-ring spectrum representing G-equivariant K-theory. In addition we construct an E_∞-model for the G-equivariant Atiyah-Bott-Shapiro orientation $MSpin^c \to K$.

1. INTRODUCTION

About twenty-five years ago May, Quinn and Ray introduced the concept of E_∞-ring spectra [16, Chapter IV]. The definition was motiviated by the fact that there is no way to construct an internal smash product on the category of ordinary spectra and functions between them in such a way that the smash product would equip this category of spectra with a symmetric monoidal product. Of course there is the well-defined smash product on the homotopy category of spectra. An E_∞-structure on a commutative "homotopy ring spectrum" R or on a module M over it essentially guarantees that the homotopy multiplications $R \wedge R \to R$ and $R \wedge M \to M$ satisfy "all relevant algebraic relations". For example, E_∞-structures allow to define the smash product of two E_∞-module spectra over an E_∞-ring spectrum which then again is an E_∞-module spectrum over the E_∞-ring spectrum which then again will be an E_∞-module spectrum over the E_∞-ring spectrum.

Recently however several people suceeded in defining a symmetric monoidal smash product on certain categories of spectra. Of course, for doing so one needs to put some extra structure on the spectra. In particular one should mention here the category of S-modules invented by Elmendorf, Kriz, Mandell and May, and the category of symmetric spectra invented by Smith. The two approaches both have their specific advantages and disadvantages. The definition of a commutative symmetric ring spectrum actually is quite easy to give. However, if one is interested in the associated model category structure on the category of symmetric spectra or the category of module spectra over a commutative symmetric ring spectrum then the corresponding theory becomes quite complicated. Moreover, a priori it is not clear how one should extend the definition of symmetric structure to the G-equivariant setting for an arbitrary compact Lie group G. There is an obvious generalization of the notion of a symmetric spectrum: one can replace the family of symmetric groups in the definition of a symmetric spectrum by the corresponding family of orthogonal groups. This resolves the two major disadvantages stated above: most of the difficulties one has with the model category structure on the category of symmetric spectra go away, and there is an obvious way for introducing an equivariant setting [14].

Let G be a compact Lie group. A nice way to define an equivariant orthogonal spectrum is provided by the so-called $\mathscr{I}$-FSPs, introduced in [14]. The letter $\mathscr{I}$ here stands for a certain symmetric monoidal category built out of finite dimensional real G-inner product spaces. The category $\mathscr{I}$ is a symmetric monoidal G-category as is the category $\mathscr{T}_*$ consisting of the well-pointed compactly generated weak Hausdorff G-spaces and all (not necessarily G-equivariant) maps between them.[1] Taking one point compactifications of finite dimensional real G-inner product spaces provides a strong symmetric monoidal functor $S : \mathscr{I} \to \mathscr{T}_*$, and an $\mathscr{I}$-FSP by definition is a pair $(\mathbb{X}, \eta)$ consisting of a lax symmetric monoidal functor $\mathbb{X} : \mathscr{I} \to \mathscr{T}_*$ and a monoidal natural transformation $\eta : S \to \mathbb{X}$ (see Section 2). An $\mathscr{I}$-FSP can be thought of as an G-equivariant FSP in the sense of Bökstedt which is just defined on the image $S(\mathscr{I}) \subset \mathscr{T}_*$. The concept of an $\mathscr{I}$-FSP is quite closely related to the old concept of E_∞-structure: given an $\mathscr{I}$-FSP one obtains a corresponding E_∞-ring spectrum, essentially given by a restriction of the underlying functor that defines the $\mathscr{I}$-FSP (see 2.2 and c.f. [12, VII, Proposition 2.4]).

In this paper we construct two $\mathscr{I}$-FSPs, $\mathbb{K}$ and $\mathbb{MSpin}^c$, as well as a map of $\mathscr{I}$-FSPs $\alpha : \mathbb{MSpin}^c \to \mathbb{K}$. The first $\mathscr{I}$-FSP $\mathbb{K}$ represents equivariant periodic (complex) K-theory as defined in [19], the second one represents equivariant $Spin^c$-cobordism and the map $\alpha : \mathbb{MSpin}^c \to \mathbb{K}$ represents an equivariant version of the Atiyah-Bott-Shapiro orientation $MSpin^c \to K$. In particular this implies (using the old E_∞-terminology)

Theorem

(A) *For any compact Lie group equivariant periodic complex topological K-theory can be represented by an equivariant E_∞-ring spectrum.*

(B) *For any compact Lie group an equivariant version of the Atiyah-Bott-Shapiro orientation $MSpin^c \to K$ can be realized as a map of equivariant E_∞-ring spectra.*

Until recently it was not even known if the non-equivariant periodic K-theory spectrum can be represented by an E_∞-ring spectrum. It has been known for a long time though that non-equivariant connective K-theory can be represented by an E_∞-ring spectrum [16, VIII, Theorem 2.1], and that the periodic K-theory spectrum K can be obtained from the connective K-theory spectrum k by localizing with respect to the Bott class in $\pi_2(k)$. However

---

[1]Here we adopted the notation used in [14]; it is important in the set-up of the theory to carefully distinguish between the category $\mathscr{T}_*$ and the subcategory $\mathscr{T}_*^G$ consisting if all G-equivariant maps.

one did not know that one can perform localizations within the category of E_∞-ring spectra. The latter now can be shown using the theory of S-modules, so we can obtain an E_∞-model for K this way. A similar approach can be used to show that equivariant periodic K-theory can be represented through an equivariant E_∞-spectrum in case G is a finite group (see [20]), but the methods do not generalize to the case where G is an arbitrary compact Lie group. For $G = 1$ an explicit geometric construction of an E_∞-model for the periodic K-theory spectrum also can be obtained from [9], where we constructed a symmetric ring spectrum spectrum representing periodic real K-theory. Where the second statement of the theorem is concerned, here we are not even aware of any previously known explicit E_∞-model for the Atiyah-Bott-Shapiro orientation, even in the case where $G = 1$.

The construction of the $\mathscr{I}$-FSP $\mathbb{K}$ is in fact related to the definition of the symmetric ring spectrum KO representing non-equivariant KO-theory given in [9]. There we have used Clifford linear Fredholm operators to build the spaces $KO(n)$ for the symmetric ring spectrum KO. One technical difficulty in this approach was to introduce appropriate base points. In this paper we show that more recent concepts of operator theory can be used in a straight-forward manner to construct an $\mathscr{I}$-FSP $\mathbb{K}$ which represents (periodic) equi-variant K-theory. The appropriate concept is the so-called homomorphism picture for equivariant KK-theory. In the homomorphism picture ([21],[17]) the K-homology group of a G-C^*-algebra is identified with the set of path components of a space of $*$-homomorphism between certain C^*-algebras. It follows that all spaces which are involved in our definition of $\mathbb{K}$ will be spaces of $*$-homomorphisms, hence they all naturally come equipped with a base point, given by the trivial $*$-homomorphism which sends all elements to zero. Another advantage of this approach is that it is quite clear how one can modify the definition of the $\mathscr{I}$-FSP $\mathbb{K}$ in order to obtain E_∞-ring spectra $\mathbb{K}A$ representing the equivariant K-homology groups $K_*^G(A)$ for commutative G-C^*-algebras A.

In this paper we are working with equivariant periodic complex topological K-theory. However all constructions and proofs go through equally well for equivariant periodic real topological K-theory, with the obvious modifications which correspond to the different period of Bott periodicity in the real set-ting. It follows that our constructions also can be used to show that equivari-ant periodic real topological K-theory can be represented by an equivariant E_∞-spectrum, and that an equivariant version of the Atiyah-Bott-Shapiro orientation $MSpin \to KO$ can be modelled by a map of equivariant E_∞-ring spectra.

The paper is organized as follows. In Section 2 we recall the definition of an $\mathscr{I}$-FSP and we show how an $\mathscr{I}$-FSP $\mathbb{X}$ determines a multiplicative equi-variant cohomology theory $\mathbb{X}^*$ on the category $\mathscr{T}_*^G$, which is the subcategory of $\mathcal{T}_*$ consisting of the equivariant maps. In the following two sections we construct the $\mathcal{I}$-FSP $\mathbb{K}$ which represents equivariant K-theory. By the latter we mean that there is a natural isomorphism $\Phi^* : \mathbb{K}^* \to \widetilde{K}_G^*$ of (reduced) multiplicative equivariant cohomology theories on $\mathcal{K}_*^G$, the full subcategory of $\mathcal{T}_*^G$ generated by the compact spaces in $\mathcal{T}_*^G$. In Section 5 we construct the corresponding natural isomorphism Φ^*, using well-known concepts of equivariant KK-theory. The construction of the $\mathcal{I}$-FSP $\mathbb{M}\mathrm{Spin}^c$ as well as the map $\alpha : \mathbb{M}\mathrm{Spin}^c \to \mathbb{K}$ are discussed in Section 6.

CONVENTION: Throughout the paper we will work in the category of compactly generated weak Hausdorff spaces. If a certain construction (like taking mapping spaces) does not produce a compactly generated space at the first place we think of it as being hit by the Kelley functor k without notice.

2. SYMMETRIC MONOIDAL CATEGORIES AND *FSP*'S

A *symmetric monoidal topological G-category* is a topological G-category $\mathcal{C}$ together with a continuous G-functor $\oplus_\mathcal{C} : \mathcal{C} \times \mathcal{C} \to \mathcal{C}$ and a unit object $u_\mathcal{C}$ such that the functor $\oplus_\mathcal{C}$ is associative, commutative and unital up to coherent natural isomorphism.[2] Given two symmetric monoidal topological G-categorgies $\mathcal{C}$ and $\mathcal{D}$, a *lax symmetric monoidal G-functor* $\mathcal{C} \to \mathcal{D}$ consists of a continuous G-functor $\mathbb{X} : \mathcal{C} \to \mathcal{D}$ together with a natural transformation of functors from $\mathcal{C} \times \mathcal{C}$ to $\mathcal{D}$ consisting of equivariant maps

$$\mu = \mu_{a_1,a_2} : \mathbb{X}(a_1) \oplus_\mathcal{D} \mathbb{X}(a_2) \to \mathbb{X}(a_1 \oplus_\mathcal{C} a_2), \quad a_1, a_2 \in \mathcal{C}$$

and an equivariant map $\lambda : u_\mathcal{D} \to \mathbb{X}(u_\mathcal{C})$ such that all coherence diagrams relating associativity, commutativity and unit isomorphisms of $\mathcal{C}$ and $\mathcal{D}$ are commutative. A lax symmetric monoidal G-functor $\mathbb{X} = (\mathbb{X}, \mu, \lambda)$ is called a *strong symmetric monoidal G-functor* if μ and λ are isomorphisms. A natural transformation of lax symmetric monoidal G-functors $\alpha : \mathbb{X} \to \mathbb{Y}$ is called *monoidal* if the following diagrams commute

$$
\begin{array}{ccc}
\mathbb{X}(a_1) \oplus_\mathcal{D} \mathbb{X}(a_2) & \xrightarrow{\alpha \oplus_\mathcal{D} \alpha} & \mathbb{Y}(a_1) \oplus_\mathcal{D} \mathbb{Y}(a_2) \\
\mu_\mathbb{X} \downarrow & & \downarrow \mu_\mathbb{Y} \\
\mathbb{X}(a_1 \oplus_\mathcal{C} a_2) & \xrightarrow{\alpha} & \mathbb{Y}(a_1 \oplus_\mathcal{C} a_2)
\end{array}
\qquad
\begin{array}{ccc}
 & u_\mathcal{D} & \\
\lambda_\mathbb{X} \swarrow & & \searrow \lambda_\mathbb{Y} \\
\mathbb{X}(u_\mathcal{C}) & \xrightarrow{\alpha} & \mathbb{Y}(u_\mathcal{C}).
\end{array}
$$

Note that if $\mathbb{X}_1, \mathbb{X}_2 : \mathcal{C} \to \mathcal{D}$ are lax symmetric monoidal G-functors, then we can define the product of the two functors $\mathbb{X}_1 \oplus_\mathcal{D} \mathbb{X}_2 : \mathcal{C} \to \mathcal{D}$, given by $(\mathbb{X}_1 \oplus_\mathcal{D} \mathbb{X}_2)(a) = \mathbb{X}_1(a) \oplus_\mathcal{D} \mathbb{X}_2(a)$, which naturally is a lax symmetric monoidal G-functor.

Example 2.1. Let $\mathcal{C} = (\mathcal{C}, \oplus_\mathcal{C}, u_\mathcal{C})$ and $\mathcal{D} = (\mathcal{D}, \oplus_\mathcal{D}, u_\mathcal{D})$ be symmetric monoidal topological G-categories. Then there always is the *trivial* strong

---

[2]See [13, VII §1, §7] for a precise definition of the latter.

symmetric monoidal G-functor called $u_{\mathscr{D}} : \mathscr{C} \to \mathscr{D}$, denoted as the unit of $\mathscr{D}$. It associates to any object $a \in \mathscr{C}$ the unit $u_{\mathscr{D}}$ and to any morphism $a \to b$ in $\mathscr{C}$ the identity on $u_{\mathscr{D}}$. The corresponding natural transformation $\mathscr{C} \times \mathscr{C} \to \mathscr{D}$ for any pair of objects $(a_1, a_2) \in \mathscr{C}^2$ is given through the canonical map $u_{\mathscr{D}} \oplus_{\mathscr{D}} u_{\mathscr{D}} \to u_{\mathscr{D}}$, and the corresponding morphism λ is the identity on $u_{\mathscr{D}}$. Note also, given any lax symmetric monoidal G-functor $\mathbb{X} : \mathscr{C} \to \mathscr{D}$ then there is a canonical monoidal natural isomorphism of lax symmetric monoidal G-functors $\mathbb{X} \to \mathbb{X} \oplus_{\mathscr{D}} u_{\mathscr{D}}$. This terminology will be used in the following sections.

Recall that a *G-universe U* is a sum of countably many copies of each real G-inner product space in some set of irreducible representations of G that includes the trivial representation $\mathbb{R}$. U is called *complete* if it contains all irreducible representations of G up to isomorphism. A finite dimensional G-invariant linear subspace of U is called an *indexing space*. The subspace relation gives the set of indexing spaces the structure of a directed set, which we will denote $\mathscr{D} = \mathscr{D}(U)$ in the sequel. Let $\mathscr{I} = \mathscr{I}(U)$ be the topological G-category whose objects are all real G-inner product spaces that are equivariantly isomorphic to indexing spaces in U, and whose morphisms are the linear isometric isomorphisms equipped with the G-action given by conjugation. $\mathscr{I}$ becomes a symmetric monoidal topological G-category under taking direct sums of inner product spaces in $\mathscr{I}$. From this point of view we want to regard the direct sum on $\mathscr{I}$ as a functor $\oplus : \mathscr{I} \times \mathscr{I} \to \mathscr{I}$, and without loss of generality we will assume that the unit is the trivial subspace $0 \subset U$.

Next let us consider the topological G-category $\mathscr{T}_*$ of well-pointed compactly generated weak Hausdorff G-spaces and all (not necessarily G-equivariant) continuous maps between them. The topology on the morphism sets shall be given by the compact-open topology. The category has the structure of a symmetric monoidal topological G-category given through the smash product of pointed G-spaces with unit the space S^0 equipped with the trivial G-action. Let $S : \mathscr{I} \to \mathscr{T}_*$ be the strong symmetric monoidal G-functor that maps an inner product space $V \in \mathscr{I}$ to its one point compactification S^V.

Definition 2.2. A commutative *$\mathscr{I}$-FSP* is a lax symmetric monoidal G-functor $\mathbb{X} : \mathscr{I} \to \mathscr{T}_*$ together with a monoidal natural transformation $\eta : S \to \mathbb{X}$ of lax symmetric monoidal G-functors.

Let $\eta : S \to \mathbb{X}$ be a commutative *$\mathscr{I}$-FSP*. For any pair of indexing spaces V_1, V_2 with $V_1 \perp V_2$ and any pair of inner product spaces $W, Z \in \mathscr{I}$ the natural transformation η provides maps

$$(2.3) \quad \sigma : S^{V_1 \oplus W} \wedge \mathbb{X}(V_2 \oplus Z) \xrightarrow{\eta_{V_1 \oplus W} \wedge Id} \mathbb{X}(V_1 \oplus W) \wedge \mathbb{X}(V_2 \oplus Z)$$
$$\xrightarrow{\mu_{\mathbb{X}}} \mathbb{X}(V_1 \oplus W \oplus V_2 \oplus Z) \xrightarrow{\cong} \mathbb{X}(V_1 + V_2 \oplus W \oplus Z).$$

Here, the last map[3] in the composition is given by applying the functor $\mathbb{X}$ to the canonical map

$$V_1 \oplus W \oplus V_2 \oplus Z \to V_1 + V_2 \oplus W \oplus Z.$$

The maps σ define a coordinate-free prespectrum in the sense of Lewis-May-Steinberger (cf. [12]).

As usual, a prespectrum defines a corresponding equivariant cohomology theory on $\mathscr{T}_*^G$, the subcategory of $\mathscr{T}_*$ consisting of the G-equivariant maps. Given inner product spaces $W, Z \in \mathscr{I}$ we obtain groups $\mathbb{X}^{W,Z}$ for pointed spaces $X \in \mathscr{T}_*$ by defining

$$(2.4) \qquad \mathbb{X}^{W,Z}(X) = \operatorname*{colim}_{V \in \mathscr{D}} \, [S^{V \oplus W} \wedge X, \mathbb{X}(V \oplus Z)]_G.$$

Here, $[\,,\,]_G$ stands for pointed G-homotopy classes of pointed G-maps, and the colimites are taken over the indexing spaces $V \in \mathscr{D}$ using the suspension maps (2.3). The bigraded groups (2.4) can be used (using standard techniques) to define a cohomology theory $\mathbb{X}^*$ whose grading is over the real representation ring $RO(G)$.

Given inner product spaces $W_1, Z_1, W_2, Z_2 \in \mathscr{I}$ and pointed spaces $X_1, X_2 \in \mathscr{T}_*$ the multiplicative structure of $\mathbb{X}$ provides a natural pairing

$$(2.5) \qquad \cup \; : \; \mathbb{X}^{W_1,Z_1}(X_1) \times \mathbb{X}^{W_2,Z_2}(X_2) \longrightarrow \mathbb{X}^{W_1 \oplus W_2, Z_1 \oplus Z_2}(X_1 \wedge X_2).$$

More explicitly, if G-equivariant pointed maps

$$\varphi_1 : S^{V_1 \oplus W_1} \wedge X_1 \;\to\; \mathbb{X}(V_1 \oplus Z_1),$$
$$\varphi_2 : S^{V_2 \oplus W_2} \wedge X_2 \;\to\; \mathbb{X}(V_2 \oplus Z_2)$$

represent classes $x_1 \in \mathbb{X}^{W_1,Z_1}(X_1)$ and $x_2 \in \mathbb{X}^{W_2,Z_2}(X_2)$, and $V_1 \perp V_2$ in U, then the product $x_1 \cup x_2 \in \mathbb{X}^{W_1 \oplus W_2, Z_1 \oplus Z_2}(X_1 \wedge X_2)$ is represented by the map

$$S^{V_1 + V_2 \oplus W_1 \oplus W_2} \wedge X_1 \wedge X_2 \longrightarrow \left(S^{V_1 \oplus W_1} \wedge X_1\right) \wedge \left(S^{V_2 \oplus W_2} \wedge X_2\right)$$
$$\xrightarrow{\varphi_1 \wedge \varphi_2} \mathbb{X}(V_1 \oplus Z_1) \wedge \mathbb{X}(V_2 \oplus Z_2) \xrightarrow{\mu_{\mathbb{X}}} \mathbb{X}(V_1 \oplus Z_1 \oplus V_2 \oplus Z_2)$$
$$\to \mathbb{X}(V_1 + V_2 \oplus Z_1 \oplus Z_2),$$

where the first map is induced by the various natural isomorphisms and the last one is obtained by applying $\mathbb{X}$ to the canonical map $V_1 \oplus Z_1 \oplus V_2 \oplus Z_2 \to V_1 + V_2 \oplus Z_1 \oplus Z_2$. As every pair of equivalence classes $x_1 \in \mathbb{X}^{W_1,Z_1}, x_2 \in \mathbb{X}^{W_2,Z_2}$ can be represented by maps φ_1 and φ_2 as above (i.e. with $V_1 \perp V_2$) this can actually be used as the definition of the pairing (2.5). The natural pairing (2.5) induces a corresponding natural pairing for the $RO(G)$-graded

---

[3]The appearance of the last isomorphism in the composition (2.3) is motivated by the fact that $V_1 + V_2$ is an indexing space while $V_1 \oplus V_2$ is not. The issue will become relevant when we define $\mathbb{X}^{W,Z}(X)$ in (2.4).

cohomology groups $\mathbb{X}^*$ which equips the $RO(G)$-graded cohomology theory $\mathbb{X}^*$ with a multiplicative structure.

3. CLIFFORD-ALGEBRAS AND COMPACT OPERATORS

Recall that a $\mathbb{Z}/2$-graded G-C^*-algebra A is a $\mathbb{Z}/2$-graded C^*-algebra A with a grading preserving covariant representation of G on A (see [5, 10.1]), i.e. in particular, A regarded as a topological space carries a continuous G-action. Let us denote by $\mathscr{A}$ the topological G-category of $\mathbb{Z}/2$-graded G-C^*-algebras with morphisms the grading preserving $*$-homomorphisms. The topology on its morphism sets $Hom(A, B)$ shall be the (compactly generated) compact-open topology. The $\mathbb{Z}/2$-graded tensor product $\otimes : \mathscr{A} \times \mathscr{A} \to \mathscr{A}$ with unit $\mathbb{C}$ gives $\mathscr{A}$ the structure of a symmetric monoidal topological G-category.[4] The unit for $\otimes$ is the C^*-algebra $\mathbb{C}$ regarded as a $\mathbb{Z}/2$-graded C^*-algebra using the trivial $\mathbb{Z}/2$-grading and the trivial G-action.

Definition 3.1. Given an inner product space $V \in \mathscr{I}$ we define the Clifford algebra $\mathbb{C}l_V$ to be the quotient of the complex tensor algebra on V modulo the ideal generated by the elements $v \otimes v - \langle v, v \rangle \cdot 1$, $v \in V$. The Clifford algebras naturally come with a $\mathbb{Z}/2$-grading $\mathbb{C}l_V = \mathbb{C}l_V^{even} + \mathbb{C}l_V^{odd}$ induced from the $\mathbb{N}_0$-grading of the complex tensor algebra generated by V. Moreover there are canonical isomorphisms $\mathbb{C}l_V \otimes \mathbb{C}l_W \cong \mathbb{C}l_{V \oplus W}$, and we have the obvious isomorphism $\mathbb{C} \to \mathbb{C}l_0$ sending the unit in $\mathbb{C}$ to the unit in $\mathbb{C}l_0$. Finally $\mathbb{C}l_V$ is canonically isomorphic as a vector space to the exterior algebra $\Lambda^*(V \otimes \mathbb{C})$. Using this isomorphism left multiplication on $\mathbb{C}l_V$ defines a $*$-algebra monomorphism $\mathbb{C}l_V \to End(\Lambda^*(V \otimes \mathbb{C}))$. Using the standard scalar product product on $\Lambda^* V \otimes \mathbb{C}$ we interpret $End(\Lambda^*(V \otimes \mathbb{C}))$ as the C^*-algebra of bounded operators of a (finite dimensional) Hilbert space. The inclusion $\mathbb{C}l_V \subset End(\Lambda^*(V \otimes \mathbb{C}))$ then induces a C^*-norm on $\mathbb{C}l_V$, which gives $\mathbb{C}l_V$ the structure of a $\mathbb{Z}/2$-graded G-C^*-algebra. Altogether we get a strong symmetric monoidal G-functor $\mathbb{C}l : \mathscr{I} \to \mathscr{A}$ by sending V to $\mathbb{C}l_V$.

Definition 3.2. Given an inner product space $V \in \mathscr{I}$ let $L^2(V)$ denote the L^2-completion of the pre-Hilbert space of functions of compact support on V with values in $\mathbb{C}$. The space $L^2(V)$ comes equipped with a natural $\mathbb{Z}/2$-grading which corresponds to the decomposition into even and odd functions on V. For two inner product spaces $V, W \in \mathscr{I}$ there is the canonical G-invariant isomorphism

$$(3.3) \qquad L^2(V) \otimes L^2(W) \xrightarrow{\cong} L^2(V \oplus W), \ f \otimes g \mapsto f \cdot g$$

---

[4]On the algebraic tensor product of two C^*-algebras there is typically more than one meaningful definition of a C^*-norm, and the algebraic tensor product typically is not complete with respect to any of these C^*-norms. Thus, in order to define a tensor product which is a C^*-algebra one specifies a corresponding choice of a C^*-norm and defines the tensor product to be the completion of the algebraic tensor product with respect to it. The C^*-norm that we want to use is the so-called spatial C^*-norm, see [22, App. T] for details.

where the tensor product here is the graded tensor product of Hilbert spaces.

We now associate to V the $\mathbb{Z}/2$-graded G-C^*-algebra of compact operators on $L^2(V)$

$$\mathcal{K}_V = \mathcal{K}(L^2(V)).$$

The canonical maps (3.3) induce corresponding canonical isomorphisms

$$\mathcal{K}_V \otimes \mathcal{K}_W \xrightarrow{\cong} \mathcal{K}_{V \oplus W}.$$

Moreover we have the obvious map $\mathbb{C} \to \mathcal{K}_0$ sending the unit $1 \in \mathbb{C}$ to the unit in $\mathcal{K}_0$. This gives $\mathcal{K}$ the structure of a strong symmetric monoidal G-functor $\mathscr{I} \to \mathscr{A}$.

Now consider the $\mathbb{Z}/2$-graded G-C^*-algebra $\mathcal{S}$ of continuous functions on $\mathbb{R}$ vanishing at infinity, equipped with the grading given by the even and odd functions therein and the trivial G-action. Note that there is one and only one non trivial grading preserving $*$-homomorphism $ev_0 : \mathcal{S} \to \mathbb{C}$. It is given by evaluating a function $u \in \mathcal{S}$ at $0 \in \mathbb{R}$, i.e. $ev_0(u) = u(0)$.

There is a comultiplication map $\Delta : \mathcal{S} \longrightarrow \mathcal{S} \otimes \mathcal{S}$. To present it we consider the isomorphism of linear spaces $\mathcal{S} \otimes \mathcal{S} \cong C_0(\mathbb{R}) \otimes C_0(\mathbb{R}) \cong C_0(\mathbb{R}^2)$, where $C_0(\mathbb{R})$ and $C_0(\mathbb{R}^2)$ denote the C^*-algebras of continuous functions vanishing at infinity on $\mathbb{R}$ and $\mathbb{R}^2$ respectively. We also use the fact that the C^*-algebra $\mathcal{S}$ is generated by the functions $u(t) = (1+t^2)^{-1}$ and $v(t) = t(1+t^2)^{-1}$. The map Δ then is given by

$$\Delta(u)(x,y) = \frac{1}{1+x^2+y^2} \quad \text{and} \quad \Delta(v)(x,y) = \frac{x+y}{1+x^2+y^2}.$$

One easily checks that Δ is coassociative and cocommutative. The following lemma is an immediate consequence of the coassociativity and the cocommutativity of the comultiplication map.

Below we want to consider the spaces $Hom(\mathcal{S}, A)$ for $\mathbb{Z}/2$-graded G-C^*-algebras A. Without further assumptions on A one maybe should not expect this homomorphism space to be well-pointed. Thus we temporarily will work with the category $\mathscr{T}_*'$ of pointed compactly generated weak Hausdorff spaces and all continuous maps between them.

Lemma 3.4. *The functor $Hom(\mathcal{S}, ?) : \mathscr{A} \to \mathscr{T}_*'$ sending a $\mathbb{Z}/2$-graded G-C^*-algebra A to the pointed G-space $Hom(\mathcal{S}, A)$ together with the natural transformation $* : \mathscr{A} \times \mathscr{A} \to \mathscr{T}_*'$ given by the family of maps*

$$* : Hom(\mathcal{S}, A) \wedge Hom(\mathcal{S}, B) \longrightarrow Hom(\mathcal{S} \otimes \mathcal{S}, A \otimes B) \xrightarrow{\Delta^*} Hom(\mathcal{S}, A \otimes B)$$

and the pointed isomorphism $S^0 \cong Hom(\mathcal{S}, \mathbb{C})$ is a lax symmetric monoidal G-functor.
(The unlabeled map in the composition is given by taking the tensor product of $*$-homomorphisms.)

Corollary 3.5. *Let $\mathscr{C}$ be a symmetric monoidal topological G-category.*

(1) *Let $\mathbb{X} = (\mathbb{X}, \mu, \lambda) : \mathscr{C} \to \mathscr{A}$ be a lax symmetric monoidal G-functor. Then the G-functor $\widehat{\mathbb{X}} : \mathscr{C} \to \mathscr{T}'_*$, defined by*

$$\widehat{\mathbb{X}} : a \longmapsto Hom(\mathcal{S}, \mathbb{X}(a)), \quad a \in \mathscr{C},$$

together with the natural transformation of G-functors $\mathscr{C} \times \mathscr{C} \to \mathscr{T}'_*$

$$Hom(\mathcal{S}, \mathbb{X}(a)) \wedge Hom(\mathcal{S}, \mathbb{X}(b)) \xrightarrow{*} Hom(\mathcal{S}, \mathbb{X}(a) \otimes \mathbb{X}(b)) \xrightarrow{\mu_{a,b_*}} Hom(\mathcal{S}, \mathbb{X}(a \oplus b))$$

and the canonical map $S^0 \xrightarrow{\cong} Hom(\mathcal{S}, \mathbb{C}) \xrightarrow{\lambda_*} Hom(\mathcal{S}, \mathbb{X}(u_{\mathscr{C}}))$ is a lax symmetric monoidal G-functor $\widehat{\mathbb{X}}$ from $\mathscr{C}$ to $\mathscr{T}'_*$.

(2) *Let $\mathbb{X}_1, \mathbb{X}_2 : \mathscr{C} \to \mathscr{A}$ be lax symmetric monoidal G-functors, then the maps*

$$* : Hom(\mathcal{S}, \mathbb{X}_1(a)) \wedge Hom(\mathcal{S}, \mathbb{X}_2(a)) \longrightarrow Hom(\mathcal{S}, \mathbb{X}_1(a) \otimes \mathbb{X}_2(a))$$

define a monoidal natural transformation $* : \widehat{\mathbb{X}_1} \wedge \widehat{\mathbb{X}_2} \to \widehat{\mathbb{X}_1 \otimes \mathbb{X}_2}$.

Definition 3.6. Let $\mathbb{K}$ be the lax symmetric monoidal G-functor $\mathbb{K} = \widehat{\mathcal{Cl} \otimes \mathcal{K}}$, where $\widehat{\mathcal{Cl} \otimes \mathcal{K}}$ is defined as in Corollary 3.5. In particular we have

$$\mathbb{K}(V) = Hom(\mathcal{S}, \mathcal{Cl}_V \otimes \mathcal{K}_V) \quad \text{for all } V \in \mathscr{I}.$$

One can show that these homomorphism spaces are in fact all well-pointed; thus $\mathbb{K}$ is a lax symmetric monoidal G-functor from $\mathscr{I}$ to $\mathscr{T}_*$.

4. THE K-THEORY SPECTRUM

One goal of this section is to complete the definition of of the $\mathscr{I}$-FSP $\mathbb{K}$, i.e. in this section we will present the relevant monoidal natural transformation $\eta : S \to \mathbb{K}$. Moreover we will state the first main theorem and give some comments on the proof.

As already mentioned earlier the C^*-algebra $\mathcal{S}$ is generated by the two functions $u(t) = (1 + t^2)^{-1}$ and $v(t) = t(1 + t^2)^{-1}$. Thus a $*$-homomorphism out of $\mathcal{S}$ is specified by saying where these two elements go to. Now let W be a finite dimensional G-inner product space and let $w \in W$. The element w then can be regarded as an element of the Clifford algebra $\mathbb{Cl}_W$ by means of the natural inclusion $W \subset \mathbb{Cl}_W$. We can specify a $*$-homomorphism $fc(w) : \mathcal{S} \to \mathbb{Cl}_W$ by requiring that $fc(w)(u) = (1 + |w|^2)^{-1}$ and $fc(w)(v) = w \cdot (1 + |w|^2)^{-1}$. Here the letters fc stand for "functional calculus". We shall think of $fc(w)$ as been given by substituting the variable t by w. If w becomes big in norm then the norm of $fc(w)$ tends to zero.

Definition 4.1. Let V be an inner product space in $\mathscr{I}$. Associating to $v \in V$ the $*$-homomorphism $fc(v)$ induces a continuous map $fc_V : V \to Hom(\mathcal{S}, \mathbb{Cl}_V)$. Since the norm of $fc(v)$ tends to zero when v goes to infinity it follows that the map fc_V has a continuous pointed extension

$$\beta_V : S^V \longrightarrow Hom(\mathcal{S}, \mathbb{Cl}_V).$$

For inner product spaces $V, W \in \mathscr{I}$ and $v \in V, w \in W$ the element $v + w \in \mathbb{C}l_{V \oplus W}$ is the image of $v \otimes 1 + 1 \otimes w \in \mathbb{C}l_V \otimes \mathbb{C}l_W$ under the canonical isomorphism $\mathbb{C}l_V \otimes \mathbb{C}l_W \cong \mathbb{C}l_{V \oplus W}$. A direct calculation shows that $fc(v + w)$ coincides with the composition

$$\mathcal{S} \xrightarrow{\Delta} \mathcal{S} \otimes \mathcal{S} \xrightarrow{fc(v) \otimes fc(w)} \mathbb{C}l_V \otimes \mathbb{C}l_W \cong \mathbb{C}l_{V \oplus W}.$$

We define a monoidal natural transformation $\beta : S \to \widehat{\mathbb{C}l}$ through the maps $\beta_V : S^V \to Hom(\mathcal{S}, \mathbb{C}l_V)$ which send a vector $v \in V \subset S^V$ to $\beta_V(v) = fc(v)$.

Definition 4.2. Given an inner product space $V \in \mathscr{I}$, let $P_V \in \mathcal{K}_V$ denote the projection onto $\mathbb{C}i_V$, the linear subspace in $L^2(V)$ generated by the function $i_V \in L^2(V)$ which sends a vector v to $\exp(-\langle v, v \rangle)$. We then define the element $p_V^0 \in Hom(\mathcal{S}, \mathcal{K}_V)$ to be the composition of the non-trivial $*$-homomorphism $ev_0 : \mathcal{S} \to \mathbb{C}$ with the $*$-homomorphism $\mathbb{C} \to \mathcal{K}_V$ which sends the unit in $\mathbb{C}$ to the projection P_V. The family $(p_V^0)_{\{V \in \mathscr{I}\}}$ is multiplicative, since $i_{V \oplus W} = i_V \cdot i_W$ for inner product spaces $V, W \in \mathscr{I}$. Now recall from the second section that the unit S^0 of the symmetric monoidal category $\mathscr{T}_*$ gives rise to a lax monoidal functor $S^0 : \mathscr{I} \to \mathscr{T}_*$, which sends any inner product space $V \in \mathscr{I}$ to $S_V^0 = S^0$. The family $(p_V^0)_{\{V \in \mathscr{I}\}}$ then defines a lax symmetric monoidal natural transformation $p : S^0 \to \widehat{\mathcal{K}}$. More explicitly this transformation is given by the collection of pointed maps $p_V : S_V^0 \to \widehat{\mathcal{K}}_V = Hom(\mathcal{S}, \mathcal{K}_V)$ which send the point different from the base-point in S_V^0 to the $*$-homomorphism $p_V^0 \in Hom(\mathcal{S}, \mathcal{K}_V)$.

Definition 4.3. We define $\eta : S \to \mathcal{K}$ to be the monoidal natural transformation given by the composition

$$\eta : S \xrightarrow{\cong} S \wedge S^0 \xrightarrow{\beta \wedge p} \widehat{\mathbb{C}l} \wedge \widehat{\mathcal{K}} \xrightarrow{*} \widehat{\mathbb{C}l \otimes \mathcal{K}} = \mathbb{K},$$

i.e. for an inner product space $V \in \mathscr{I}$ the map η_V sends $v \in V \subset S^V$ to the $*$-homomorphism $fc(v) * p_V^0 : \mathcal{S} \to \mathbb{C}l_V \otimes \mathcal{K}_V$.

Recall from the second section that a monoidal natural transformation as $\eta : S \to \mathbb{K}$ gives rise to an equivariant spectrum in the sense of Lewis-May-Steinberger [12]. Hence η defines a multiplicative G-equivariant cohomology theory on $\mathscr{T}_*^G$, the category of well-pointed weak Hausdorff G-spaces and equivariant pointed maps between them.

Let $\mathscr{K}^G$ denote the category of (unpointed) compact G-spaces and equivariant maps. In the following $K_G^*(X)$ shall stand for unreduced equivariant K-cohomology for spaces $X \in \mathscr{K}^G$. That is, $K_G^0(X)$ is the Grothendieck group of isomorphism classes of G-vector bundles over X. For any pair of inner product spaces $V, W \in \mathscr{I}$ one has the groups

$$K_G^{V,W}(X) = \tilde{K}_G^0(S^V \wedge S^W \wedge X_+),$$

where X_+ as usual denotes the one point compactification of X. The groups can be used (using standard techniques) to define an $RO(G)$-graded cohomology theory (c.f. Section 2); the relevant suspension isomorphisms

$$\sigma : K^0(X) \longrightarrow \widetilde{K}_G^{0,V}(S^V \wedge X_+) = \widetilde{K}_G^0(S^{V \oplus V} \wedge X_+)$$

are given by multiplication with the so-called Bott element $b_V^{top} \in \widetilde{K}_G^0(S^{V \oplus V})$. For the definition of the Bott class we refer to (6.8) below.

Theorem 4.4. *Let U be a complete universe. Then the $\mathscr{I}$-FSP $(\mathbb{K}, \eta)$, defined by 3.6 and 4.3, represents equivariant K-theory, i.e. there is a natural isomorphism of unreduced $RO(G)$-graded multiplicative (equivariant) cohomology theories on $\mathscr{K}^G$*

$$\Phi^*(X) : \mathbb{K}^*(X_+) \longrightarrow K_G^*(X).$$

Note that the natural isomorphism $\Phi^*(X)$ of the previous theorem induces a corresponding natural isomorphism $\widetilde{\Phi}^*(X) : \mathbb{K}^*(X) \to \widetilde{K}_G^*(X)$ of reduced $RO(G)$-graded multiplicative (equivariant) cohomology theories on the category of well-pointed compact spaces $\mathscr{K}_*^G$. The definition of the natural isomorphism Φ^* as well as a rigorous proof of the theorem will be given in the following section using elements of Kasparovs KK-theory. In the remaining part of this section we comment a little on some ideas that go into the proof so that the statement becomes plausible even without knowing too much about Kasparov K-theory and operator theory respectively.

Assume ℓ is an evenly graded G-Hilbert space and P is a G-fixed compact operator which also is a projection. Then $P(\ell)$ is a finite dimensional G-representation. Conversely, if $V \subset \ell$ is a finite-dimensional G-invariant linear subspace, then we have an isomorphism $\ell \cong V \oplus V^\perp$, and we can define a compact projection P_V by

$$P_V : \ell \to \ell, \quad P_V(h) = \begin{cases} h & \text{for } h \in V \\ 0 & \text{for } h \in V^\perp \end{cases}.$$

The space[5] of all G-invariant projections in $\mathcal{K}(\ell)$ is homeomorphic to the space of $*$-homomorphisms $Hom(\mathbb{C}, \mathcal{K}(\ell))$: the homeomorphism is given by mapping a G-invariant compact projection P to the unique $*$-homomorphism f_P which maps the unit $1 \in \mathbb{C}$ to P. Given a compact G-space X an equivariant continuous map $f : X \to Hom(\mathbb{C}, \mathcal{K}(\ell))$ defines a corresponding family of compact projections $\{P_{f(x)}\}_{x \in X}$, which defines a G-vector bundle E_f over X with fiber $E_{fx} = P_{f(x)}(\ell)$ over a point $x \in X$. By construction the vector bundle E_f comes equipped with a G-equivariant linear embedding $E_f \subset X \times \ell$ which is a map over X. Conversely, given a G vector bundle E and a G-equivariant linear embedding $j : E \subset X \times \ell$ over X one defines a corresponding G-equivariant map $X \to Hom(\mathbb{C}, \mathcal{K}(\ell))$ by mapping a point $x \in X$ to the

---

[5]For the topology we take the norm topology.

$*$-homomorphism $f_{P(x)}$ where $P(x)$ is the compact projection which projects to the embedded fiber $j(E_x) \subset \ell$.

If ℓ is universal in the sense that any G-vector bundle over any compact space X can be G-equivariantly embedded into $X \times \ell$ (as a map over X) then $Hom(\mathbb{C}, \mathcal{K}(\ell))$ represents the functor $X \mapsto Vect_G(X)$, i.e. we have a natural isomorphism

$$Vect_G(X) \cong [X_+, Hom(\mathbb{C}, \mathcal{K}(\ell))]_G,$$

where $[\ ,\ ]_G$ stands for pointed G-homotopy classes of pointed continuous G-equivariant maps.

Now let H be the $\mathbb{Z}/2$-graded G-vectorspace with $H^{even} = H^{odd} = \ell$ for a universal ungraded G-Hilbert space ℓ.[6] Let $\iota : \mathcal{K}(\ell) \to \mathcal{K}(H)$ be the inclusion which is induced by the inclusion $\ell = H^{even} \subset H$, and let $ev_0 : \mathcal{S} \to \mathbb{C}$ be the $*$-homomorphism which evaluates a function $u \in \mathcal{S}$ at the point $0 \in \mathbb{R}$. Then one has a commutative diagram as follows

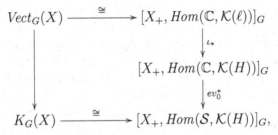

where the left vertical arrow is group completion. The fact that the space $Hom(\mathcal{S}, \mathcal{K}(H))$ represents K-theory is a well-known result in operator theory. Roughly speaking an element $f \in Hom(\mathcal{S}, \mathcal{K}(H))$ can be used to define[7] an odd self-adjoint G-equivariant Fredholm operator $\mathcal{F}_f$ on a sub-G-Hilbert space $\mathcal{E}_f \subset H$. The corresponding element in $K_G^0(\{pt\})$ which is represented by $[f] \in \pi_0(Hom(\mathcal{S}, \mathcal{K}(H)))$ then is given by the $\mathbb{Z}/2$-graded index of $\mathcal{F}_f$. More generally one can define an index map $[X_+, Hom(\mathcal{S}, \mathcal{K}(H))]_G \to K_G^0(X)$ which then is an inverse to the bottom isomorphism in the above diagram. This index map is similiar in spirit to the one given in [2, Appendix].

If V is a finite dimensional G-inner product space there is an analogous interpretation for the space $Hom(\mathcal{S}, \mathbb{C}l_V \otimes \mathcal{K}(H))$. Elements in $\mathbb{C}l_V \otimes \mathcal{K}(H)$ correspond to $\mathbb{C}l_V$-right linear compact operators on the G-Hilbert space $\mathbb{C}l_V \otimes H$ which we regard as a right $\mathbb{C}l_V$-module by means of right multiplication on the first factor. An element $f \in Hom(\mathcal{S}, \mathbb{C}l_V \otimes \mathcal{K}(H))$ now can be used to define an odd self-adjoint $\mathbb{C}l_V$-right linear G-equivariant Fredholm operator $\mathcal{F}_f$ on a sub-G-Hilbertspace $\mathcal{E}_f \subset \mathbb{C}l_V \otimes H$, so one can think of $Hom(\mathcal{S}, \mathbb{C}l_V \otimes \mathcal{K}(H))$ as a space of $\mathbb{C}l_V$-linear Fredholm operators. Both

---

[6]Then H is universal in the sense defined on page 102.

[7]To see how this works see 5.2: take $A = \mathbb{C}$ and use Theorem 5.2 to identify $KK_G^0(\{pt\})$ with $K_G^0(\{pt\})$; moreover use the fact that a Hilbert-$\mathbb{C}$-module is nothing but a $\mathbb{Z}/2$-graded G-Hilbert space.

spaces, $Hom(\mathcal{S}, \mathbb{C}l_V \otimes \mathcal{K}(H))$ as well as the honest space of odd self-adjoint $\mathbb{C}l_V$-right linear G-equivariant Fredholm operators on $\mathbb{C}l_V \otimes H$ are V-th deloopings of a space representing equivariant K-theory.[8] The statement for the space of odd self-adjoint $\mathbb{C}l_V$-right linear G-equivariant Fredholm operator is more classical and essentially due to Atiyah and Singer who proved the statement in the non-equivariant setting [4, Theorem A(k), Prop. 5.3], but see [9, Theorem 3.3] for a formulation which fits better with the present context.

Given a $\mathbb{Z}/2$-graded G-Hilbert space we call it universal in the following, if it is isomorphic to H above. It follows from Lemma 5.5 below that there are indexing spaces W for which $L^2(W)$ is universal, and that these indexing spaces form a cofinal diagram in the diagram of all indexing spaces.

We now want to connect the above to Theorem 4.4. Recall that by definition we have

$$\mathbb{K}^0(X) = \operatorname*{colim}_{V \in \mathscr{D}} [S^V \wedge X_+, Hom(\mathcal{S}, \mathbb{C}l_V \otimes \mathcal{K}_V)]_G.$$

Computing this colimit involves homomorphisms which fit as bottom rows in commutative diagrams of the following form (with the remaining maps defined as below)

$$
\begin{array}{ccc}
[X_+, Hom(\mathcal{S}, \mathcal{K}_V)]_G & \longrightarrow & [X_+, Hom(\mathcal{S}, \mathcal{K}_{V+W})]_G \\
\downarrow & & \downarrow \\
[S^V \wedge X_+, Hom(\mathcal{S}, \mathbb{C}l_V \otimes \mathcal{K}_V)]_G & [S^{V+W} \wedge X_+, Hom(\mathcal{S}, \mathbb{C}l_{V+W} \otimes \mathcal{K}_{V+W})]_G.
\end{array}
$$

We may assume that $L^2(V)$ is universal, and since $L^2(V) \subset L^2(V + W)$ then so is $L^2(V + W)$. Therefore the two groups on the top of the above diagram are both canonically isomorphic to $K_G^0(X)$. The map between them is the one which is induced by the natural inclusion $\mathcal{K}(j) : \mathcal{K}_V \subset \mathcal{K}_{V+W}$ and induces an isomorphism (in fact the identity)[9] on $K_G^0(X)$. The left vertical arrow in the diagram is induced by the map which maps a pointed function $f : X_+ \to Hom(\mathcal{S}, \mathcal{K}_V)$ to the function $\beta_V * f : S^V \wedge X_+ \to Hom(\mathcal{S}, \mathbb{C}l_V \otimes \mathcal{K}_V)$ given by $(\beta_V * f)(v, x) = \beta_V(v) * f(x)$; while the vertical right arrow is induced by the corresponding map which maps a pointed function $g : X_+ \to Hom(\mathcal{S}, \mathcal{K}_{V+W})$ to the function $\beta_{V+W} * g$. The statement of Theorem 4.4 then follows from the fact that the vertical arrows are isomorphisms, which follows from the

---

[8]The topology on the spaces of Fredholm operators considered here is the norm topology. We would like to emphasize that the natural G-action on the set of Fredholm operators in general is not compatible with the norm topology, i.e. the G-action does not give the spaces of Fredholm operators the structure of a G-space.

[9]To see that assume that an element in $[X_+, Hom(\mathcal{S}, \mathcal{K}_V)]_G$ is represented by $ev_0 \circ f$ for some map $f : X \to Hom(\mathbb{C}, \mathcal{K}(L^2(W)))$. As explained above, the corresponding element in $K_G^0(X)$ then is represented by the G-vector bundle E_f. The element in $K_G^0(X)$ which is represented by $j \circ ev_0 \circ f$ then is $E_{j \circ f}$ which is isomorphic to E_f since j is an embedding. Hence $[E_f] = [E_{j \circ f}] \in K_G^0(X)$.

fact that the vertical arrows are the operator-theoretic descriptions of Bott periodicity isomorphisms within the theory of Clifford algebras.

5. ELEMENTS OF EQUIVARIANT KK-THEORY

In this section we will define the natural homomorphism $\Phi^*(X) : \mathbb{K}^*(X_+) \to K^*_G(X)$ of Theorem 4.4 and see that it is an isomorphism which then proves the theorem. For doing this we need to recall some facts about equivariant Kasparov K-theory first.

5.1. The KK-homology groups.

Given a $\mathbb{Z}/2$-graded G-algebra A there is the notion of a *Hilbert-A-module* (see [5, 13.1 & 20.1.1]). In particular, a Hilbert-A-module is a $\mathbb{Z}/2$-graded Banach space with a right A-action, an A-sesquilinear A-valued scalar product and a continuous representation of G on it which is compatible with the $\mathbb{Z}/2$-grading, the A-action and the A-valued scalar product. The concept of a Hilbert-A-module generalizes the notion of a Hilbert space in the sense that a Hilbert-$\mathbb{C}$-module is nothing but a $\mathbb{Z}/2$-graded Hilbert space with a unitary representation of G on it, which preserves the grading. Given a Hilbert-A-module $\mathcal{E}$ one defines the corresponding $\mathbb{Z}/2$-graded C^*-algebras $\mathbb{B}(\mathcal{E})$, consisting of the so-called *(adjointable) bounded operators* on $\mathcal{E}$, and $\mathbb{K}(\mathcal{E})$, consisting of the *(adjointable) compact operators* on $\mathcal{E}$ ([5, 13.2.1 & 20.1.1]). By definition, $\mathbb{B}(\mathcal{E})$ is contained in the subspace of A-linear bounded linear operators $\mathcal{B}(\mathcal{E})_A$, and $\mathbb{K}(\mathcal{E})$ is a subspace of the A-linear compact operators $\mathcal{K}(\mathcal{E})_A$. If A is finite dimensional then we have equality, i.e. $\mathbb{B}(\mathcal{E}) = \mathcal{B}(\mathcal{E})_A$ and $\mathbb{K}(\mathcal{E}) = \mathcal{K}(\mathcal{E})_A$. Note that $\mathbb{K}(\mathcal{E})$ is a G-C^*-algebra while $\mathbb{B}(\mathcal{E})$ in general is not (cf. [5, 20.1]).

For the following let A be σ-unital. Moreover let us regard $[0, 1]$ as a G-space with trivial G-action, and let $A[0, 1]$ be the G-C^*-algebra consisting of the functions on $[0, 1]$ with values in A. If $\mathcal{E}$ is a Hilbert-A-module let $\mathcal{E}[0, 1]$ denote the Hilbert-$A[0, 1]$-module consisting of the functions from the unit interval into $\mathcal{E}$. Moreover, for any $t \in [0, 1]$ we have the evaluation $*$-homomorphism $ev_t : A[0, 1] \to A$, $ev_t(f) = f(t)$, and given a Hilbert-$A[0, 1]$-module $\mathcal{E}$ we use the evaluation $*$-homomorphisms to define the *evaluations* of $\mathcal{E}$ at t by $\mathcal{E}_t = \mathcal{E} \otimes_{A[0,1]} A$. For bounded operators $F \in \mathbb{B}(\mathcal{E})$ the *evaluation* at t is given by $F_t = F \otimes_{A[0,1]} Id \in \mathbb{B}(\mathcal{E}_t)$.

Definition 5.1. Let A be a $\mathbb{Z}/2$-graded σ-unital G-C^*-algebra. Then we define a *Kasparov-A-cycle* to be a pair $(\mathcal{E}, \mathcal{F})$ consisting of a countably generated Hilbert-A-module $\mathcal{E}$ and a G-invariant self-adjoint bounded operator $\mathcal{F} \in \mathbb{B}(\mathcal{E})$ satisfying $\mathcal{F}^2 - Id \in \mathbb{K}(\mathcal{E})$ and $||\mathcal{F}|| \leq 1$. Two Kasparov-A-cycles $(\mathcal{E}_0, \mathcal{F}_0), (\mathcal{E}_1, \mathcal{F}_1)$ are called *isomorphic (or unitary equivalent)* if there is an isomorphism (i.e. a G-equivariant grading preserving unitary) $U : \mathcal{E}_0 \to \mathcal{E}_1$ such that $\mathcal{F}_1 = U\mathcal{F}_0 U^*$. The two cycles are called *homotopic* if there is a Kasparov-$A[0, 1]$-cycle $(\mathcal{E}, \mathcal{F})$ whose evaluations at 0 and 1 are $(\mathcal{E}_0, \mathcal{F}_0)$ and

$(\mathcal{E}_1, \mathcal{F}_1)$. The corresponding Kasparov cycle $(\mathcal{E}, \mathcal{F})$ then is called a *Kasparov homotopy* between $(\mathcal{E}_0, \mathcal{F}_0)$ and $(\mathcal{E}_1, \mathcal{F}_1)$. Homotopy induces an equivalence relation on the set of isomorphism classes of Kasparov-A-cycles, and the direct sum of Kasparov-A-cycles induces a group structure on this set of homotopy classes. Our notation for this group is $KK_0^G(A)$.

Equivariant KK-theory is related to equivariant K-theory by means of the equivariant Swan isomorphism.

Theorem 5.2 (Equivariant Swan Theorem, (cf. [18, 2.3])). *Let X be a compact G-space and let $C(X)$ denote the trivially graded G-C^*-algebra of continuous functions on X with values in $\mathbb{C}$. Associating to a G-vector bundle with a G-invariant inner product over X the corresponding Hilbert-$C(X)$-module of continuous sections $\Gamma(E)$ induces an isomorphism*

$$(5.3) \qquad \psi_X : K_G^0(X) \cong KK_0^G(C(X)).$$

5.2. The homomorphism picture. Given a countably generated Hilbert-A-module $\mathcal{E}$ we want to think about the G-fixpoint set $Hom(\mathcal{S}, \mathbb{K}(\mathcal{E}))^G$ as a space of Kasparov-A-cycles as follows. Let us be given an equivariant $*$-homomorphism $f : \mathcal{S} \to \mathbb{K}(\mathcal{E})$. Then we define the Hilbert-A-module $\mathcal{E}_f$ to be the completion

$$\mathcal{E}_f = \overline{span\ f(\mathcal{S})\mathcal{E}}.$$

As all operators $K \in f(\mathcal{S})$ are G-invariant $\mathcal{E}_f$ is G-invariant, and $\mathcal{E}_f$ canonically has the structure of a Hilbert-A-module. Moreover we have $\mathbb{K}(\mathcal{E}_f) = \overline{f(\mathcal{S})\mathbb{K}(\mathcal{E})f(\mathcal{S})}$. Thus we obtain a $*$-homomorphism $f_r : \mathcal{S} \to \mathbb{K}(\mathcal{E}_f)$ by restricting the range of the $*$-homomorphism $f : \mathcal{S} \to \mathbb{K}(\mathcal{E})$. The $*$-homomorphism f_r uniquely extends to a unital $*$-homomorphism of the multiplier algebras $f_r : C_b(\mathbb{R}) \to \mathbb{B}(\mathcal{E}_f)$ (cf. [11, Proposition 2.1]). The G-invariant operator $\mathcal{F}_f$ then is given by

$$\mathcal{F}_f = f_r(q) \in \mathbb{B}(\mathcal{E}_f),$$

with $q \in C_b(\mathbb{R})$ given by $q(s) = s/\sqrt{1+s^2}$. It follows from this construction that $(\mathcal{E}_f, \mathcal{F}_f)$ is a Kasparov-A-cycle. Note that the pair $(\mathcal{E}_f, \mathcal{F}_f)$ is a function of f. We can associate to $(\mathcal{E}_f, \mathcal{F}_f)$ its equivalence class in $KK_0^G(A)$, which leads to a map $kc : Hom(\mathcal{S}, \mathbb{K}(\mathcal{E}))^G \longrightarrow KK_0^G(A)$ which factors through taking path components on the left side, i.e. we have a corresponding map

$$kc_* : [Hom(\mathcal{S}, \mathbb{K}(\mathcal{E}))]_G \longrightarrow KK_0^G(A),$$

where $[\, , \,]_G$ denotes pointed G-homotopy classes of pointed G-maps. Using the description above one sees that an inclusion $i : \mathcal{E} \subset \mathcal{E}'$ yields a commutative diagram

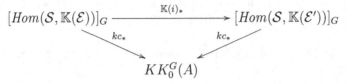

In special cases kc_* is an isomorphism as we shall recall next. For the following let $\widehat{H}$ be a fixed choice of a separable $\mathbb{Z}/2$-graded Hilbert space satisfying $\dim \widehat{H}^{even} = \dim \widehat{H}^{odd} = \infty$. The standard $\mathbb{Z}/2$-graded G Hilbert space then is defined to be $L^2(G) \otimes \widehat{H}$. A separable $\mathbb{Z}/2$-graded G-Hilbert space H is called *universal* if it is isomorphic to $L^2(G) \otimes \widehat{H}$.

Theorem 5.4. *If H is a universal $\mathbb{Z}/2$-graded G-Hilbert space then for any σ-unital G-C^*-algebra A the map*

$$kc_* : [Hom(\mathcal{S}, \mathbb{K}(A \otimes H))]_G \longrightarrow KK_0^G(A)$$

is an isomorphism.

Proof. The theorem is folklore in operator theory; although we are not aware of any written account for the equivariant setting. The non-equivariant statement is proved in [21]. $\qquad\Box$

Lemma 5.5. *Let W be a G-representation equipped with a G-invariant inner product. Assume that there is a G-equivariant smooth embedding of G into W. Then the $\mathbb{Z}/2$-graded G-Hilbert space $L^2(W \oplus \mathbb{R})$ is universal.*

Proof. By the equivariant Kasparov stabilization theorem [10, §1, 12] it suffices to show that there is an embedding of the standard $\mathbb{Z}/2$-graded G-Hilbert space $L^2(G) \otimes \widehat{H}$ as a direct summand into $L^2(W \oplus \mathbb{R})$. We have $L^2(W \oplus \mathbb{R}) \cong L^2(W) \otimes L^2(\mathbb{R})$ and $\widehat{H} \cong L^2(\mathbb{R})$. Thus it suffices to show that $L^2(G)$ is contained in $L^2(W)$ as a direct summand (ignoring the $\mathbb{Z}/2$-gradings). A corresponding direct summand can be constructed from a smooth embedding of G into W and a choice of a G-invariant tubular neighborhood of the embedding. $\qquad\Box$

In the following we will say that a G-representation W is *full* if $L^2(W)$ is universal. For later reference note that if the universe U is complete then the full indexing spaces are cofinal in the diagram of all indexing spaces.

5.3. The exterior Kasparov product and Bott periodicity. If $\mathcal{E}_1$ and $\mathcal{E}_2$ are countably generated Hilbert-modules over G-C^*-algebras A_1 and A_2 respectively let $\mathcal{E}_1 \otimes \mathcal{E}_2$ denote the Hilbert module tensor product (cf. [5, 13.5]). In particular, $\mathcal{E}_1 \otimes \mathcal{E}_2$ is a Hilbert-$A_1 \otimes A_2$-module. The canonical map $\mathbb{K}(\mathcal{E}_1) \otimes \mathbb{K}(\mathcal{E}_2) \to \mathbb{K}(\mathcal{E}_1 \otimes \mathcal{E}_2)$ then is an isomorphism, which we will use in the sequel without notice. If we are given Kasparov-A_i-cycles $(\mathcal{E}_i, \mathcal{F}_i)$ for $i = 1, 2$ then one can define a corresponding Kasparov-$A_1 \otimes A_2$-cycle $(\mathcal{E}_1 \otimes \mathcal{E}_2, \mathcal{F}_1 \# \mathcal{F}_2)$. The definition of $\mathcal{F}_1 \# \mathcal{F}_2$ typically involves various choices (c.f. [5, 18.2]), however the resulting operators for any two sets of choices just differ by a Kasparov homotopy and one obtains a lax symmetric monoidal pairing, the exterior Kasparov product

$$KK_0^G(A_1) \times KK_0^G(A_2) \xrightarrow{\otimes} KK_0^G(A_1 \otimes A_2).$$

Theorem 5.6 (Exterior Kasparov product, e.g. see [17, p. 221]). *Let $\mathcal{E}_1$ and $\mathcal{E}_2$ be countably generated Hilbert modules over $\mathbb{Z}/2$-graded σ-unital G-C^*-algebras A_1 and A_2. Then the $*$-homomorphism $\mathbb{K}(\mathcal{E}_1) \otimes \mathbb{K}(\mathcal{E}_2) \to \mathbb{K}(\mathcal{E}_1 \otimes \mathcal{E}_2)$ induces a commutative diagram*

$$
\begin{array}{ccc}
Hom(\mathcal{S}, \mathbb{K}(\mathcal{E}_1))^G \times Hom(\mathcal{S}, \mathbb{K}(\mathcal{E}_2))^G & \xrightarrow{\;*\;} & Hom(\mathcal{S}, \mathbb{K}(\mathcal{E}_1 \otimes \mathcal{E}_2))^G \\
\downarrow{\scriptstyle kc \times kc} & & \downarrow{\scriptstyle kc} \\
KK_0^G(A_1) \times KK_0^G(A_2) & \xrightarrow{\;\otimes\;} & KK_0^G(A_1 \otimes A_2).
\end{array}
$$

For more details concerning the Kasparov product we refer the reader to [5, VIII.18].

Theorem 5.7 (Bott periodicity in KK-theory, c.f. [10, §5]). *Let V be a G-inner product space and let A be a σ-unital $\mathbb{Z}/2$-graded G-C^*-algebra. Then there is a natural (in A) isomorphism*

$$KK_0^G(A) \cong KK_0^G(C_0(S^V) \otimes \mathbb{C}l_V \otimes A),$$

where $C_0(S^V)$ denotes the trivially graded G-C^*-algebra consisting of continuous functions on S^V vanishing at infinity. The isomorphism is given by exterior multiplication with a specific element, the so-called Bott element $b_V \in KK_0^G(C_0(S^V) \otimes \mathbb{C}l_V)$.

5.4. The exponential law. For spaces $X, Y \in \mathscr{T}_*$ let $C_0(X, Y) \in \mathscr{T}_*$ denote the pointed space of pointed maps from X to Y equipped with the compact-open topology. Given a G-C^*-algebra we also regard it as an object in $\mathscr{T}_*$, the basepoint being $0 \in A$. In case $X \in \mathscr{K}_*$, the full subcategory of $\mathscr{T}_*$ whose objects are the pointed compact G-spaces, the corresponding space $C_0(X, A)$ naturally has the structure of a G-C^*-algebra. The C^*-algebra $C_0(X, A)$ can also be identified with the tensor product of $C_0(X) = C_0(X, \mathbb{C})$ and A since the canonical map $C_0(X) \otimes_{alg} A \to C_0(X, A)$, $f \otimes a \mapsto f \cdot a$ becomes an isomorphism after completing the algebraic tensor product $C_0(X) \otimes_{alg} A$.

Lemma 5.8. *Given a G-space $X \in \mathscr{K}$ and a G-C^* algebra A then there is a pointed G-homeomorphism*

$$(5.9) \qquad \nu : C_0(X, Hom(\mathcal{S}, A)) \;\cong\; Hom(\mathcal{S}, C_0(X) \otimes A).$$

The G-homeomorphism is natural in X and A.

Proof. Using the canonical isomorphism $C_0(X) \otimes A \cong C_0(X, A)$ this is a consequence of the fact that the exponential map $C_0(X \wedge Y, Z) \to C_0(X, C_0(Y, Z))$ is a pointed G-homeomorphism for spaces $X, Y, Z \in \mathscr{T}_*$.   □

Any G-C^*-algebra A can be regarded as a Hilbert module over itself, and there is the canonical isomorphism $\iota : A \to \mathbb{K}(A)$, which sends an element $a \in A$ to the bounded operator that multiplies with a from the left. In case B is a G-C^*-algebra and $\mathcal{E}$ a Hilbert-B-module we thus have a canonical isomorphism $\iota : A \otimes \mathbb{K}(\mathcal{E}) \cong \mathbb{K}(A) \otimes \mathbb{K}(\mathcal{E}) \cong \mathbb{K}(A \otimes \mathcal{E})$. We use this isomorphism for $A = C_0(X)$ to define the composition

$$ad = \iota_* \circ \nu : C_0(X, Hom(\mathcal{S}, \mathbb{K}(\mathcal{E}))) \to Hom(\mathcal{S}, C_0(X) \otimes \mathbb{K}(\mathcal{E}))$$

$$\to Hom(\mathcal{S}, \mathbb{K}(C_0(X) \otimes \mathcal{E})).$$

Let A be a $\mathbb{Z}/2$-graded G-C^*-algebra, let $\mathcal{E}$ be a countably generated Hilbert-A-module, and let $X \in \mathscr{K}_*$ such that $C_0(X)$ is σ-unital. Given an equivariant map $f : X \to Hom(\mathcal{S}, \mathbb{K}(\mathcal{E}))$ we associate to it the Kasparov-$C_0(X) \otimes A$-cycle $(\mathcal{E}_f, \mathcal{F}_f) = (\mathcal{E}_{ad(f)}, \mathcal{F}_{ad(f)})$. This allows us to extend the definition of kc above to a natural map

$$kc : C_0(X, Hom(\mathcal{S}, \mathbb{K}(\mathcal{E})))^G \longrightarrow KK_0^G(C_0(X) \otimes A)$$

by defining $kc(f) = [(\mathcal{E}_f, \mathcal{F}_f)]$. If we are given a path $w : [0, 1] \to C_0(X, Hom(\mathcal{S}, \mathbb{K}(\mathcal{E})))^G$ this corresponds to an element $h \in C_0(X \wedge [0, 1]_+, Hom(\mathcal{S}, \mathbb{K}(\mathcal{E})))^G$, and $(\mathcal{E}_h, \mathcal{F}_h)$ is a Kasparov homotopy between $\mathcal{E}_{w(0)}$ and $\mathcal{E}_{w(1)}$. Hence kc factors through taking homotopy classes, i.e. we obtain an induced map

(5.10)

$$kc_* : [X, Hom(\mathcal{S}, \mathbb{K}(\mathcal{E}))]_G = [X, Hom(\mathcal{S}, \mathbb{K}(\mathcal{E}))]_G \to KK_0^G(C_0(X) \otimes A).$$

5.5. **Proof of Theorem 4.4.** We now come back to our special situation. For any inner product space $V \in \mathscr{I}$ we can identify $\mathbb{C}l_V$ with the adjointable compact operators $\mathbb{K}(\mathbb{C}l_V)$, if we regard $\mathbb{C}l_V$ as a Hilbert-$\mathbb{C}l_V$-module using the action of $\mathbb{C}l_V$ given by right multiplication. The identification $\mathbb{C}l_V \cong \mathbb{K}(\mathbb{C}l_V)$ is given by identifying an element $v \in \mathbb{C}l_V$ with the operator which multiplies with v from the left. Moreover, if we regard $\mathbb{C}l_V \otimes L^2(V)$ as a Hilbert-$\mathbb{C}l_V$-module with $\mathbb{C}l_V$ acting by right multiplication on the first factor and trivially on $L^2(V)$, then we have a canonical isomorphism $\mathbb{K}(\mathbb{C}l_V \otimes L^2(V)) \cong \mathbb{C}l_V \otimes \mathcal{K}_V$.

Recall from Section 3 that we constructed natural transformations

$$p : S^0 \to \widehat{\mathcal{K}} \quad \text{and} \quad \beta : S \to \widehat{\mathbb{C}l} \quad \text{and} \quad \eta = \beta * p : S \to \mathbb{K} = \widehat{\mathbb{C}l \otimes \mathcal{K}}.$$

By construction, if we apply kc to the map $p_V : S_V^0 \to Hom(\mathcal{S}, \mathcal{K}_V)$ we obtain $kc(p_V) = [\mathbb{C}i_V, 0] = [\mathbb{C}, 0]$, which represents the unit $1 \in KK_0^G(\mathbb{C})$. If we apply kc to the map $\beta_V : S^V \to Hom(\mathcal{S}, \mathbb{C}l_V)$ introduced in Definition 4.2 then we obtain the Bott elements

$$b_V = kc(\beta_V) \in KK_0^G(C_0(S^V) \otimes \mathbb{C}l_V).$$

It follows from Theorem 5.6 that

$$kc(\eta_V) = kc(\beta_V) \otimes kc(p_V) = b_V \otimes 1 = b_V.$$

Let $X \in \mathcal{K}$, let $V \in \mathcal{D}$ be an indexing space. Using the canonical isomorphisms $C_0(S^V) \otimes C_0(X_+) \cong C_0(S^V \wedge X_+)$ we obtain commutative diagrams

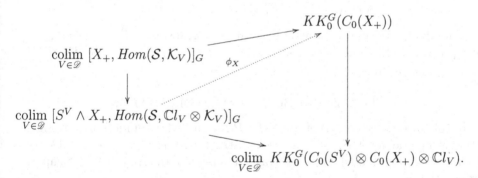

Taking colimits over the indexing spaces V we obtain commutative diagrams

It follows from Bott periodicity that each of the right vertical arrows in the two diagrams above is an isomorphism. In particular this allows us to define the homomorphism ϕ_X as indicated in the diagram. Since the full[10] indexing spaces are cofinal in the diagram of all indexing spaces (cf. Lemma 5.5) it follows from Lemma 5.8 and Theorem 5.4 that ϕ_X is an isomorphism. Note that the lower left corner, by definition, is $\mathbb{K}^0(X_+)$.

We now define $\Phi^0(X) : \mathbb{K}^0(X) \to K_G^0(X)$ as the composition

(5.11)

$$\Phi^0(X) : \mathbb{K}^0(X_+) \xrightarrow{\phi_X} KK_0^G(C_0(X_+)) \cong KK_0^G(C(X)) \xrightarrow{\psi_X^{-1}} K_G^0(X).$$

[10]For the definition of full see page 102

The definition of $\Phi^0(X)$ has a unique extension to a family of isomorphisms

$$(5.12) \qquad \Phi^{V,W}(X) : \mathbb{K}^{V,W}(X_+) \to K_G^{V,W}(X)$$

for inner product spaces $V, W \in \mathscr{I}$ such that $\Phi^{V,0}(X) = \widetilde{\Phi}^0(S^V \wedge X_+)$ for all V and such that the diagrams

$$
\begin{array}{ccc}
\mathbb{K}^0(X_+) & \xrightarrow{\;\;\Phi^0(X)\;\;} & K_G^0(X) \\
\Big\downarrow{\sigma} & & \Big\downarrow{\otimes b_W^{top}} \\
\mathbb{K}^{0,W}(S^W \wedge X_+) & \xrightarrow{\widetilde{\Phi}^{0,W}(S^W \wedge X_+)} & \widetilde{K}_G^{0,W}(S^W \wedge X_+)
\end{array}
$$

commute for all W. It follows that the natural isomorphisms $\Phi^{*,*}$ can be used to define a corresponding natural isomorphism of $RO(G)$-graded cohomology theories $\Phi^*(X) : \mathbb{K}^*(X_+) \to K_G^*(X)$. It also is not hard to see that Φ^* is compatible with the multiplicative structures of the two cohomology theories. This completes the proof of Theorem 4.4.

Remark 5.13. One can extend the definition of the KK-homology groups to a setting where the corresponding groups are labeled by pairs of inner product spaces $V, W \in \mathscr{I}$ by defining

$$KK_{V,W}^G(A) = KK_0^G(C_0(S^V) \otimes \mathbb{C}l_W \otimes A).$$

The methods above also provide natural isomorphisms $\phi_X^{V,W} : \mathbb{K}^{V,W}(X_+) \cong KK_{V,W}^G(C_0(X_+))$. The isomorphisms $\Phi^{V,W}$ of (5.12) also can be described using the isomorphisms $\phi_X^{V,W}$.

6. THE ATIYAH-BOTT-SHAPIRO ORIENTATION

In this section we construct an $\mathscr{I}$-FSP $M Spin^c$ whose underlying prespectrum is a model for $MSpin^c$. We also construct a map $MSpin^c \to \mathbb{K}$ and show that it provides a model for an equivariant version of the Atiyah-Bott-Shapiro orientation.

Given an inner product space $V \in \mathscr{I}$ we consider the group $Pin_V^c \subset \mathbb{C}l_V$ which is generated by the elements in the unit sphere $S(V) \subset V \subset \mathbb{C}l_V$ and the elements of the unit circle $S^1 \subset \mathbb{C} \cdot 1 \subset \mathbb{C}l_V$. Note that Pin_V^c is contained in the units of $\mathbb{C}l_V$ since for any $v \in S(V)$ we have $v^2 = 1 \in \mathbb{C}l_V$, and obviously any $\lambda \in S^1$ is invertible. The $\mathbb{Z}/2$-grading of $\mathbb{C}l_V$ induces a homomorphism $\xi : Pin_V^c \to \mathbb{Z}/2$. The homomorphism ξ is called the *grading homomorphism*. We now consider the twisted conjugation action of Pin_V^c on $\mathbb{C}l_V$; for elements $v \in Pin_V^c$ and $w \in \mathbb{C}l_V$ it is given by $v : w \mapsto \xi(v)vwv^{-1}$. One easily checks that the twisted conjugation action leaves $V \subset \mathbb{C}l_V$ invariant. For each $e \in Pin_V^c$ the restricted action $c_e : V \to V$ is an isometry. It is well-known that associating c_e to e defines a surjective group homomorphism $\varrho_V : Pin_V^c \to O_V$ with kernel S^1. On the other hand any isometry $S : V \to V$

uniquely extends to an algebra automorphism $\mathbb{Cl}_S : \mathbb{Cl}_V \to \mathbb{Cl}_V$. Hence the twisted conjugation action of e on $\mathbb{Cl}_V$ coincides with $\mathbb{Cl}_{\varrho_V(e)}$.

Given an inner product space $V \in \mathscr{I}$ let us write $B(\mathbb{Cl}_V \otimes L^2(V))$ for the space consisting of the bounded operators on $\mathbb{Cl}_V \otimes L^2(V)$, equipped with the strong $*$-topology, and let $\mathbb{U}_V \subset B(\mathbb{Cl}_V \otimes L^2(V))$ be the group generated by the odd unitary operators and the scalars. By construction $\mathbb{U}_V$ comes equipped with a corresponding grading homomorphism $\xi : \mathbb{U}_V \to \mathbb{Z}/2$. The group $\mathbb{U}_V$ acts on $B(\mathbb{Cl}_V \otimes L^2(V))$ by twisted conjugation, and this action factors through an effective action of the projective group $\mathbb{PU}_V = \mathbb{U}_V/S^1$. More precisely, given a unitary operator $U \in \mathbb{U}_V$ and a bounded operator $F \in B(\mathbb{Cl}_V \otimes L^2(V))$ we define $\overline{U} \in \mathbb{PU}_V$ to be the equivalence class of U in $\mathbb{PU}_V$, and $\overline{U}F = \xi(U)UFU^*$. Note that the restriction of the action to the subset $\mathbb{K}(\mathbb{Cl}_V \otimes L^2(V))$ consisting of the compact operators in $B(\mathbb{Cl}_V \otimes L^2(V))$ is compatible with the norm topology in the sense that $\mathbb{PU}_V$ is in fact acting on $\mathcal{K}_V = \mathbb{K}(\mathbb{Cl}_V \otimes L^2(V))$.

Given an element $e \in Pin_V^c$ we can associate to it the unitary operator $U_e \in B(\mathbb{Cl}_V \otimes L^2(V))$ given by

$$U_e(v \otimes f) = ev \otimes (f \circ \varrho_V(e)^{-1}), \quad v \in \mathbb{Cl}_V, f \in L^2(V).$$

This induces a homomorphism $\iota_V : Pin_V^c \to \mathbb{U}_V$. Since the kernel of ϱ_V maps onto $S^1 \subset \mathbb{U}_V$ the homomorphism ι_V induces a homomorphism $j_V : O_V \to \mathbb{PU}_V$. We use j_V to regard $\mathbb{K}(\mathbb{Cl}_V \otimes L^2(V))$ as an O_V-space. Using the canoncial isomorphism $\mathbb{K}(\mathbb{Cl}_V \otimes L^2(V)) \cong \mathbb{Cl}_V \otimes \mathcal{K}_V$ the action of an isometry $S \in O_V$ is given by $S(v \otimes k) = \mathbb{Cl}_S v \otimes \mathcal{K}_S k$ for $v \otimes k \in \mathbb{Cl}_V \otimes \mathcal{K}_V$.

Lemma 6.1. *The map* $\eta_V : S^V \to Hom(\mathcal{S}, \mathbb{K}(\mathbb{Cl}_V \otimes L^2(V)))$, *mentioned in Section 4, is O_V-equivariant.*

Proof. Note that it follows directly from the definition of β_V (see Definition 4.1) that for any orthogonal transformation $S \in O_V$ we have $S(\beta_V(v)) = \beta_V(Sv)$ for all $v \in V$. The same applies to the map $p_V : S_V^0 \to Hom(\mathcal{S}, \mathbb{K}(L^2(V)))$. Hence $\eta_V = \beta * p_V$ is O_V-equivariant. $\square$

It follows that the maps $\widetilde{\alpha_V} : (\mathbb{U}_V)_+ \wedge S^V \longrightarrow Hom(\mathcal{S}, \mathbb{K}(\mathbb{Cl}_V \otimes L^2(V)))$ given by $\widetilde{\alpha_V}(U, v) = \xi(U)U\eta_V(v)U^*$ induce corresponding maps

$$(6.2) \qquad \alpha_V : \mathbb{PU}_{V+} \wedge_{O_V} S^V \longrightarrow Hom(\mathcal{S}, \mathbb{K}(\mathbb{Cl}_V \otimes L^2(V))).$$

Definition 6.3. Let $\mathrm{MSpin}^c : \mathscr{I} \to \mathscr{T}_*$ be the G-functor given by

$$(6.4) \qquad \mathrm{MSpin}_V^c = \mathbb{PU}_{V+} \wedge_{O_V} S^V.$$

Let $\mu : \mathrm{MSpin}^c \wedge \mathrm{MSpin}^c \to \mathrm{MSpin}^c$ be the natural transformation of G-functors $\mathscr{I} \times \mathscr{I} \to \mathscr{T}_*$ given by the maps $\mu_{V,W} : \mathrm{MSpin}_V^c \wedge \mathrm{MSpin}_W^c \to \mathrm{MSpin}_{V \oplus W}^c$ which are induced by the obvious natural maps $\mathbb{U}_{V+} \wedge \mathbb{U}_{W+} \to \mathbb{U}_{V \oplus W+}$ and $S^V \wedge S^W \to S^{V \oplus W}$ respectively. Finally let $\lambda : S^0 \to \mathrm{MSpin}_0^c$

be the unique pointed homeomorphism from S^0 to $\mathrm{MSpin}^c_0 = \{\overline{Id_{L^2(0)}}\}_+ \wedge S^0 \cong S^0$. Then the triple $(\mathrm{MSpin}^c, \mu, \lambda)$ defines a lax symmetric monoidal G-functor, and the natural transformation of lax monoidal G-functors $\eta : S \to \mathrm{MSpin}^c$ given by the maps

$$(6.5) \quad \eta_V : S^V \cong O_{V+} \wedge_{O_V} S^V \xrightarrow{\jmath_{V}+\wedge_{O(V)}Id} \mathbb{P}\mathbb{U}_{V+} \wedge_{O_V} S^V = \mathrm{MSpin}^c_V$$

gives MSpin^c the structure of an $\mathscr{I}\text{-}FSP$. Moreover the maps α_V introduced in (6.2) define a natural transformation $\alpha : \mathrm{MSpin}^c \to \mathbb{K}$ of lax symmetric monoidal G-functors.

Theorem 6.6. *The $\mathscr{I}\text{-}FSP$ (MSpin^c, η) given by (6.4) and (6.5) provides a model for the G-equivariant $Spin^c$-bordism spectrum.*

Proof. Let $V \in \mathscr{I}$ be an inner product space. We have the following diagram of groups with exact rows

$$
\begin{array}{ccccccccc}
1 & \longrightarrow & S^1 & \longrightarrow & Pin^c_V & \xrightarrow{\varrho_V} & O_V & \longrightarrow & 1 \\
& & \downarrow{\scriptstyle\cong} & & \downarrow{\scriptstyle\iota_V} & & \downarrow{\scriptstyle\jmath_V} & & \\
1 & \longrightarrow & S^1 & \longrightarrow & \mathbb{U}_V & \longrightarrow & \mathbb{P}\mathbb{U}_V & \longrightarrow & 1.
\end{array}
$$

It follows that we have a homeomorphism $\mathbb{P}\mathbb{U}_V/O_V \cong \mathbb{U}_V/Pin^c_V$.

Now let us write $\mathbb{U}_V^{even}$ for the even part of $\mathbb{U}_V$ and recall that $Spin^c_V$, by definition, is the even part of Pin^c_V. Assume that $V \neq 0$, so that $\xi : Pin^c_V \to \mathbb{Z}/2$ is onto. We then have a commutative diagram of groups with exact rows

$$
\begin{array}{ccccccccc}
1 & \longrightarrow & Spin^c_V & \longrightarrow & Pin^c_V & \longrightarrow & \mathbb{Z}/2 & \longrightarrow & 1 \\
& & \downarrow & & \downarrow & & \downarrow{\scriptstyle=} & & \\
1 & \longrightarrow & \mathbb{U}_V^{even} & \longrightarrow & \mathbb{U}_V & \longrightarrow & \mathbb{Z}/2 & \longrightarrow & 1.
\end{array}
$$

It follows that we have a G-homeomorphism

$$\mathbb{P}\mathbb{U}_V/O_V \cong \mathbb{U}_V/Pin^c_V \cong \mathbb{U}_V^{even}/Spin^c_V.$$

If V in addition has the property that $L^2(V)^{even}$ is stable in the sense that $L^2(V)^{even} \cong L^2(V)^{even} \otimes \ell$ with ℓ an infinite dimensional separable Hilbert space with trivial G-action, then $\mathbb{U}_V^{even}$ is weakly G-contractible. Hence the space

$$\mathbb{P}\mathbb{U}_V/O_V \cong \mathbb{U}_V^{even}/Spin^c_V$$

is a model for $BSpin^c_V$. Note that the two assumptions on V are satisfied if V contains a copy of the trivial representation, and indexing spaces with this property are cofinal in the diagram of all indexing spaces.

Given two such inner product spaces V_1, V_2 the homomorphism

$$\mu_{V_1,V_2} : Spin^c_{V_1} \times Spin^c_{V_2} \to Spin^c_{V_1 \oplus V_2}$$

induces a commutative diagram

$$
\begin{CD}
\mathbb{PU}_{V_1}/O_{V_1} \times \mathbb{PU}_{V_2}/O_{V_2} @>>> \mathbb{PU}_{W_1 \oplus V_2}/O_{V_1 \oplus V_2} \\
@V{\cong}VV @VV{\cong}V \\
\mathbb{U}^{even}_{V_1}/Spin^c_{V_1} \times \mathbb{U}^{even}_{V_2}/Spin^c_{V_2} @>>> \mathbb{U}^{even}_{V_1 \oplus V_2}/Spin^c_{V_1 \oplus V_2}.
\end{CD}
$$

It follows that the natural maps

$$\nu_{V_1,V_2} : \mathbb{PU}_{V_1}/O_{V_1} \times \mathbb{PU}_{V_2}/O_{V_2} \to \mathbb{PU}_{W_1 \oplus V_2}/O_{V_1 \oplus V_2}$$

provide a model for $B\mu_{V_1,V_2}$.

Finally for any V satisfying the above two properties the natural map of G-vector bundles

$$\mathbb{U}^{even}_V \times_{Spin^c_V} V \longrightarrow \mathbb{PU}_V \times_{O_V} V = E_V,$$

where $Spin^c_V$ acts on V in the standard way (i.e. via ϱ_V), is an isomorphism. Hence $\mathrm{MSpin}^c_V = \mathbb{PU}_{V+} \wedge_{O_V} S^V$, which is homeomorphic to the Thom space of E_V, is a model for $MSpin^c_V$.

The G-vector bundles E_V also fit together in the sense that $E_{V_1} \oplus E_{V_2}$ is a pullback of $E_{V_1 \oplus V_2}$ along ν_{V_1,V_2}. Hence the maps $\mu_{V_1,V_2} : \mathrm{MSpin}^c_{V_1} \wedge \mathrm{MSpin}^c_{V_2} \to \mathrm{MSpin}^c_{V_1 \oplus V_2}$ provide a model for the map

$$MSpin^c(V_1) \wedge MSpin^c(V_2) \to MSpin^c(V_1 \oplus V_2).$$

Since the indexing spaces V with the two required properties are cofinal in the diagram of all indexing spaces it follows that MSpin^c provides a model for $MSpin^c$. $\square$

Definition 6.7. Motivated by the results just obtained we define for an inner product space $V \in \mathscr{I}$

$$
\begin{aligned}
\mathbb{ESpin}^c_V &= \mathbb{U}^{even}_V \\
\mathbb{BSpin}^c_V &= \mathbb{U}^{even}_V/Spin^c_V \\
\gamma_{Spin^c_V} &= \mathbb{ESpin}^c_V \times_{Spin^c_V} V.
\end{aligned}
$$

Note that $\mathbb{ESpin}^c_V$ and $\mathbb{BSpin}^c_V$ are in general not models for the spaces $ESpin^c_V$ and $BSpin^c_V$ respectively, unless Pin^c_V has a G-invariant odd element and $L^2(V)^{even}$ is stable in the sense considered in the proof of the previous theorem.

The remaining part of this section is devoted to prove that $\alpha : \mathrm{MSpin}^c \to \mathbb{K}$ represents an equivariant version of the classical Atiyah-Bott-Shapiro orientation. In order to give the more precise statement (Theorem 6.9 below) we need to recall the definition of the latter.

Let $W = V \oplus V$ for some G-inner product space V. There is a standard right $\mathbb{C}l_W$-action on the exterior algebra $\Lambda^*(V \otimes \mathbb{C})$. The action is given as follows. Let us consider the isomorphism $W = V \oplus V \to V \otimes \mathbb{C}$ which sends $w = (v'_0, v'_1) \in W$ to $v' = v'_0 + iv'_1$. For $z \in \Lambda^*(V \otimes \mathbb{C})$ let

$$z \cdot w = z \wedge v' + z \lrcorner\, v',$$

where $z \lrcorner\, v'$ for an element $z = v_1 \wedge \cdots \wedge v_p \in \Lambda^p(V \otimes \mathbb{C})$ with $v_i \in V \otimes \mathbb{C}$ is given by the formula

$$z \lrcorner\, v' = \sum_{i=1}^{p} (-1)^{p-i} \langle v', v_i \rangle v_1 \wedge \cdots \wedge v_{i-1} \wedge v_{i+1} \wedge \cdots \wedge v_p.$$

One easily checks that the action of w is $\mathbb{C}$-linear (i.e. $(\lambda z) \cdot w = \lambda(z \cdot w)$, $\lambda \in \mathbb{C}$) and that for any $z \in \Lambda^*(V \otimes \mathbb{C})$ one has $z \cdot w \cdot w = z \cdot \langle w, w \rangle$. On the other hand $\Lambda^*(V \otimes \mathbb{C})$ is a complex vector space. Hence the action of the elements $w \in W$ induces a right action of $\mathbb{C}l_W$ on $\Lambda^*(V \otimes \mathbb{C})$. In the following let us write $\mathbb{S}_W$ for the right $\mathbb{C}l_W$-module $\Lambda^*(V \otimes \mathbb{C})$, and let us write $\mathbb{S}_W^t$ for the left $\mathbb{C}l_W$-module, which coincides with $\mathbb{S}_W$, but where the elements of $\mathbb{C}l_W$ act by their adjoints. We then have a $\mathbb{C}l_W$-bimodule isomorphism

$$\mathbb{C}l_W \cong \mathbb{S}_W^t \otimes \mathbb{S}_W.$$

Since $\dim_{\mathbb{C}} \mathbb{S}_W = 2^{\dim_{\mathbb{R}} V}$ it follows that $\mathbb{S}_W$ is an irreducible $\mathbb{Z}/2$-graded right $\mathbb{C}l_W$-module, and that $\mathbb{S}_W^t$ is an irreducible $\mathbb{Z}/2$-graded left $\mathbb{C}l_W$-module.

Let E' be a G-vector bundle over a compact space X with a $Spin_W^c$-structure. Using the $Spin_W^c$-structure bundle $Spin_{E'}^c$ of E' we have a canonical isomorphism

$$Spin_{E'}^c \times_{Spin_W^c} W \cong E'.$$

Let

$$
\begin{aligned}
\mathbb{C}l_{E'} &= Spin_{E'}^c \times_{Spin_W^c} \mathbb{C}l_W, \\
\mathbb{S}_{E'}^t &= Spin_{E'}^c \times_{Spin_W^c} \mathbb{S}_W, \\
\mathbb{S}_{E'}^{Cl} &= Spin_{E'}^c \times_{Spin_W^c} \mathbb{C}l_W.
\end{aligned}
$$

Here the $Spin_W^c$ action on the fiber in the first case is given by conjugation, in the second and the third case it is given by left multiplication. The bundle $\mathbb{C}l_{E'}$ is denoted the *Clifford bundle* associated to E'. The other two bundles are called the *spinor bundle* of E' and the *Clifford spinor bundle* of E' respectively. The latter two are module bundles over $\mathbb{C}l_{E'}$ by means of the action given by $[s, w] \cdot [s, z] = [s, wz]$ for $s \in Spin_{E'}^c$, $w \in \mathbb{C}l_W$, and $z \in \mathbb{S}_W^t$ or $z \in \mathbb{C}l_W$ respectively. Note that E' itself is canonically contained in $\mathbb{C}l_{E'}$. Hence an element e' in the fiber E'_x over some $x \in X$ gives rise to endomorphisms $e' \cdot : (\mathbb{S}_{E'}^t)_x \to (\mathbb{S}_{E'}^t)_x$ and $e' \cdot : (\mathbb{S}_{E'}^{Cl})_x \to (\mathbb{S}_{E'}^{Cl})_x$. In case e is of norm 1, the endomorphism is a unitary isomorphism.

Now let E be a G-vector bundle equipped with a $Spin_V^c$-structure for some inner product space $V \in \mathscr{I}$. Let $E' = V \oplus E$. Let $D(E')$ and $S(E')$ denote the disk and the sphere bundle of E' respectively. The *Atiyah-Bott-Shapiro orientation* of E is a cohomology class $u_E \in \widetilde{K}_G^{0,V}(Th(E))$. We define it by saying what its image is under the sequence of isomorphisms

$$\widetilde{K}_G^V(Th(E)) = \widetilde{K}_G^0(S^V \wedge Th(E)) \cong \widetilde{K}_G^0(D(E')/S(E')) \cong K_G^0(D(E'), S(E')).$$

Recall that elements in relative K-theory for a pair of compact spaces (X, A) can be described by pairs $(\widetilde{E}, \widetilde{\varphi})$ consisting of a $\mathbb{Z}/2$-graded G-vector bundle $\widetilde{E} = \widetilde{E}^{even} \oplus \widetilde{E}^{odd}$ over X and an isomorphism $\widetilde{\varphi} : \widetilde{E}|_A^{even} \to \widetilde{E}|_A^{odd}$. The image of the Atiyah-Bott-Shapiro orientation u_E by definition is represented by the pair $[\mathscr{S}_{E'}^t, \varphi']$, where $\mathscr{S}_{E'}^t = \pi^* \mathbb{S}_{E'}^t$ is the pullback of $\mathbb{S}_{E'}^t$ along $\pi : D(E') \to X$, and where $\varphi' : \mathscr{S}_{E'}^t|_{S(E')} \to \mathscr{S}_{E'}^t|_{S(E')}$ is the isomorphism which in a fiber over a point $e' \in S(E')$ is given by multiplication with e'. For the case where E just is an inner product space $E = V \in \mathscr{I}$ the Atiyah-Bott-Shapiro orientation is the Bott periodicity element introduced in Section 4

$$(6.8) \qquad b_V^{top} = u_V \in \widetilde{K}_G^{0,V}(S^V) = \widetilde{K}_G^0(S^V \wedge S^V).$$

Theorem 6.9. *The natural transformation of $\mathscr{I}$-FSPs $\alpha : \mathrm{MSpin}^c \to \mathbb{K}$, defined in Definition 6.3, represents an equivariant version of the classical Atiyah-Bott-Shapiro orientation, i.e. in particular: if G is the trivial group then the map $\alpha : \mathrm{MSpin}^c \to \mathbb{K}$ induces the classical Atiyah-Bott-Shapiro orientation homomorphism $MSpin_*^c \to K_*$ on the level of homotopy groups.*

Taking for granted Theorem 6.6 a technically precise statement is given as follows. If E is a G-vector bundle with structure group $Spin_V^c$ over some compact space X for some inner product space $V \in \mathscr{I}$, with a classifying map $c : E \to \gamma_{Spin_V^c}$, then the isomorphism

$$\widetilde{\Phi}^{0,V}(Th(E)) : \mathbb{K}^{0,V}(Th(E)) \to \widetilde{K}_G^{0,V}(Th(E)),$$

which is the reduced version of the natural isomorphism $\Phi^{0,V}(Th(E))$ introduced in (5.12), maps the class $\{f_E\} \in \mathbb{K}^{0,V}(Th(E))$ which is represented by the composition

$$(6.10) \quad f_E = \alpha_V \circ Th(c) : Th(E) \to Th(\gamma_{Spin_V^c}) = \mathrm{MSpin}_V^c \xrightarrow{\alpha_V} \mathbb{K}_V$$

to the Atiyah-Bott-Shapiro orientation $u_E \in \widetilde{K}_G^{0,V}(Th(E))$.

It is easy to see that the natural transformation $\alpha : \mathrm{MSpin}^c \to \mathbb{K}$ defines an orientation in the sense of stable homotopy theory. That it is a model for the Atiyah-Bott-Shapiro orientation essentially follows from Theorem 6.6 and the fact that the definition of the natural transformation α is modelled on an operator-theoretic description of the classical Atiyah-Bott-Shapiro.

Sketch of Proof. Let $W = V \oplus V$ and regard the irreducible $\mathbb{Z}/2$-graded $\mathbb{C}l_W$-module $\mathbb{S}_W$ as a Hilbert-$\mathbb{C}l_W$-module. Let $(\mathbb{S}_W, 0)$ be the Kasparov-$\mathbb{C}l_W$-cycle consisting of the irreducible $\mathbb{Z}/2$-graded $\mathbb{C}l_W$-module $\mathbb{S}_W$ and the zero operator on it. The Kasparov cycle represents a generator $[\mathbb{S}_W, 0] \in$

$KK_0^G(\mathbb{C}l_W) \cong \mathbb{Z}$. Performing the exterior product with this generator induces an isomorphism

$$\otimes[\mathbb{S}_W, 0] : KK_0^G(A) \longrightarrow KK_0^G(A \otimes \mathbb{C}l_W)$$

for any σ-unital $\mathbb{Z}/2$-graded G-C^*-algebra A. Given a Kasparov-A-cycle $(\mathcal{E}, \mathcal{F})$ the image in $KK_0^G(A \otimes \mathbb{C}l_W)$ is represented by the Kasparov cycle $(\mathcal{E} \otimes \mathbb{S}_W, \mathcal{F} \otimes Id)$.

On the other hand there is a version of the Swan isomorphism (5.3) which applies to pointed compact spaces Y

$$\widetilde{\psi}_Y : \widetilde{K}_G^0(Y) \to KK_0^G(C_0(Y)).$$

Similarly there is a pointed version $\widetilde{\phi}_Y^{0,V} : \mathbb{K}^{0,V}(Y) \to KK_0^G(C_0(Y) \otimes \mathbb{C}l_V)$ of the isomorphism $\phi_Y^{0,V}$ that we introduced in Remark 5.13. The isomorphism $\widetilde{\Phi}^{0,V}(Y) : \mathbb{K}^{0,V}(Y) \to \widetilde{K}_G^{0,V}(Y)$, which by definition is the reduced version of the natural isomorphism $\Phi^{0,V}(Y)$ introduced in (5.12), then fits into the commutative diagram

$$
\begin{array}{ccc}
\mathbb{K}^{0,V}(Y) & \xrightarrow{\;\;\widetilde{\Phi}^{0,V}(Y)\;\;} & \widetilde{K}_G^{0,V}(Y) \\
\downarrow{\scriptstyle \widetilde{\phi}_Y^{0,V}} & & \downarrow{\scriptstyle \widetilde{\psi}_{S^V \wedge Y}} \\
KK_0^G(C_0(Y) \otimes \mathbb{C}l_V) & & KK_0^G(C_0(S^V) \otimes C_0(Y)) \\
\downarrow{\scriptstyle \otimes b_V} & & \downarrow{\scriptstyle \otimes[\mathbb{S}_W, 0]} \\
KK_0^G(C_0(Y) \otimes \mathbb{C}l_V \otimes C_0(S^V) \otimes \mathbb{C}l_V) & & \\
\downarrow{\scriptstyle KK_0^G(Id \otimes \tau \otimes Id)} & & \\
KK_0^G(C_0(Y) \otimes C_0(S^V) \otimes \mathbb{C}l_V \otimes \mathbb{C}l_V) & & KK_0^G(C_0(S^V) \otimes C_0(Y) \otimes \mathbb{C}l_W)
\end{array}
$$

$$KK_0^G(\tau \otimes Id \otimes Id)$$

We now compare

$$a = \widetilde{\psi}_{S^V \wedge Th(E)}(u_E) \in KK_0^G(C_0(S^V) \otimes C_0(Th(E)))$$

with the element

$$b \in KK_0^G(C_0(S^V) \otimes C_0(Th(E)) \otimes \mathbb{C}l_W)$$

which we define to be the image of $\{f_E\} \in \mathbb{K}^{0,V}(Th(E))$ under the composition in the above diagram (for $Y = Th(E)$) which starts in the upper left corner, passes through the lower left corner and ends in the lower right corner.

Therefore consider the diagram of isomorphisms

$$a \in KK_0^G(C_0(S^V) \otimes C_0(Th(E))) \longrightarrow KK_0^G(C_0(D(E')/S(E')))$$

$$\downarrow \otimes [\mathbb{S}_W, 0] \qquad\qquad\qquad\qquad \downarrow \otimes [\mathbb{S}_W, 0]$$

$$b \in KK_0^G(C_0(S^V) \otimes C_0(Th(E)) \otimes \mathbb{C}l_W) \to KK_0^G(C_0(D(E')/S(E')) \otimes \mathbb{C}l_W)$$

The image of a in $KK_0^G(C_0(D(E')/S(E')))$ is represented by the Kasparov cycle

$$(\mathcal{E}^t, \mathcal{F}^t) = (\Gamma_0(\mathscr{S}_{E'}^t), L), \text{ with } L : \sigma \longmapsto (L(\sigma) : e' \mapsto e'\sigma(e')),$$

while the image of b in $KK_0^G(C_0(D(E')/S(E')) \otimes \mathbb{C}l_W)$ is represented by

$$(\mathcal{E}^{\mathbb{C}l}, \mathcal{F}^{\mathbb{C}l}) = (\Gamma_0(\mathscr{S}_{E'}^{\mathbb{C}l}), L), \text{ with } L : \sigma \longmapsto (L(\sigma) : e' \mapsto e'\sigma(e')).$$

Hence $(\mathcal{E}^t \otimes \mathbb{S}_W, \mathcal{F}^t \otimes Id) \cong (\mathcal{E}^{\mathbb{C}l}, \mathcal{F}^{\mathbb{C}l})$. Therefore $a \otimes [\mathbb{S}_W, 0] = b$, hence $\widetilde{\Phi}^{0,V}(\{f_E\}) = u_E$. $\qquad\square$

References

[1] Adams, J.F.: *"Stable Homotopy and Generalised Homology"*, Chicago Lecture Notes in Mathematics , University of Chicago Press (1974)

[2] Atiyah, M.F.: *K-Theory*, Benjamin (1967)

[3] Atiyah, M.F., Bott R. and Shapiro, A: *"Clifford Modules"*, Topology 3 (1964), 3–38

[4] Atiyah, M.F. and Singer, I.M.: *"Index theory for skew-adjoint Fredholm operators"*, Publ. Math. IHES 37 (1969), 305–326

[5] Blackadar, B.: *"K-theory for operator algebras"*, M.S.R.I. Monographs, volume 5, second edition, Cambridge University Press (1998)

[6] Dixmier, J. and Doudy, A.: *"Champs continus d'espaces hilbertiens et de C*-algèbres"*, Bull. Soc. Math. France 91 (1963), 227–284

[7] Greenlees, J.P.C.: *K-homology of universal sapces and local cohomology of the representation ring*, Topology 32 (1993), 295–308

[8] Joachim, M.: *"The twisted Atiyah orientation and manifolds whose universal cover is spin"*, Ph.D. Thesis, University of Notre Dame (1997)

[9] Joachim, M.: *"A symmetric ring spectrum representing KO-theory"*, Topology 40 (2001), 299–308

[10] Kasparov, G.G.: *"The operator K-functor and extensions of C*-algebras"*, Math. USSR Izvestija 16 (1981), 513–572

[11] Lance, E.C.:*"Hilbert C*-modules"*, LMS Lecture Note Series, volume 210, Cambridge University Press (1995)

[12] Lewis, L.G, May, J.P., and Steinberger, M.: *"Equivariant stable homotopy theory"*, Lecture Notes in Mathematics, volume 1213, Springer (1986)

[13] MacLane, S.: *"Categories for the working mathematician"*, Graduate texts in Mathematics, volume 5, Springer (1971)

[14] Mandell, M.A., May, J.P.: *"Equivariant orthogonal spectra and S-modules"*, Memoirs of the AMS, volume 755, AMS (2002)

[15] Mandell, M.A., May, J.P., Schwede, S., Shipley, B.: *"Diagram spaces, diagram spectra and S-modules, and FSP's"*, Proc. London Math. Soc. (3) 82 (2001), 441–512

[16] May, J.P.: *"E_∞ Ring Spaces and E_∞ Ring Spectra"*, Lecture Notes in Mathematics, volume 577, Springer (1977)

114

[17] Meyer, R.: *"Equivariant Kasparov theory and generalized homomorphisms"*, K-theory 21 (2000), 201–228

[18] Phillips, N.C.: *"Equivariant K-theory and freeness of group actions on C*-algebras"*, Lecture Notes in Mathematics, volume 1274, Springer (1987)

[19] Segal, G.B.: *"Equivariant K-theory"*, Publ. Math. I.H.E.S. 34 (1968), 129–151

[20] Shimakawa, K.: *"Note on the Equivariant K-Theory Spectrum"*, Publ. RIMS, Kyoto Univ. 29 (1993), 449–453

[21] Trout, J.:*"On graded K-theory, Elliptic Operators, and the Functional Calculus"*, Illinios Journal of Math. 44 (2000), 294–309

[22] Wegge-Olsen, N.E.: *"K-theory and C*-algebras"*, Oxford University Press (1993)

MATHEMATISCHES INSTITUT, FACHBEREICH MATHEMATIK UND INFORMATIK, EINSTEINSTR. 62, 48149 MÜNSTER, GERMANY

email-address: joachim@math.uni-muenster.de

(CO-)HOMOLOGY THEORIES FOR COMMUTATIVE (S-)ALGEBRAS

MARIA BASTERRA AND BIRGIT RICHTER

The aim of this paper is to give an overview of some of the existing homology theories for commutative (S-)algebras. We do not claim any originality; nor do we pretend to give a complete account. But the results in that field are widely spread in the literature, so for someone who does not actually work in that subject, it can be difficult to trace all the relationships between the different homology theories. The theories we aim to compare are

- topological André-Quillen homology
- Gamma homology
- stable homotopy of Γ-modules
- stable homotopy of algebraic theories
- the André-Quillen cohomology groups which arise as obstruction groups in the Goerss-Hopkins approach

As a comparison between stable homotopy of Γ-modules and stable homotopy of algebraic theories is not explicitly given in the literature, we will give a proof of Theorem 2.1 which says that both homotopy theories are isomorphic when they are applied to augmented commutative algebras. This result is well-known to experts.

The comparison results provided by Mike Mandell [M] and Basterra-McCarthy [B-McC] can be cobbled together to prove that Gamma cohomology and the André-Quillen cohomology groups in the Goerss-Hopkins approach coincide.

Acknowledgements We are grateful for help we got from Paul Goerss, Mike Mandell and Stefan Schwede: Paul insisted on the isomorphism which is now the subject of Theorem 2.6, Stefan initiated the whole project and the study of the relationship stated in Theorem 2.1; the proof of the isomorphism came out of a discussion with him. And last but not least, Mike patiently answered all our questions concerning André-Quillen cohomology of E_∞ algebras.

1. DIFFERENT COHOMOLOGY THEORIES FOR COMMUTATIVE (S-)ALGEBRAS

We will briefly describe the definitions of the above mentioned homology theories, their range and the domain, on which they coincide. Here range means, that some of them are defined for genuine commutative algebras whereas others are homology theories for commutative S-algebras à la [EKMM]. These reduce to homology theories for algebras by considering Eilenberg-MacLane spectra of commutative rings.

1.1. André-Quillen homology, AQ.

To start with, we should mention the algebraic predecessor of these theories, namely André-Quillen homology of commutative algebras. The standard references for this homology theory are [A], [Q2, Q1], and [W]. For a pointed model category, Quillen defined the notion of homology of objects: he considers the subcategory of abelian objects in that model category. If the inclusion of this subcategory in the whole category has a left adjoint – called abelianization– then the homology of an object is the left derived functor of abelianization. That is, one takes an object, considers a cofibrant resolution and applies the abelianization functor to that resolution.

Let k be a commutative ring with unit. For a commutative (simplicial) k-algebra A this means to take a free simplicial resolution $P_* \to A$, to apply the module of Kähler differentials to P_*, and then define André-Quillen homology of A with respect to the ground ring k and coefficients in an A-module M to be

$$\mathsf{AQ}_*(A|k; M) := \pi_*(\Omega^1_{P_*|k} \otimes_{P_*} M).$$

The module $\Omega^1_{P_*|k}$ is called the *cotangent complex* of A over k and is denoted by $\mathbf{L}_{A|k}$; this is well-defined, because the homotopy groups of $\Omega^1_{P_*|k}$ do not depend on the resolution.

For A as above let I denote the kernel of the multiplication map $I :=$ $\ker(A \otimes_k A \to A)$. Then the module of Kähler differentials has an alternative description: the ideal I has an induced multiplication and the Kähler differentials $\Omega^1_{A|k}$ are isomorphic to I/I^2. The quotient I/I^2 is the module of indecomposables in I and is often denoted by $Q(I)$.

André-Quillen homology vanishes in positive degrees for smooth algebras: if A is smooth over k, then $\mathsf{AQ}_*(A|k; M) = 0$ for all $* > 0$ and $\mathsf{AQ}_0(A|k; M) \cong \Omega^1_{A|k} \otimes_A M$. In particular, if A is étale over k, then André-Quillen homology vanishes in all degrees.

Let $\Lambda^q(V)$ denote the q-th exterior power on a module V. Quillen [Q2, 8.1] constructs a spectral sequence

$$(1.1) \qquad E^2_{p,q} = H_p(\Lambda^q \mathbf{L}_{A|k}) \Longrightarrow \mathrm{Tor}^{A \otimes_k A}_{p+q}(A, A)$$

which starts with André-Quillen homology and its higher versions $H_*(\Lambda^q \mathbf{L}_{A|k})$ and converges to Hochschild homology for commutative algebras A which are k-flat.

The properties, which make André-Quillen homology actually a homology theory, are a transitivity long exact sequence, i.e., for a triple of algebras $A \to B \to C$ the following sequence is long exact:

$$\begin{aligned} \cdots \longrightarrow \; & \mathsf{AQ}_n(B|A; M) \to \mathsf{AQ}_n(C|A; M) \to \mathsf{AQ}_n(C|B; M) \\ \longrightarrow \; & \mathsf{AQ}_{n-1}(B|A; M) \to \cdots \end{aligned}$$

In addition, there is a flat-base change property: if A and B are two commutative k-algebras and B is k-flat, then André-Quillen homology does not see the difference between $A \otimes_k B$ relative to B and A relative to k

$$\mathsf{AQ}_*(A \otimes_k B | B; M) \cong \mathsf{AQ}_*(A|k; M)$$

for all $A \otimes_k B$-modules M. Similarly, if $\mathsf{Tor}_*^k(A, B) = 0$ for $* > 0$, then André-Quillen homology of $A \otimes_k B$ relative to k splits as

$$\mathsf{AQ}_*(A \otimes_k B | k; M) \cong \mathsf{AQ}_*(A|k; M) \oplus \mathsf{AQ}_*(B|k; M).$$

The zeroth André-Quillen homology gives the module of Kähler differentials; the zeroth cohomology is therefore the module of derivations. The first André-Quillen cohomology of A with coefficients in M classifies 'infinitesimal extensions'. To be more precise, $\mathsf{AQ}^1(A|k; M)$ classifies surjections of k-algebras $\pi : E \twoheadrightarrow A$ such that the kernel of π is isomorphic to M as an A-module. Here M is considered as a trivial algebra $M^2 = 0$, and the kernel of π gets its A-module structure from the inclusion into E, i.e., $\pi(e)m = em$ for $e \in E$ and m in the kernel of π.

1.2. Topological André-Quillen homology, TAQ.

1.2. Topological André-Quillen homology, TAQ. Several authors (Waldhausen, McClure and Hunter, Kriz, and Robinson among others) initiated the study of a corresponding theory in the category of E_∞ ring spectra before the necessary foundations where in place. The construction of the category of commutative S-algebras in [EKMM], a model category equivalent to the category of E_∞ ring spectra, allowed M. Basterra to mimic the construction of Kähler differentials and define topological André-Quillen (co)-homology. We give a brief account of this theory in the following section. For a more extensive description see [La] or the original account [B].

Let A be a commutative S-algebra and let B be an A-algebra over A. Then one can build the pullback in the category of A-modules

$$
\begin{array}{ccc}
I_A(B) & \longrightarrow & * \\
\downarrow & & \downarrow \\
B & \longrightarrow & A
\end{array}
$$

and $I_A(B)$ is called the *augmentation ideal of B*. Similarly, for a non-unital A-algebra C (like $I_A(B)$), the multiplication allows us to construct the pushout in the category of A-modules

$$
\begin{array}{ccc}
C \wedge_A C & \longrightarrow & * \\
\downarrow & & \downarrow \\
C & \longrightarrow & Q(C).
\end{array}
$$

We call the outcome $Q(C)$ the *module of indecomposables*.

For every commutative A-algebra B, the smash product $B \wedge_A B$ is naturally augmented over B via the multiplication map. The *topological André-Quillen homology of B with respect to A* is defined to be

$$\mathsf{TAQ}(B|A) := (LQ)(RI_B)(B \wedge_A^L B).$$

Here L stands for left and R for right derived functor.

For any B-module M, the topological André-Quillen homology groups of B with coefficients in M are the homotopy groups of the TAQ spectrum

$$\mathsf{TAQ}_*(B|A; M) := \pi_*(\mathsf{TAQ}(B|A) \wedge_B M).$$

Topological André-Quillen homology has properties similar to algebraic André-Quillen homology. For any triple of cofibrant commutative S-algebras we have a transitivity long exact sequence and there is a 'cofibrant base change' property [B, 4.2,4.3 & 4.6].

For a connective commutative S-algebra A, it is shown in [B, Theorem 8.1(Kriz)] that the usual Postnikov tower of A can be refined to a Postnikov towers consisting of commutative S-algebras, such that the k-invariants live in topological André-Quillen cohomology. This result can be used for an obstruction theory for commutative S-algebra structures: assume for a connective S-module A that there is a commutative S-algebra structure on the n-th Postnikov stage. Then this structure can be lifted to a commutative S-algebra structure on the $(n + 1)$-st stage, if the k-invariant for that stage can be lifted to a k-invariant in topological André-Quillen cohomology.

Recall that given an E_∞ space X, the suspension spectrum of X_+, the space obtained by adjoining a disjoint point to X, is an S-module with an E_∞-ring structure coming from the H-space structure. Hence, $S \wedge (X_+) = \Sigma^\infty(X_+)$ is a commutative S-algebra. In work in progress, M. Basterra and M. Mandell show that its cotangent complex is equivalent to $\underline{X}$, the S-module associated to the spectrum obtained from X using an *infinite loop space machine* (see [EKMM, VII.3]).

More generally, given an augmented commutative A-algebra B there is a reduced version of TAQ with $\widetilde{\mathsf{TAQ}}(B|A) = (LQ_A)(RI_A)(B)$. Then, for an E_∞ space X,

$$\widetilde{\mathsf{TAQ}}(A \wedge X_+|A) \cong A \wedge \underline{X}.$$

The authors use this fact and the weak equivalence of E_∞-ring spectra $MU \wedge MU \to MU \wedge BU_+$ provided by the Thom isomorphism to calculate:

$$\mathsf{TAQ}(MU|S) \cong \widetilde{\mathsf{TAQ}}(MU \wedge BU_+|MU) \cong MU \wedge bu$$

i.e., the cotangent complex of MU, the complex cobordism S-algebra, is the connective complex K-theory module.

In [Mi], Minasian constructed a spectral sequence similar to the spectral sequence (1.1) in the algebraic setting. It is of the form

$$E_1^{s,t} \cong \pi_{t-s}\left(\left(\bigwedge_{i=1}^{s-1}\Sigma\mathsf{TAQ}(A|S)\right)_{h\Sigma_{s-1}}\right) \qquad \text{for} \quad t \geq s \geq 0.$$

This spectral sequence converges to the reduced topological Hochschild homology of A. With the help of this spectral sequence, Minasian could prove [Mi, Corollary 2.8] that for a connective cofibrant S-algebra A topological André-Quillen homology vanishes, if and only if the reduced topological Hochschild homology of A is trivial. Here, it is crucial to assume, that A is connective (compare with the discussion at the end of the paper).

In [McC-Mi] McCarthy and Minasian developed a notion of TAQ-*smooth* and THH-*smooth* commutative S-algebras and they could prove an analogue of the Hochschild-Kostant-Rosenberg theorem for usual Hochschild-homology, which states that Hochschild homology of smooth algebras is isomorphic to the exterior powers of the modules of Kähler differentials.

The counterpart in the context of S-algebras of this theorem [McC-Mi, Theorem 6.1] says, that for a THH-smooth R-algebra A in the category of connective S-algebras, there is a natural equivalence of A-algebras

$$\mathbf{P}_A(\Sigma\mathsf{TAQ}(A|R)) \simeq \mathsf{THH}(A|R)$$

between the free commutative A-algebra on $\Sigma\mathsf{TAQ}(A|R)$ and topological Hochschild homology of A.

With the help of TAQ, one can distinguish certain classes of commutative S-algebras. For instance, the algebraic notion of étaleness can be transferred to étaleness for commutative S-algebras. John Rognes, Randy McCarthy and others use the notion of TAQ-étale maps of S-algebras – maps $A \to B$ of S-algebras with the property that $\mathsf{TAQ}(B|A; B) \sim *$ – and THH-étale maps of S-algebras – maps, such that the reduced topological Hochschild homology $\widetilde{\mathsf{THH}}(B|A; B)$ is trivial – to transfer statements of classical algebra to the theory of commutative spectra. In particular, Rognes applies these and other notions in his work on Galois theory of commutative S-algebras.

1.3. **Gamma homology, HΓ.** In the mid 90's, Alan Robinson and Sarah Whitehouse developed a homology theory for E_∞ algebras, called Gamma homology (HΓ). A published account of this work is [Ro-Wh]. The general definition of Gamma homology is quite involved: they construct an analog of the cotangent complex in the case of E_∞-algebras: if A is a k-algebra over some E_∞ operad $\mathcal{C}$ and M is an A-module, then the *realization* of these data is defined as the cofibre $\mathcal{K}(A|k; M)$ of $|\mathcal{M}(A|k; M)|' \to \mathcal{M}_2$ where $|\mathcal{M}(A|k; M)|'$ is a quotient of $\bigoplus_{n \geq 2} \mathcal{C}_{n+1}\otimes_{\Sigma_n} A^{\otimes n}\otimes M$; $|\mathcal{M}|'$ has a natural filtration by taking the k-th filtration to be everything that is the quotient of $\bigoplus_{2 \leq n \leq k} \mathcal{C}_{n+1}\otimes A^{\otimes n}\otimes$

M under the action of the symmetric groups and the other identifications defined in [Ro-Wh, 2.8 (1),(2)]. The part $\mathcal{M}_2$ is the bottom filtration piece.

For an E_∞ subalgebra A of B the cotangent complex is defined to be the cofibre of

$$\mathcal{K}(A|k; M) \to \mathcal{K}(B|k; M)$$

and Gamma homology is the homology of the cotangent complex. They provide a transitivity long exact sequence [Ro-Wh, 3.4] for a triple of inclusions of E_∞ algebras $A \hookrightarrow B \hookrightarrow C$ and there is also a variant of Gamma homology for cyclic E_∞-algebras [Ro-Wh, 2.9]. In the special example of commutative algebras (viewed as E_∞ algebras) there are several concrete chain complex models for Gamma homology. Sarah Whitehouse gave one model in her thesis [Wh], and Alan Robinson uses a quasi-isomorphic one in [Ro1]. We will briefly give the description of the latter (compare [Ro1, 2.5]).

For a commutative k-algebra A and an A-module M the complex for Gamma homology, $C\Gamma$ is the total complex of a bicomplex $\Xi_{*,*}$, which in bidegree (p, q) consists of

$$\Xi_{p,q} = \mathrm{Lie}_{q+1}^* \otimes k[\Sigma_{q+1}]^{\otimes p} \otimes M \otimes A^{\otimes q+1}.$$

Here, all tensor products are taken with respect to the ground ring k. The k-module Lie_n^* is the dual of the n-th part of the operad for Lie-algebras, i.e., Lie_n (without the dualization) is the free k-module on all Lie words on n generators $x_1, \ldots, x_n$ which contain each x_i exactly once; this is a left-Σ_n-module, where the action of $\sigma \in \Sigma_n$ on a word of length n is given by the sign-action and the permutation of the variables $x_1, \ldots, x_n$.

The horizontal differential is just the differential in the two-sided bar construction of the symmetric group, using the right action of Σ_{q+1} on Lie_{q+1}^* and the left-action on $M \otimes A^{\otimes q+1}$ by permuting the tensor factors in $A^{\otimes q+1}$. The vertical differential uses an action of certain standard surjection on Lie_{q+1}^*. For the precise definition see [Ro1, 2.2–2.5]. In order to get a homotopy invariant definition one should either insist that the algebra A is k-flat or assume that A is replaced by a simplicial flat resolution and the complex $\Xi_{*,*}$ is applied to that.

In the case of commutative algebras, Gamma homology vanishes on étale extensions. There is a transitivity long exact sequence for a triple $A \to B \to C$ of algebras and there is a flat-base change theorem. Gamma homology agrees with André-Quillen homology for algebras over the rational numbers and in general, Gamma homology in degree zero gives the first Hochschild homology group. The zeroth Gamma cohomology is the module of derivations and the first Gamma cohomology group is the module of 'infinitesimal extensions', i.e., it is isomorphic to the first André-Quillen cohomology group.

Some calculations of Gamma homology can be found in [Ri-Ro]. In particular, for smooth algebras, for group rings and for truncated polynomial algebras, there are explicit formulae for Gamma homology.

For a commutative ring spectrum E, Gamma cohomology groups of the algebra of cooperations E_*E contain information about the obstructions for refining the given multipication on the ring spectrum to an E_∞ ring structure. Alan Robinson established this obstruction theory in [Ro1]; an overview can be found in this volume [Ro2]. For instance, the existence of the unique E_∞ structures on the Lubin-Tate spectra E_n [Ro1, Ri-Ro], on real and complex K-theory, on the Adams summand and on the I_n-adic completion of the Johnson-Wilson spectra $\widehat{E(n)}$ [B-R] can be proven this way.

1.4. Stable homotopy of Γ-modules, π_*^{st}. Let Γ denote the skeleton of the category of finite pointed sets with set of objects $[n] = \{0, \ldots, n\}$ with 0 as base point.

There is a well-known way to associate a spectrum to any covariant functor F from Γ to some pointed category $\mathcal{C}$ which has a forgetful functor the category Sets$_*$ of pointed sets. Let us call such an F a left Γ-object in $\mathcal{C}$. Such a functor F can be prolonged to a functor from pointed simplicial sets to simplicial $\mathcal{C}$-objects by approximating an arbitrary pointed set by finite pointed sets and by applying F degreewise: for a pointed simplicial set X_* let $F(X_*)$ be $F(X_n)$ in simplicial degree n.

For two pointed simplicial spaces X_* and Y_*, and for a left Γ-object F in $\mathcal{C}$ we obtain a map $X_* \wedge F(Y_*) \to F(X_* \wedge Y_*)$: each element $x \in X_n$ defines a morphism $x : Y_n \to (X_* \wedge Y_*)_n$ by sending an element y in Y_n to $x(y) := [(x, y)]$, i.e., to the equivalence class of (x, y) in the smash-product. This yields the desired transformation $X_* \wedge F(Y_*) \to F(X_* \wedge Y_*)$ by naturality of F. In particular, we obtain maps

$$\mathbb{S}^1 \wedge F(\mathbb{S}^n) \to F(\mathbb{S}^{n+1})$$

such that the sequence $(F(\mathbb{S}^n))_{n \geqslant 0}$ becomes a spectrum and we denote the stable homotopy groups of that spectrum by $\pi_*^{st}(F)$.

For a commutative ring with unit k, a left Γ-module is a functor from Γ to the category of k-modules. Teimuraz Pirashvili showed in [P, Prop.2.2], that the groups $\pi_*^{st}(F)$ are isomorphic to the derived functors $\mathrm{Tor}_*^\Gamma(t, F)$ of the tensor product $t \otimes_\Gamma F$. Here t is a contravariant functor from Γ to k-modules, which is given by $t[n] = \mathrm{Hom}_{\mathrm{Sets}_*}([n], k)$. The Tor-groups in turn have been identified with the homology groups of the first layers in the Goodwillie tower for F in [Ri2, Theorem 4.5].

If one considers the particular case of the left Γ-module $\mathcal{L}(A|k; M)$ which is given by $[n] \mapsto M \otimes A^{\otimes n}$, for any commutative k-algebra A and any A-module M, then we obtain stable homotopy groups associated to an algebra and a module. Here a map of finite pointed sets $f : [n] \to [m]$ induces multiplication in A, insertion of the unit or the action of A on M: the map f sends an element $a_0 \otimes a_1 \otimes \cdots \otimes a_n \in M \otimes A^{\otimes n}$ to $b_0 \otimes b_1 \otimes \cdots \otimes b_m \in M \otimes A^{\otimes m}$ where each b_i is a product $\prod_{f(j)=i} a_j$, where we interpret this to be the unit of A whenever $f^{-1}(i) = \varnothing$.

There is a visible relationship to Hochschild homology: recall the usual specific chain complex for Hochschild homology. The underlying simplicial set looks like

$$M \rightrightarrows M \otimes A \underset{\leftarrow}{\overset{\leftarrow}{\rightleftarrows}} M \otimes A \otimes A \quad \cdots$$

Here the face maps induce the multiplication in the algebra A respectively the action of A on M.

Taking the simplicial model for the 1-sphere which consists of $n+1$ elements in degree n it is visible, that Hochschild homology of A with coefficients in M is the homotopy of $\mathcal{L}(A|k; M)$ evaluated at $\mathbb{S}^1$. This gives a stabilization map from Hochschild homology to stable homotopy of $\mathcal{L}(A|k; M)$

$$\pi_* \mathcal{L}(A|k; M)(\mathbb{S}^1) \to \operatorname{colim}_n \pi_{*+n} \mathcal{L}(A|k; M)(\mathbb{S}^{n+1}) = \pi_{*-1}^{st}(\mathcal{L}(A|k; M)).$$

As stable homotopy splits tensor products of Γ-modules into sums (see [P, 4.2]) in the following way

$$\pi_*^{st}(F \otimes G) \cong \pi_*^{st}(F) \otimes G[0] \oplus F[0] \otimes \pi_*^{st}(G)$$

we obtain, that stable homotopy of $\mathcal{L}(k[x_i, i \in I]|k; k)$ for an arbitrary indexing set I is isomorphic to $\bigoplus_I \pi_*^{st}(\mathcal{L}(k[x]|k; k))$; therefore, stable homotopy of a free simplicial resolution of an algebra gives as many copies of $\pi_*^{st}(\mathcal{L}(k[x]|k; k))$ as there are generators in the resolving algebra. This additivity property leads to an Atiyah-Hirzebruch spectral sequence (compare [Ri1]) for stable homotopy of augmented commutative algebra, which is of the form

$$E_{*,*}^2 = \mathsf{AQ}_*(A|k; \pi_*^{st}(\mathcal{L}(k[x]|k; k))) \Rightarrow \pi_*^{st}(\mathcal{L}(A|k; k)).$$

The algebra $k[x]$ is the free commutative algebra on one generator and might be interpreted as the 'base point' in this context.

We will meet this spectral sequence again in Schwede's [Sch2] stable homotopy of the algebraic theory of augmented commutative algebras.

1.5. Stable homotopy of algebraic theories, $\pi_*^{\mathcal{T}}$. We will describe this approach by Stefan Schwede in some detail, because we will later give a proof of Theorem 2.1, which compares stable homotopy of the algebraic theory of augmented commutative k-algebras to stable homotopy of the functor $\mathcal{L}(-|k; k)$. Note that our category Γ is denoted by Γ^{op} in [Sch2].

The model for the category of connective spectra used in this approach is the symmetric monoidal category of Γ-spaces, i.e., functors from Γ to the category sSets of simplicial sets which send [0] to a one-point simplicial set. The monoidal structure is given by a smash-product whose definition and properties can be found in [Ly].

Start with a pointed simplicial algebraic theory. This is a pointed simplicial category $\mathcal{T}$ which has the same objects as the category of finite pointed sets Γ and which has a functor from Γ^{op} to $\mathcal{T}$ which preserves products and is the identity on objects. Note that the object $[n]$ is the n-fold product of the object $[1]$ in the category Γ^{op}.

If you do not feel comfortable with algebraic theories, then think of the morphisms from $[n]$ to $[1]$ as all possible n-ary operations in the theory, i.e., in our example of augmented commutative k-algebras. Such a morphism gives an operation from A^n to A for every such algebra A. For any theory $\mathcal{T}$, $\mathcal{T}$-algebras are product-preserving simplicial functors from $\mathcal{T}$ to the category sSets$_*$ of pointed simplicial sets. Therefore these functors are determined by their value on $[1]$. For the theory of augmented commutative k-algebras, a functor $G : \mathcal{T} \to$ sSets$_*$ corresponds to an algebra A as above by $G[1] \cong A$.

Schwede establishes in Theorem [Sch2, 3.1] a simplicial model category of $\mathcal{T}$-algebras. The simplicial structure allows to talk about suspensions of objects: for any $\mathcal{T}$-algebra A, the suspension ΣA is the geometric realization of the simplicial object that sends the simplicial object $\{0 < \ldots < m\}$ to the m-fold coproduct $\coprod_m A$ of A.

Spectra of $\mathcal{T}$-algebras can now be defined by the suspension functor as sequences of $\mathcal{T}$-algebras (A_n) together with maps $\rho_n^A : \Sigma A_n \to A_{n+1}$. Maps of spectra $f : (A_n) \to (B_n)$ are strict maps in the sense that $\rho_n^B \circ \Sigma(f_n) = f_{n+1} \circ \rho_n^A$.

Theorem [Sch2, 4.3] states that the category of spectra of $\mathcal{T}$-algebras, called $\mathcal{S}p(\mathcal{T})$, is a closed simplicial model category. To any theory $\mathcal{T}$, one can associate a monoid in the symmetric monoidal category of Γ-spaces, T^s, such that there is an equivalence between the homotopy category of modules over T^s and the homotopy category of connective spectra (cf. [Sch2, 4.4]).

Stable homotopy of $\mathcal{T}$-algebras can be defined as the homotopy groups of the suspension spectrum of any $\mathcal{T}$-algebra

$$\pi_*^{\mathcal{T}}(A) := \pi_*^{st}(\Sigma^\infty(A)).$$

Having a nice model category around, it makes also sense to talk about Quillen homology which is defined ([Sch2, 5.1]) as:

$$H_*(A) := \pi_*(X_{ab}^c); \qquad H_*(A; M) := \pi_*(M \otimes_{T_{ab}} X_{ab}^c).$$

Here $(-)^c$ is the cofibrant replacement, $(-)_{ab}$ denotes the abelianization of a $\mathcal{T}$-algebra, and M is a right simplicial module over a certain simplicial ring T_{ab}. In a similar way as connective spectra of $\mathcal{T}$-algebras are equivalent to T^s-modules, the category of abelian objects in $\mathcal{T}$-algebras is equivalent to modules over T_{ab}.

This simplicial ring T_{ab} can be described in a more explicit way: there is a *linearization functor* L (see [Sch2, 5.2]) from Γ-spaces to simplicial abelian groups. Let $\overline{\mathbb{Z}}[S_*]$ denote the free abelian group of the pointed simplicial set S_* with the relation that the base point is equivalent to zero. The linearization takes a Γ-space F and assigns

$$L(F) = \mathrm{coker}((p_1)_* + (p_2)_* - \nabla_* : \overline{\mathbb{Z}}[F[2]] \to \overline{\mathbb{Z}}[F[1]])$$

to it. Here the p_1 and p_2 are the projections

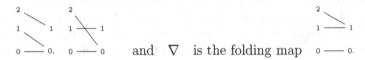

and ∇ is the folding map

The simplicial ring T_{ab} is isomorphic to $L(T^s)$; in particular, if the theory is discrete, then the description of T_{ab} as the linearization of T^s shows that T_{ab} reduces to $\pi_0(T^s)$ in that case (cf. [Sch2, 5.2]).

The suspension spectrum of a $\mathcal{T}$-algebra can be identified with a different spectrum, which is closer related to the stabilization process for Γ-spaces. In [Sch2, 5.1] an alternative to the suspension spectrum is described: define the functor

$$\widetilde{\Sigma^\infty}(A) : \Gamma \longrightarrow \mathcal{T}\text{-algebras}$$

by $\widetilde{\Sigma^\infty}(A)[n] := \coprod_n A$. Surjective maps of finite pointed sets induce folding maps or the projection of components and injective maps of finite pointed sets induce inclusion maps on the coproduct. This functor has the following properties

- The spectrum associated to the Γ-space $\widetilde{\Sigma^\infty}(A)$ is equivalent to the suspension spectrum of A.
- The abelianization of an arbitrary $\mathcal{T}$-algebra is isomorphic to the linearization $L(\widetilde{\Sigma^\infty}(A))$.

Schwede constructs a universal coefficient spectral sequence and an Atiyah-Hirzebruch spectral sequence. The latter has the following shape:

$$E^2_{p,q} = H_p(A; \pi_q W) \Rightarrow W_{p+q}(A).$$

Here W is a right T^s-module and W-homology is defined to be the homotopy of the derived smash product of W with the suspension spectrum of A, $W \wedge^L_{T^s} \Sigma^\infty A$.

In particular, for $W = T^s$ we obtain a spectral sequence which starts with André-Quillen homology of A with coefficients in the homotopy groups of T^s converging to the stable homotopy of A

$$E^2_{p,q} = H_p(A, \pi_q^T(T^s)) \Rightarrow \pi^T_{p+q}(A).$$

2. Comparison results

As promised, we will describe the relationship between the different homology theories for commutative $(S\text{-})$algebras. Except for the first theorem, we will not give proofs of the comparison results, because these can be found in the literature.

We will start with the two homology theories arising from Γ-spaces which have their range of definition in purely algebraic objects:

Theorem 2.1. *Stable homotopy of the Γ-module $\mathcal{L}(A|k;k)$ of an augmented unital commutative k-algebra A is isomorphic to stable homotopy of A, $\pi_*^{\mathcal{T}}(A)$ for the theory $\mathcal{T}$ of commutative augmented k-algebras.*

Proof. We will identify the two Γ-spaces which give stable homotopy of algebraic theories on the one hand and stable homotopy of $\mathcal{L}(A|k;k)$ on the other hand. So let A be an arbitrary augmented commutative k-algebra. The model $\widetilde{\Sigma^\infty}(A)$ of the suspension spectrum looks as follows: the object $[n] \in \Gamma$ is sent to the n-fold coproduct $\coprod_n A$ of A. In the category of commutative algebras, this is the same as the n-fold tensor product of A with itself, $A^{\otimes n}$.

Order-preserving injective maps of finite pointed sets induce the insertion of units on both functors. Let us distinguish surjective maps of finite pointed sets with the property that the preimage of the basepoint zero is only zero from all other surjective maps. Maps of the first kind induce the folding map on the coproduct (which is multiplication), and maps of the second kind involve the projection of components in the coproduct to the basepoint, which is the ring k. Consequently, in the first case elements in A are just multiplied whereas in the other case there is an additional action of A on k by the augmentation.

The Γ-module $\mathcal{L}(A|k;k)$ sends the object $[n]$ to the n-fold tensor product $A^{\otimes n}$ and from the definition of $\mathcal{L}$ in part 1.4 it follows that maps of finite pointed sets induce the same maps on this Γ-module. Therefore the two Γ-spaces are isomorphic and the defining spectra for stable homotopy in both cases agree. $\qquad\square$

Corollary 2.2. *The Atiyah-Hirzebruch spectral sequence for stable homotopy of the algebraic theory of augmented commutative k-algebras coincides with the one for stable homotopy of the functor $\mathcal{L}(-|k;k)$.*

Proof. Stable homotopy of T^s for the theory of augmented commutative k-algebras is isomorphic to the singular k-homology of the Eilenberg-MacLane spectrum of the integers, because T^s is stably equivalent to $Hk \wedge^L H\mathbb{Z}$ (see [Sch2, 7.9]). The result [Ri1, 3.1] (or [Ri-Ro, 3.2]) identifies $Hk_*H\mathbb{Z}$ with stable homotopy of $\mathcal{L}(k[x]|k;k)$, so there is an isomorphism on the level of E^2-terms.

This is not only an additive isomorphism but will lead to an isomorphism of spectral sequences. Let us denote the linearization functor from the category Γ to the category of k-modules which sends a set $[n]$ to the free module k^n by ℓ (in order to distinguish it from the functor L used above). The identification of $\pi_*^{st}(\mathcal{L}(k[x]|k;k))$ with $Hk_*H\mathbb{Z}$ in [Ri1] uses the fact, that the functor $\mathcal{L}(k[x]|k;k)$ can be identified with the linearization functor composed with the infinite symmetric product functor Sym^* from k-modules to k-modules. Stable homotopy of any such composed functor $G \circ \ell$ is isomorphic to the stable derived functors L_*^{st} of Eilenberg and MacLane. See for instance Betley's paper [Be] for a proof of this last claim.

Schwede proves in [Sch2, 7.9] an equivalence between T^s and the composite functor $Hk \circ \mathsf{Sym}$, where Sym is the infinite symmetric product functor on pointed spaces. Therefore we get a natural stable weak equivalence of Γ-spaces and the claim follows. $\qquad\qquad\qquad\qquad\qquad\qquad\qquad\qquad\qquad\square$

The second comparison result relates Gamma homology, a homology theory for commutative rings, which at first sight has nothing to do with functors from finite pointed sets to modules, to stable homotopy of Γ-modules. The proof of Theorem 1 in [P-R] uses an enlargement of the domain of definition for Gamma homology to all Γ-modules. See also [Ro2] for a proof.

Theorem 2.3. [P-R] *Gamma homology of any commutative k-algebra A with coefficients in an A-module M is isomorphic to stable homotopy of the Γ-module $\mathcal{L}(A|k; M)$.*

The second result obtained by Basterra and McCarthy compares topological André-Quillen homology – a homology theory for genuine S-algebras – with Gamma homology – a homology theory for algebras.

Theorem 2.4. [B-McC, 4.2] *Gamma homology is isomorphic to* TAQ *of the corresponding Eilenberg-MacLane spectra for flat algebras, i.e., if A is k-flat, then*

$$\mathsf{TAQ}_*(H(A)|H(k); H(A)) \cong \mathrm{H}\Gamma_*(A|k; A).$$

Using the 'hyperhomology' spectral sequence from [EKMM, 4.1] for the $H(A)$-modules $\mathsf{TAQ}(H(A)|H(k); H(A))$ and $H(M)$ for an A-module M

$$E_{p,q}^2 = \mathrm{Tor}_{p,q}^A(\mathsf{TAQ}_*(H(A)|H(k); H(A)), M)$$
$$\Rightarrow \mathrm{Tor}_{p+q}^{H(A)}(\mathsf{TAQ}(H(A)|H(k); H(A)), H(M)) = \mathsf{TAQ}_{p+q}(H(A)|H(k); H(M))$$

on the one hand and the corresponding spectral sequence for modules on the other hand for the chain complex $C\Gamma_*(A|k; M) = C\Gamma_*(A|k; A) \otimes_A M$, we can extend this isomorphism. The theorem above yields an isomorphism on the level of spectral sequences and we obtain that

$$(2.1) \qquad\qquad \mathsf{TAQ}_*(H(A)|H(k); H(M)) \cong \mathrm{H}\Gamma_*(A|k; M)$$

for k-flat A. Similarly, the corresponding spectral sequences for Ext-groups ensure, that

$$(2.2) \qquad\qquad \mathsf{TAQ}^*(H(A)|H(k); H(M)) \cong \mathrm{H}\Gamma^*(A|k; M)$$

for k-projective A.

In the flat case we obtain an equivalence of all these theories

$$\mathsf{TAQ}_*(H(A)|H(k); H(A)) \cong \mathrm{H}\Gamma_*(A|k; A) \cong \pi_*^{st}(\mathcal{L}(A|k; A))$$

and for A an augmented k-flat algebra we get isomorphisms between all these homology theories:

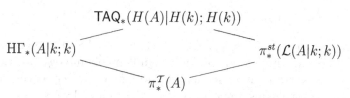

$$\mathsf{TAQ}_*(H(A)|H(k); H(k))$$

$$\mathsf{H\Gamma}_*(A|k; k) \qquad\qquad \pi_*^{st}(\mathcal{L}(A|k; k))$$

$$\pi_*^{\mathcal{T}}(A)$$

The last comparison theorem which we will mention is a result by Mike Mandell. He relates topological André-Quillen cohomology of spectra to TAQ in a differential graded resp. simplicial setting of E_∞-algebras.

Let k be again an arbitrary commutative ring with unit. Mandell defines in [M, 1.1] André-Quillen (co)homology for E_∞-differential graded k-algebras – which we will call $\mathsf{AQ}^*_{dgE_\infty}$ – and for simplicial E_∞-algebras – here denoted by $\mathsf{AQ}^*_{sE_\infty}$.

Theorem 2.5.

(1) [M, 1.8] *The normalization functor N from simplicial k-modules to differential graded k-modules transforms $\mathsf{AQ}^*_{sE_\infty}$ into André-Quillen homology of differential graded E_∞-algebras: for any simplicial E_∞ k-algebra A and any A-module M there is a natural isomorphism*

$$\mathsf{AQ}^*_{sE_\infty}(A|k; M) \cong \mathsf{AQ}^*_{dgE_\infty}(N(A)|k; N(M)).$$

This isomorphism can be extended to simplicial E_∞-algebras relative to another algebra: if $f : A \to B$ is a map of simplicial E_∞-algebras, then

$$\mathsf{AQ}^*_{sE_\infty}(B|A; M) \cong \mathsf{AQ}^*_{dgE_\infty}(N(B)|N(A); N(M)).$$

If the homotopy groups of the module M are concentrated in non-positive degrees then André-Quillen cohomology with coefficients in M resp. in $N(M)$ is concentrated in non-negative degrees.

(2) [M, 7.8–7.10] *Let R be a connective and cofibrant commutative S-algebra. There is a functor Ξ from the category of E_∞ R-algebras to differential graded E_∞-algebras and there is a functor $\mathbf{R}$ from the homotopy category of modules over $\Xi(A)$, for A an E_∞-algebra over R, to the homotopy category of R-modules such that*

$$\mathsf{TAQ}^*(A|R; \mathbf{R}(M)) \cong \mathsf{AQ}^*_{dgE_\infty}(\Xi(A)|\Xi(R); M).$$

A similar result applies to any map $f : A \to B$ of E_∞-R-algebras:

$$\mathsf{TAQ}^*(B|A; \mathbf{R}(M)) \cong \mathsf{AQ}^*_{dgE_\infty}(\Xi(B)|\Xi(A); M).$$

In the cases of coefficients in an Eilenberg-MacLane spectrum, the isomorphism specializes to something very concrete: let A be a connective E_∞-algebra over R and let N be a module over $\pi_0(A)$. Then for any

$\Xi(A)$-*module* M *with* $H_0(M) \cong N$ *and trivial other homology groups we obtain*

$$\mathsf{TAQ}^*(A|R; H(N)) \cong \mathsf{AQ}^*_{\mathrm{dg}E_\infty}(\Xi(A)|\Xi(R); M).$$

The above isomorphisms preserves more structure than the mere additive module structure: all three kinds of André-Quillen cohomology mentioned in the theorem possess transitivity sequences and long exact sequences for short exact and the isomorphisms respects them ([M, 1.9 and 13.2]).

The identification of topological André-Quillen cohomology of spectra with a cohomology theory for differential graded objects made it for instance possible to find a concrete example for an S-algebra, which is TAQ-étale but not THH-étale (see [McC-Mi]). This example (and its chain model – which is just the cochain algebra on the n-th Eilenberg-MacLane space on the field with p elements for $n > 1$) are necessarily not connective, because Minasian's work in [Mi] proves that both notions coincide for connective commutative S-algebras.

Paul Goerss and Mike Hopkins develop an obstruction theory for the existence of E_∞-structures on ring spectra (see [GH2]). The obstruction groups that arise in that context are André-Quillen cohomology groups of algebras over simplicial E_∞ algebras. More precisely, the obstructions for E_∞ structures on a commutative ring spectrum E live in

$$\mathsf{AQ}^*(E_*E|E_*; E_*)$$

where AQ means that one views the graded commutative commutative algebra E_*E of cooperations as a constant simplicial E_∞ algebra.

It is a natural question to ask, what the relationship is between these obstruction groups and the ones developed by Alan Robinson (see [Ro2] and [Ro1]). In his approach, the obstruction groups live in Gamma cohomology of the algebra of cooperations

$$\mathsf{H}\Gamma^*(E_*E|E_*; E_*).$$

In the following we sketch an argument, why the obstruction groups in the two approaches are actually isomorphic. Let k be a commutative ring with unit, let A be a unital commutative k-algebra which is projective as a k-module and let M be an A-module. The rough idea of the proof is to combine Mike Mandell's results [M] with the comparison result in [B-McC] to obtain the desired isomorphism.

The Goerss-Hopkins groups do not actually depend on the choice of a simplicial E_∞ operad, neither do the simplicial André-Quillen groups in Mandell's work. André-Quillen cohomology in both contexts is defined via a cofibrant resolution in the category of simplicial E_∞ algebras. Here the used model categories (in [M, 3.3] resp. [GH2, 4.1]) agree: the weak equivalences are given by maps which induce an isomorphism on homotopy groups and the fibrations are maps which lead to surjective maps in positive degrees after normalization.

Therefore we obtain

$$\mathsf{AQ}^*(A|k; M) \cong \mathsf{AQ}^*_{sE_\infty}(A|k; M)$$

where the first groups denote the Goerss-Hopkins groups and the latter Mandell's groups. As A, k and M are viewed as constant simplicial objects, the result 2.5 yields an isomorphism of these groups with André-Quillen cohomology groups in the category of differential graded E_∞ algebras:

$$\mathsf{AQ}^*_{sE_\infty}(A|k; M) \cong \mathsf{AQ}^*_{dgE_\infty}(A|k; M).$$

These cohomology groups have a relationship with topological André-Quillen cohomology of Eilenberg-MacLane spectra in the following way.

First of all, in the case of constant coefficients M, the functor $\mathbf{R}$ from differential modules over A into modules over HA reduces to

$$\mathbf{R}(M) \simeq H(M).$$

An argument for this can be found in [M, 7.10]. So the cohomology groups on the level of E_∞ ring spectra

$$\mathsf{TAQ}^*(H(A)|H(k); H(M))$$

are isomorphic to $\mathsf{AQ}^*_{dgE_\infty}(\Xi(H(A))|\Xi(H(k)); M)$ and we have to compare these groups to $\mathsf{AQ}^*_{dgE_\infty}(A|k; M)$.

As the algebra $\Xi(H(A))$ is connected, there is a natural map to $H_0(\Xi(H(A))) = A$. This map

$$\varphi : \Xi(H(A)) \longrightarrow A$$

is a map of differential graded E_∞ algebras and is the unique map which gives the inverse of the isomorphism

$$A = \pi_0 H(A) \cong H_0(H(A)) = H_0(\Xi(H(A)))$$

on homology. The functor Ξ is a composition $C_* \circ \Gamma$, where Γ is a CW approximation functor in the category of E_∞ $H(k)$-algebras and C_* is a cellular chains functor. By construction [M, 10.3] there is a canonical weak equivalence

$$\gamma : \Gamma(H(A)) \longrightarrow H(A).$$

The cellular chain functor constructed in [M, §9] does not change the homology which for Eilenberg-MacLane spectra gives the ordinary homotopy groups. Therefore φ is a weak equivalence of E_∞ algebras:

$$H_*\Xi(H(A)) = \pi_*\Gamma(H(A)) \xrightarrow{\simeq} \pi_*H(A) = A.$$

Topological André-Quillen cohomology of commutative $H(k)$-algebras in the category of E_∞ $H(k)$-algebras is isomorphic to usual topological André-Quillen cohomology of commutative $H(k)$-algebras. Taking all these steps together, the Goerss-Hopkins groups $\mathsf{AQ}^*(A|k; M)$ are isomorphic to $\mathsf{TAQ}^*(H(A)|H(k); H(M))$.

Using the comparison result from [B-McC] and adapting it to cohomology as above (2.2) we get an isomorphism of the latter to Gamma cohomology. In fact, for the comparison result we do not need A to be projective over k. The comparison of Gamma homology and topological André-Quillen homology works for flat algebras. To transfer this to cohomology, we just have to have that the universal coefficient spectral sequence collapses. So for such commutative k-algebras A and A-modules M we obtain:

Theorem 2.6. *The Goerss-Hopkins André-Quillen cohomology groups* $AQ^*(A|k; M)$ *are isomorphic to Alan Robinson's Gamma cohomology groups* $H\Gamma^*(A|k; M)$.

References

[A]   M. André, *Homologie des algèbres commutatives*, Die Grundlehren der mathematischen Wissenschaften, **206**, Springer Berlin-New York (1974) xv+341 pp.

[B-R]   A. Baker & B. Richter, Γ-*cohomology of rings of numerical polynomials and* E_∞ *structures on K-theory*, preprint (2003), available at the arXiv

[B]   M. Basterra, *André-Quillen cohomology of commutative S-algebras*, J. Pure Appl. Algebra **144** 2 (1999), 111–143.

[B-McC] M. Basterra & R. McCarthy, Γ-*homology, topological André-Quillen homology and stabilization*, Topology Appl. **121** 3 (2002), 551–566.

[Be]   S. Betley, *Stable derived functors, the Steenrod algebra and homological algebra in the category of functors*, Fund. Math. **168** (2001), 279–293.

[EKMM]   A.D. Elmendorf, I. Kriz, M.A. Mandell & J.P. May, *Rings, modules, and algebras in stable homotopy theory*, with an appendix by M. Cole, Mathematical Surveys and Monographs, **47**, AMS, Providence, RI (1997)

[GH1]   P.G. Goerss & M.J. Hopkins, *André-Quillen (co-)homology for simplicial algebras over simplicial operads*, Une dégustation topologique: homotopy theory in the Swiss Alps (Arolla, 1999), Contemp. Math. **265**, Amer. Math. Soc., Providence, RI, (2000), 41–85,

[GH2]   P.G. Goerss & M.J. Hopkins, *Moduli Spaces of Commutative Ring Spectra*, this volume

[La]   A. Lazarev, *Cohomology theories for highly structured ring spectra*, this volume

[L]   J. L. Loday, *Cyclic homology*, Appendix E by María O. Ronco. Second edition. Chapter 13 by the author in collaboration with Teimuraz Pirashvili, *Grundlehren der Mathematischen Wissenschaften* **301** Springer-Verlag, Berlin (1998) xx+513 pp.

[Ly]   M. Lydakis, *Smash-products and Γ-spaces*, Math. Proc. Camb. Phil. Soc. **126** (1999) 311–328.

[M]   M.A. Mandell, *Topological André-Quillen cohomology and E_∞ André-Quillen cohomology*, Advances in Mathematics **177** no. 2, (2003) 227–279.

[McC-Mi] R. McCarthy & V. Minasian, *HKR theorem for smooth S-algebras*, Journal of Pure and Applied Algebra **185** (2003) 239–258.

[Mi]   V. Minasian, *André-Quillen spectral sequence for THH*, Topology Appl. **129** (2003), 273–280.

[P]   T. Pirashvili, *Hodge decomposition for higher order Hochschild homology*, Ann. Sci. École Norm. Sup. (4) **33** 2 (2000), 151–179

[P-R]   T. Pirashvili & B. Richter, *Robinson-Whitehouse complex and stable homotopy*, Topology **39** 3 (2000), 525–530.

[Q1]   D: Quillen, *Homotopical Algebra*, Lecture Notes in Mathematics, **43** Springer-Verlag, Berlin-New York (1967), iv+156 pp.

[Q2]   D. Quillen, *On the (co-) homology of commutative rings*, Applications of Categorical Algebra, Proc. Sympos. Pure Math., Vol. XVII, New York, 1968, AMS, Providence, R.I. (1970), 65–87.

[Ri1]   B. Richter, *An Atiyah-Hirzebruch spectral sequence for topological André-Quillen homology*, J. Pure Appl. Algebra **171** 1 (2002), 59–66.

[Ri2]   B. Richter, *Taylor towers for Γ-modules*, Ann. Inst. Fourier (Grenoble) **51** 4 (2001), 995–1023

[Ri-Ro]   B. Richter & A. Robinson, *Gamma-homology of group algebras and of polynomial algebras*, to appear in the Proceedings of the 2002 Northwestern Conference on Algebraic Topology, eds. P. Goerss, M. Mahowald & S. Priddy

[Ro1]   A. Robinson, *Gamma homology, Lie representations and E_∞ multiplications*, Invent. Math. **152** 2 (2003), 331–348

[Ro2]   A. Robinson, *Classical obstructions and S-algebras*, this volume

[Ro-Wh]   A. Robinson & S. Whitehouse, *Operads and Γ-homology of commutative rings*, Math. Proc. Cambridge Philos. Soc. **132** 2 (2002), 197–234.

[Sch1]   S. Schwede, *Spectra in model categories and applications to the algebraic cotangent complex*, J. Pure Appl. Algebra **120** 1 (1997), 77–104

[Sch2]   S. Schwede, *Stable homotopy of algebraic theories*, Topology **40** 1 (2001), 1–41

[W]   C. Weibel, *An introduction to homological algebra*, Camb. studies in advanced math. **38**, Cambridge University Press, Cambridge, (1994), xiv+450 pp.

[Wh]   S. Whitehouse, *Gamma (co)homology of commutative algebras and some related representations of the symmetric group*, Thesis, University of Warwick (1994)

DEPARTMENT OF MATHEMATICS, UNIVERSITY OF NEW HAMPSHIRE IN DURHAM, KINGSBURY HALL, DURHAM, NEW HAMPSHIRE 03824, USA
email-address: basterra@cisunix.unh.edu

MATHEMATISCHES INSTITUT DER UNIVERSITÄT BONN, BERINGSTRASSE 1, 53115 BONN, GERMANY
email-address: richter@math.uni-bonn.de

CLASSICAL OBSTRUCTIONS AND S-ALGEBRAS

ALAN ROBINSON

ABSTRACT. Classical obstruction theory can be applied to the problem of finding an S-algebra structure, or a commutative S-algebra structure, on a ring spectrum. It is shown that there is no obstruction to upgrading the homotopy unit in the ring spectrum to a strict unit in the S-algebra.

1. INTRODUCTION

A *ring spectrum* is a spectrum E equipped with a homotopy-associative multiplication map $\mu : E \wedge E \to E$ which has a two-sided homotopy unit $\eta : S \to E$. It is a *commutative ring spectrum* if μ is homotopic to $\mu\tau$, where τ interchanges factors in $E \wedge E$. Thus the (commutative) ring spectra are the (commutative) monoids in the stable homotopy category.

We should like to replace the multiplication μ by a strictly associative multiplication map in the general case; and by a strictly associative and commutative multiplication map in the case of a commutative ring spectrum. (The object E may be replaced by a weakly equivalent object in the process.) These are notions at the point set or model category level, and they make sense if the model category which we are using has a symmetric monoidal smash product. We work in the category of S-modules, which has this property. It would be possible to adapt the theory, making necessary modifications, to other symmetric monoidal model categories for stable homotopy theory or to other contexts such as differential graded objects in an abelian category.

We consider the associative case in §2. The strictly associative multiplication which we seek on the S-module E can equivalently be described as an action of the associative operad $\mathcal{M}$, given by a morphism $\varphi : \mathcal{M} \to \mathrm{End}(E)$ of topological operads. To construct the action φ we replace $\mathcal{M}$ by a suitable cofibrant resolution (in the appropriate category of topological operads). Under our assumptions regarding units in the multiplicative theories, this is the Stasheff operad $\mathcal{A}$ of associahedra. In §3 we describe in detail an obstruction theory for finding a morphism of topological operads $\varphi : \mathcal{A} \to \mathrm{End}(E)$, beginning with the map φ_2 which takes the one-point space $\mathcal{A}_2$ to the point $\mu \in \mathrm{End}(E)_2 = \mathrm{Map}(E \wedge E, E)$. The vanishing of the obstructions suffices for E to be weakly equivalent to an S-algebra. This is a refinement of the theory described in [16]; by comparing the new theory with the old, we show in 3.12 that there is no obstruction to upgrading a homotopy unit to a strict unit when the multiplication is associative.

In §§4–5 we develop the corresponding obstruction theory for refining a commutative ring spectrum to a commutative S-algebra. This runs exactly parallel to the foregoing associative theory, except that the obstructions lie in the Γ-cohomology of the Hopf algebroid E_*E instead of the Hochschild cohomology. Our results here are refinements of those for the homotopy-unital case outlined in [17].

2. Background to the associative case

Suppose that E is an S-module which is also a ring spectrum. Let $\mathcal{M}$ be the topological operad governing associative multiplications. We work here without permutations (these are "non-Σ operads"), so that every space $\mathcal{M}_n$ has a single point. We denote by $\mathrm{End}(E)$ the operad with nth space $\mathrm{Map}(E^{(n)}, E)$, where $E^{(n)} = E \wedge E \wedge \cdots \wedge E$ is the S-module smash product of n factors. The structure which we should like to have on E is a morphism of non-Σ operads $\mathcal{M} \to \mathrm{End}(E)$, as this makes E into an S-algebra. We shall show that it is sufficient to construct this when $\mathcal{M}$ is replaced by a cofibrant resolution. This apparently weaker requirement can be tackled by obstruction theory. We make essential use of properties of cofibrant operads. It is not really necessary for our purposes to formalize the model category concerned, because we only need mapping properties for specific examples where they can be verified simply and directly. However, the formalization can be done: a closed model structure on the category of operads has been described in the algebraic case by Hinich [9] and the model structure on topological operads can be defined in close analogy with [9, §6]. The fibrations (resp. weak equivalences) are the maps of operads which are fibrations (resp. weak equivalences) at each level.

The canonical cofibrant resolution of $\mathcal{M}$ is the operad $W\mathcal{M}$ of "plane trees with stumps" described in [5] and [10]. A morphism $W\mathcal{M} \to \mathrm{End}(E)$ corresponds to a multiplication on E which satisfies all higher associativity conditions and has a two-sided homotopy unit $S \to E$ satisfying all expected coherence conditions.

The operad $W\mathcal{M}$ is larger and freer than is necessary for our obstruction theory. (The situation resembles one in homological algebra, where one need not use a free resolution to calculate Tor, if a flat resolution is much simpler; nor need one use an injective resolution to calculate sheaf cohomology, as a flasque resolution will do.) The simplification here arises from the fact that it is unnecessary to investigate coherent homotopy units because strict units, which are better, are so easy to analyse. Let $\mathcal{A}$ be the Stasheff operad of associahedra, described in detail below. We note that there is a factorization $W\mathcal{M} \to \mathcal{A} \to \mathcal{M}$ of the cofibrant resolution $W\mathcal{M} \to \mathcal{M}$; in the terminology of [4] the first map corresponds to making stumps ignorable. In plainer terms, $\mathcal{A}$ represents A_∞-structures with strict unit.

We need to examine the Stasheff operad $\mathcal{A}$ in some detail. For $n = 0$ and $n = 1$ the space $\mathcal{A}_n$ is a point, corresponding in these two cases respectively to the unit element and the identity map. For $n \geq 2$ the space $\mathcal{A}_n$ is a convex affine cell of dimension $n - 2$, and the composition maps $\mathcal{A}_i \times \mathcal{A}_j \to \mathcal{A}_{i+j-1}$ are affine inclusions corresponding precisely to the inclusions of the top-dimensional faces of $\mathcal{A}_{i+j-1}$. Indeed $\mathcal{A}_2$ is a point, representing a map specifying a multiplication of two factors. Next, $\mathcal{A}_3$ is a line segment, representing an associativity homotopy between the maps represented by its two endpoints, which correspond to the substitutions of the multiplication $\mathcal{A}_2$ for either of the two factors in that multiplication. Then $\mathcal{A}_4$ is the famous Stasheff pentagon, in which the five edges correspond to substitutions of $\mathcal{A}_2$ for variables in the 3-factor multiplication $\mathcal{A}_3$, or *vice versa*. The polytope $\mathcal{A}_5$ is an affine 3-cell which has six pentagonal faces isomorphic to $\mathcal{A}_4 \times \mathcal{A}_2$ or $\mathcal{A}_2 \times \mathcal{A}_4$, and three rectangular faces isomorphic to $\mathcal{A}_3 \times \mathcal{A}_3$; and so on.

If E were an S-algebra, we should have a morphism of operads

$$\mathcal{A} \to \mathcal{M} \to \mathrm{End}(E)$$

obtained by composing the $\mathcal{M}$-action with the resolution above. This composite is, philosophically speaking, the real homotopical nub of the algebra structure. The following proposition shows that an S-algebra can be recovered from it.

Proposition 2.1. *Let E be an S-module which is also a ring spectrum. Suppose that there is a morphism of operads $\mathcal{A} \to \mathrm{End}(E)$ under which the point $\mathcal{A}_2$ is mapped to the given multiplication on E. Then E is weakly equivalent to an S-algebra.*

Proof. Using the cofibrancy of $\mathcal{A}$, we can construct an augmentation $\mathcal{A} \to \mathcal{L}$ from the operad $\mathcal{A}$ into the linear isometries operad $\mathcal{L}$, because the spaces $\mathcal{L}_n$ are contractible. Now we can apply the non-Σ variant of [8, II, Prop. 4.3] to replace the $\mathcal{A}$-spectrum E by a weakly equivalent non-Σ $\mathcal{L}$-spectrum. By [8, II, Props. 4.6 and 3.6] this yields an A_∞ ring spectrum, which can be converted (by smash product with S) into a weakly equivalent S-algebra. □

3. Obstruction theory in the A_∞ case

We now need to describe how obstruction theory can allow us to prove that the hypotheses of Proposition 2.1 can be satisfied. This will require some conditions on the homology theory represented by our ring spectrum E.

In order to simplify the algebra, we shall assume that E is homotopy commutative. We denote by R the graded coefficient ring $\pi_* E$, and by Λ the graded ring $E_* E$. Our assumption implies that these are both commutative; and Λ becomes an R-algebra by means of the homomorphism, conventionally denoted η_L, induced in homology by the unit map $\eta : S \to E$. The multiplication map on E induces an augmentation $\Lambda \to R$, so that Λ splits as a

Λ-module into $R \oplus \tilde{\Lambda}$, where $\tilde{\Lambda}$ is the quotient module Λ/R or the augmentation ideal of Λ.

Definition 3.1. The ring spectrum E satisfies the *perfect universal coefficient formula* if the following two conditions hold.

(1) The algebra Λ is R-flat. Consequently $E_*(Y \wedge E) \approx E_*Y \otimes_R \Lambda$ for every spectrum Y. By induction, the smash power $E^{(n)}$ has E-homology $\Lambda^{\otimes n}$.

(2) The natural map
$$E^*(E^{(n)}) \longrightarrow \operatorname{Hom}_R(E_*(E^{(n)}), R) \approx \operatorname{Hom}_R(\Lambda^{\otimes n}, R)$$
is an isomorphism for every n.

The first condition in 3.1 is satisfied by many ring spectra including all those representing Landweber exact homology theories. The second is more restrictive, but is true for a wide range of useful spectra (see [14]).

The second condition can be rewritten in a more convenient way. Assuming that 3.1(1) holds, the algebra Λ is the Hopf algebroid of E-homology co-operations, or dual Steenrod algebroid, and the homology of any spectrum Y is a Λ-comodule via a natural homomorphism $E_*(Y) \longrightarrow E_*(Y) \otimes_R \Lambda$. Using the cofreeness of the Λ-comodule Λ, we can write the condition in 3.1(2) as

$$3.1(3) \quad E^*(E^{(n)}) \approx \operatorname{Cohom}_\Lambda(E_*(E^{(n)}), E_*E) \approx \operatorname{Cohom}_\Lambda(\Lambda^{\otimes n}, \Lambda)$$

where Cohom denotes homomorphisms of comodules.

Standing hypothesis 3.2. We assume henceforth throughout this paper that the ring spectrum E has a perfect universal coefficient formula: that is, E satisfies Definition 3.1; and that the map $\eta : S \to E$ is a cofibration.

(It would be interesting to know whether the obstruction theory can be set up when these conditions are relaxed. There may well be a derived-category variant which works more generally.)

Hochschild complexes 3.3. We shall need two versions of the Hochschild cochain complex of Λ over R. Let $C^{**}(\Lambda|R; R)$ be the standard *unnormalized* edition (with R as coefficients): thus

$$C^{m,*}(\Lambda|R; R) \approx \operatorname{Hom}_R^*(\Lambda^{\otimes m}, R)$$

where the second grading is the internal grading in the rings. The Λ-module structure on R is given by the augmentation $\Lambda = \pi_*(E \wedge E) \to \pi_* E = R$, and the formula for the coboundary $\delta : C^{m,*}(\Lambda|R; R) \longrightarrow C^{m+1,*}(\Lambda|R; R)$ is

$$(\delta\theta)(\lambda_0 \otimes \lambda_1 \otimes \cdots \otimes \lambda_m) = \lambda_0 \cdot \theta(\lambda_1 \otimes \cdots \otimes \lambda_m)$$
$$+ \sum_{i=1}^{m}(-1)^i \theta(\lambda_0 \otimes \cdots \otimes \lambda_{i-1}\lambda_i \otimes \cdots \otimes \lambda_m)$$
$$+ (-1)^{m+1}\theta(\lambda_0 \otimes \lambda_1 \otimes \cdots \otimes \lambda_{m-1}) \cdot \lambda_m .$$

On the other hand the *normalized* Hochschild cochain complex of Λ over R is $\tilde{C}^{m,*}(\Lambda|R;R) \approx \operatorname{Hom}^*_R(\tilde{\Lambda}^{\otimes m}, R)$ where $\tilde{\Lambda}$ is Λ/R (which is isomorphic to the augmentation ideal). The formula above still defines a coboundary, making $\tilde{C}^{*,*}(\Lambda|R;R)$ a subcomplex of $C^{*,*}(\Lambda|R;R)$. By the Normalization Theorem, the inclusion is a weak equivalence, so that each complex has cohomology $HH^{**}(\Lambda|R;R)$.

Definition 3.4. An $\hat{\mathcal{A}}_n$-*structure* on the ring spectrum E is a collection of maps $\mu_m : \mathcal{A}_m \to \operatorname{End}(E)_m$ for $2 \le m \le n$, such that

   (1) the point $\mathcal{A}_2$ is mapped by μ_2 to the multiplication in E
   (2) the conditions for a morphism of operads are satisfied where defined.

The second clause in this definition means the following. Recall that the boundary of the $(m-2)$-cell $\mathcal{A}_m$ is a union of faces, each an embedded copy of $\mathcal{A}_i \times \mathcal{A}_j$ where $i + j = m + 1$. The condition is that the restriction of μ_m to each face must be the composite $c \circ (\mu_i \times \mu_j)$, where c is the corresponding composition in the operad $\operatorname{End}(E)$. Note that we are here temporarily working with operads without unit, as was done in [16]. (The notation $\hat{\mathcal{A}}$ is intended to suggest that something is omitted.) The homotopy unit is present in E, but is not part of the operad structure.

The unit condition 3.5. We have assumed that the given homotopy unit $\eta : S \to E$ is a cofibration. Hence the wedge $(S \wedge E) \vee (E \wedge S)$ is now included as a subspectrum in the smash product $E \wedge E$ by the cofibration $(\eta \wedge 1) \vee (1 \wedge \eta)$. Since E is assumed to be an S-module, this wedge is isomorphic to $E \vee E$.

Lemma 3.6. *The given multiplication* $\mu_2 : E \wedge E \to E$ *can be deformed by a homotopy to make its restriction into the folding map* $1_E \vee 1_E$.

Proof. By the homotopy extension property, it suffices to show that $\mu_2|E \vee E$ is homotopic to the folding map. However, it is not obvious that this condition is satisfied, because there is no reason why the left and right unit homotopies should agree on $S \wedge S$. Thus there appears to be an obstruction in $\pi_1 E$. This obstruction is in fact zero, by the same argument as proves that an H-space is always simple. $\square$

Thus η can be assumed to be a strict unit for μ_2.

We should like to have an $\mathcal{A}_\infty$ structure in which η is a strict unit for all the maps $\mu_n : \mathcal{A}_n \to \operatorname{End}(E)_n$. Let us consider what that means.

The Stasheff cells $\mathcal{A}_n$ are related not only by face maps but also by degeneracy maps $s_i : \mathcal{A}_n \to \mathcal{A}_{n-1}$ which are defined for $1 \le i \le n$ and are related to the principal faces of $\mathcal{A}_n$ very much as faces and degeneracies among simplices are related [20]. In terms of trees, s_i corresponds to pruning off the ith twig. In terms of operads, the s_i define the n operad compositions $\mathcal{A}_0 \times \mathcal{A}_n \to \mathcal{A}_{n-1}$ with the one-point space $\mathcal{A}_0$, thus completing the operad $\mathcal{A}$ to an operad with unit.

For $1 \leq i \leq n$ there is a cofibration $\eta_i : E^{(n-1)} \to E^{(n)}$ defined as the composite

$$E^{(n-1)} \approx E^{(i-1)} \wedge S \wedge E^{(n-i)} \xrightarrow{1^{(i-1)} \wedge \eta \wedge 1^{(n-i)}} E^{(n)}$$

We define the *large wedge* $\bigvee^n E$ to be the union of the images of the η_i for $1 \leq i \leq n$: it is the S-submodule of points with at least one factor in S.

Definition 3.7. We say that η is a *strict unit for* μ_n if the following diagram commutes for $1 \leq i \leq n$

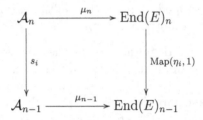

which can be interpreted by saying that a product of n factors is unaffected by a unit in the ith place among the arguments.

We note that the above condition just fixes the value of μ_n on the image of η_i for each i. The relations among the degeneracy maps s_i imply that if η is a strict unit for μ_{n-1} and for μ_n, then Definition 3.7 prescribes the adjoint $\mu'_n : \mathcal{A}_n \ltimes E^{(n)} \to E$ uniquely and coherently on $\mathcal{A}_n \ltimes \bigvee^n E$.

Definition 3.8. An $\mathcal{A}_n$-structure on the ring spectrum (E, μ, η) is a $\hat{\mathcal{A}}_n$-structure (see 3.4) such that the map $\eta : S \to E$ is a strict unit for μ_m, $2 \leq m \leq n$.

We are now ready to set up the obstruction theory in the associative case.

Theorem 3.9. *Let an $\mathcal{A}_{n-1}$ structure μ on E be given, where $n \geq 3$ and E satisfies the conditions 3.2. Then the following hold.*

(1) *There is an obstruction cocycle $\tilde{\theta}_n(\mu)$ in the normalized Hochschild cochain group $\mathrm{Hom}_R^{3-n}(\tilde{\Lambda}^{\otimes n}, R)$ which vanishes if and only if μ can be extended to an $\mathcal{A}_n$ structure on E.*

(2) *The Hochschild cohomology class $[\tilde{\theta}_n(\mu)] \in HH^{n,3-n}(\Lambda|R; R)$ is zero if and only if the underlying $\mathcal{A}_{n-2}$ structure on E can be extended to an $\mathcal{A}_n$ structure.*

Proof. To extend μ to an $\mathcal{A}_n$ structure on E, we need only construct $\mu_n : \mathcal{A}_n \to \mathrm{End}(E)_n$, or equivalently its adjoint $\mu'_n : \mathcal{A}_n \ltimes E^{(n)} \longrightarrow E$, in such a way as to be compatible with composition in the operads. There are two cases of this condition, and we must consider them separately. Preserving the compositions $\mathcal{A}_i \times \mathcal{A}_n \to \mathcal{A}_{n+i-1}$ when $i > 0$ means that μ_n is already defined on the decomposable elements of $\mathcal{A}_n$, which form the boundary of this $(n-2)$-cell. Preserving the compositions $\mathcal{A}_0 \times \mathcal{A}_n \to \mathcal{A}_{n-1}$, which is the condition of strong unitality, means that μ'_n is already defined on $\mathcal{A}_n \ltimes \bigvee^n E$. In all, the condition fixes μ'_n on $\partial\mathcal{A}_n \ltimes E^{(n)} \cup \mathcal{A}_n \ltimes \bigvee^n E$.

The obstruction to extending μ'_n over $\mathcal{A}_n \ltimes E^{(n)}$ therefore lies in the group

$$E^{-1}((\mathcal{A}_n, \partial\mathcal{A}_n) \ltimes (E^{(n)}/\overset{n}{\bigvee} E)) \, .$$

By the Künneth Theorem we know that

$$E_*(E^{(n)}/\overset{n}{\bigvee} E) \quad \approx \quad E_*((E/S)^{(n)}) \quad \approx \quad \tilde{\Lambda}^{\otimes n}$$

and the homology sequence of the pair $(E^{(n)}, \bigvee^n E)$ splits.

Furthermore $\mathcal{A}_n/\partial\mathcal{A}_n$ is an $(n-2)$-sphere, so by the universal coefficient formula the above group becomes

$$
\begin{aligned}
E^{-1}((\mathcal{A}_n, \partial\mathcal{A}_n) \ltimes (E^{(n)}/\overset{n}{\bigvee} E)) &\approx E^{-1}(S^{n-2} \wedge (E^{(n)}/\overset{n}{\bigvee} E)) \\
&\approx \operatorname{Hom}_R^{3-n}(\tilde{\Lambda}^{\otimes n}, R) \\
&\approx \tilde{C}^{n,3-n}(\Lambda | R; R) \, .
\end{aligned}
$$

Therefore the obstruction is a normalized Hochschild cochain, as claimed. We denote it by $\tilde{\theta}_n(\mu)$. If $\tilde{\theta}_n(\mu) = 0$, then the homotopy classes of extensions μ_n are enumerated by difference classes in $\tilde{C}^{n,2-n}(\Lambda | R; R)$.

Consider the effect upon the obstruction cochain $\tilde{\theta}_n(\mu)$ of changing μ_{n-1} while keeping the $\mathcal{A}_{n-2}$ structure fixed. Making this change alters the already-specified map $\mu'_n \mid (\partial\mathcal{A}_n \ltimes E^{(n)} \cup \mathcal{A}_n \ltimes \bigvee^n E)$ on $\mathcal{A}_n \ltimes \bigvee^n E$, and on certain faces of $\mathcal{A}_n$. Altering μ'_n on $\mathcal{A}_n \ltimes \bigvee^n E$ does not affect the obstruction cochain, because the homology sequence for the pair $(E^{(n)}, \bigvee^n E)$ splits. To find the effect of altering $\mu_n | \partial\mathcal{A}_n$ we consider the faces separately. On a face which is an embedded copy of $\mathcal{A}_i \times \mathcal{A}_j$, the restriction of μ_n is determined by $\mu_i \times \mu_j$. Since we are changing only μ_{n-1}, those affected are the two faces isomorphic to $\mathcal{A}_2 \times \mathcal{A}_{n-1}$ and the $n-1$ faces isomorphic to $\mathcal{A}_{n-1} \times \mathcal{A}_2$. These faces correspond precisely to the $n+1$ terms (of two kinds) in the above formula for the Hochschild coboundary.

Let us briefly explain why these faces give precisely the terms in the Hochschild coboundary formula. For the first and last terms, this is verified most easily by using formula 3.1(3). For all the other terms in the Hochschild formula, it is obvious, apart from the sign. We have to verify that the signs alternate. This is forced by the fact that, as we show at the end of this proof, the geometrical obstruction must always be a cocycle. (The details of calculation need to be slightly elaborated in the lowest case $n = 3$, but the result is the same.)

It follows that changing μ_{n-1} by a difference class $\alpha \in \tilde{C}^{n-1,3-n}(\Lambda | R; R)$ has the effect of altering the obstruction cochain by $\tilde{\theta}_n(\mu)$ by $\delta\alpha$.

Therefore the obstruction $\tilde{\theta}_n(\mu)$ can be reduced to zero by altering μ_{n-1} if and only if it is a Hochschild coboundary. To complete the proof of Theorem 3.9 we show that $\tilde{\theta}_n(\mu)$ is always a cocycle. This is true because the coboundary of $\theta_n(\mu)$ is, by the argument over faces used above, the obstruction to extending μ over the boundary of the boundary of $\mathcal{A}_{n+1}$; and this is an empty space. $\qquad\square$

To apply 3.9 recursively, we need a $\mathcal{A}_2$ structure to start the induction. This is provided by Lemma 3.6 In fact we only need the cohomology to be zero for $n \geq 4$, because the obstruction cohomology class for the existence of a $\mathcal{A}_3$-structure is always zero, as we shall see in Theorem 3.12 below. This means we can always choose the associativity homotopy μ_3 so that it is constant when one of the three factors is the unit. (It is not particularly easy to prove this by direct, bare-hands construction.)

Corollary 3.10. *If $HH^{n,3-n}(\tilde{\Lambda}|R;R) = 0$ for all $n \geq 4$, then the ring spectrum E has an $\mathcal{A}_\infty$ structure. By Proposition 2.1, E can thus be represented by an S-algebra.*

Comparison with the homotopy-unital theory 3.11. In [16] we developed a theory exactly parallel to the above, but without the strict unital condition. The given ring spectrum E has a homotopy unit, but now we do not regard the unit as part of the operad structure. In terms of trees, we allow no stumps. In this case one again builds the $\hat{\mathcal{A}}_n$-structure – that is, the higher associativity conditions – by induction on n. The obstructions lie in the unnormalized Hochschild cochain complex. The proof is a simplified version of the proof of 3.9.

A ring spectrum is homotopy associative, and therefore already has an $\mathcal{A}_3$ structure. We can therefore begin a recursive application of our homotopy-unital analogue at $n = 4$.

This result was used in [16] to show every Morava K-theory at an odd prime has an $\hat{\mathcal{A}}_\infty$ structure (indeed, has uncountably many such structures).

We therefore have two variants of the obstruction theory, which seem to be essentially equivalent: they just give rise to the normalized and non-normalized Hochschild complexes, which have the same homology. One therefore guesses that the existence of a strict unit is, homotopically speaking, no more of a restriction than the existence of a homotopy unit; a fact which is confirmed by the next theorem. (This is not too surprising. There is a close analogy with the theory of H-spaces, and it has long been known, for instance through quasifibration theory, that a connected associative H-space with homotopy unit is equivalent to a Moore loop space, which has a strict unit.)

Theorem 3.12. *Suppose E is a ring spectrum satisfying 3.2. Let $3 \leq n \leq \infty$, and suppose that E admits a $\hat{\mathcal{A}}_n$ structure. Then*

(1) E admits a $\mathcal{A}_n$-structure

(2) if $\mathcal{A}_n(E)$ (respectively $\hat{\mathcal{A}}_n(E)$) denotes the space of $\mathcal{A}_n$-structures (respectively $\hat{\mathcal{A}}_n$-structures) on E, then the forgetful map $\Phi : \mathcal{A}_n(E) \to \hat{\mathcal{A}}_n(E)$ is a weak homotopy equivalence.

Proof. We shall show that all the homotopy groups of the map Φ are trivial. Specifically, if for any $k \geq 0$ we have any maps $f : \Delta^k \to \hat{\mathcal{A}}_r(E)$ and $g : \partial\Delta^k \to \mathcal{A}_r(E)$ such that $f|\partial\Delta_k \simeq \Phi g$, then g extends to a map $\Delta^k \to \mathcal{A}_r(E)$. Since we can take $k = 0$, this will prove (1) as well as (2).

We prove the claim by setting up obstruction theory like that in 3.9, proceeding step by step up the operad. At the rth stage, we have precisely the problem of extending, over the interior of a k-cell, a deformation of $\mu'_r \mid (\mathcal{A}_r \ltimes \bigvee^r E)$ into the map prescribed by the unitality condition. (By homotopy extension, the deformation over $\mathcal{A}_r \ltimes E^{(r)}$ then follows.) The obstruction is a cocycle in the kth suspension of the mapping cone of the standard cochain map from the normalized Hochschild complex $\mathrm{Hom}_R^*(\tilde{\Lambda}^{\otimes *}, R)$ to the unnormalized one, because the E-homology of $\bigvee^r E$ is the degenerate part of $\Lambda^{\otimes r}$. Since normalization does not affect cohomology, this mapping cone is contractible; so the cocycle is a coboundary, and the extension exists. $\square$

4. BACKGROUND TO THE COMMUTATIVE CASE: GAMMA HOMOLOGY

We now aim to prove theorems exactly analogous to 3.9 and 3.12 which will handle commutativity and associativity simultaneously. This will allow us to replace a commutative ring spectrum (which is an abelian monoid object in the stable homotopy category) by a commutative S-module (which is the equivalent at the point-set level) provided that certain cohomological obstructions vanish.

Whereas Hochschild cohomology of algebras has been known for 50 years and the Stasheff operad for 40 years, the cohomology theory for commutative algebras and the cofibrant resolution of the commutative operad $\mathcal{C}$ needed here are recent developments. The cohomology theory is Γ-cohomology [18]. It is no longer very new: this homology for commutative algebras had its origins in different ideas developed independently by the author and by F. Waldhausen in the late 1980's. A related theory, called topological André-Quillen cohomology, was invented for essentially the same purpose by Basterra [2] and Kriz. The dual homology theory also arises as an instance of Schwede's stable homotopy of algebraic theories [19]. The relations among all these different approaches are surveyed by Basterra and Richter [3].

We need a commutative ground ring for our homological algebra. With a view to the application, we denote it by Λ. There is no restriction on the characteristic. From our point of view, the homology involves the Lie representations of the symmetric groups. The connection of these with geometry will become apparent in 5.6.

The Lie representations 4.1. Let $\mathcal{L}_n$ be the free Lie algebra over Λ on the set of generators $\{x_i\}_{1 \leq i \leq n}$. We denote by Lie_n the so-called *multilinear part* of $\mathcal{L}_n$. This can be described in many different ways. First, it is defined as the direct summand of $\mathcal{L}_n$ spanned by all Lie monomials containing each of the n generators exactly once. Second, it is the nth module in the Lie operad. Third, it is isomorphic to the module of all natural transformations $\Phi^{\otimes n} \to \Phi$, where Φ is the forgetful functor from Lie algebras to Λ-modules.

The symmetric group Σ_n acts upon Lie_n by permuting the n generators. The Σ_n-module thus obtained is known as the *Lie representation*. We twist it by the sign character, so that the left action of Σ_n on the abelian group Lie_n is defined by setting

$$\sigma \cdot f(x_1, \ldots, x_n) = \varepsilon(\sigma) \, f(x_{\sigma(1)}, \ldots, x_{\sigma(n)})$$

for every multilinear Lie monomial f and every $\sigma \in \Sigma_n$, where $\varepsilon(\sigma)$ is the sign of σ. Let Lie_n^* be the dual module $\mathrm{Hom}(\mathrm{Lie}_n, \Lambda)$, which is thus a right Σ_n-module.

We shall require the following properties of Lie_n (see [21, 2.3]):

(1) the left regulated Lie brackets

$$\sigma \cdot [x_1, [x_2, [x_3, \ldots, [x_{n-1}, x_n]..]]] \quad \text{for} \quad \sigma \in \Sigma_{n-1}$$

form a Λ-basis of Lie_n. Therefore

(2) the Λ-modules Lie_n and Lie_n^* are free of rank $(n-1)!$, and

(3) the restricted Σ_{n-1}-modules $\mathrm{Res}_{\Sigma_{n-1}}^{\Sigma_n}\mathrm{Lie}_n$ and $\mathrm{Res}_{\Sigma_{n-1}}^{\Sigma_n}\mathrm{Lie}_n^*$ are respectively
isomorphic to the left and right regular representations.

The Ξ-complex and stable homotopy 4.2. Let Γ be the category of finite based sets, and $[n]$ the typical object $\{0, 1, \ldots, n\}$ with 0 as the basepoint. A *left Γ-module* is a functor Φ from Γ to Λ-modules. Such a functor converts simplicial finite sets into simplicial modules. Bousfield and Friedlander [6], developing ideas of G.B. Segal, show that the homotopy groups $\pi_{n+i}\Phi(S^i)$ are independent of the simplicial model of the sphere S^i, and indeed independent of i for $i > n$. One therefore defines

$$\pi_n\Phi \quad = \quad \pi_{n+i}\Phi(S^i) \qquad \text{for } i > n.$$

This result was originally proved for the more general case of Γ-spaces, but we now specialize to Γ-modules. A *right Γ-module* is a cofunctor from Γ to Λ-modules.

Let Ω be the category of unbased finite sets $\underline{n} = \{1, 2, \ldots, n\}$, $n \geq 0$, with surjective maps as morphisms. Adding a disjoint basepoint defines an inclusion functor $\Omega \to \Gamma$ taking $\underline{n}$ to $[n]$. We regard the morphisms in the image of this functor, and other surjections, as face operators in Γ, and strict injections as degeneracy operators. There are additive categories $\Lambda\Gamma$ and $\Lambda\Omega$ with the same objects, indexed by non-negative integers, as Γ and Ω, but having as morphisms the free Λ-modules generated by the morphism-sets of Γ or Ω. We regard these additive categories as rings with many objects. One checks directly that $\Lambda\Gamma$ is a free right $\Lambda\Omega$-module. Pirashvili has shown [13] that there is a Morita equivalence between the categories $\Lambda\Gamma$-*mod* and $\Lambda\Omega$-*mod*, given by the functor which we denote

$$\Gamma \otimes_\Omega - : \Lambda\Omega\text{-}mod \longrightarrow \Lambda\Gamma\text{-}mod$$

which ought perhaps to be regarded as an abbreviation for $\Lambda\Gamma \otimes_{\Lambda\Omega} -$. Its inverse is the *cross-effect functor*

$$cr : \Lambda\Gamma\text{-}mod \longrightarrow \Lambda\Omega\text{-}mod$$

given by an idempotent in $\Lambda\Gamma$ which kills non-surjective morphisms of Γ. For the categories of right modules there is a dual situation. The Morita equivalence is given by tensoring with the Λ-dual $(\Lambda\Gamma)^*$, which is the inverse functor to the dual cross-effect $cr : mod\text{-}\Lambda\Gamma \longrightarrow mod\text{-}\Lambda\Omega$.

Let t be the right $\Lambda\Gamma$-module $\mathrm{Hom}_{\mathrm{Sets}_*}(-, \Lambda)$. We denote Tor of $\Lambda\Gamma$-modules by Tor^Γ and Tor of $\Lambda\Omega$-modules by Tor^Ω. The following is proved in [13, 2.2].

Theorem 4.3. *(Pirashvili) There is a natural isomorphism for $\Lambda\Gamma$-modules* Φ

$$\pi_* \Phi \quad \approx \quad \mathrm{Tor}^\Gamma_*(t, \Phi). \qquad \square$$

Scholium 4.4. *There are three other formulae for Tor-groups of the type arising in Theorem 4.3. Let Θ be any right Γ-module. As Morita equivalence preserves tensor products and is exact, we have $\mathrm{Tor}^\Omega(cr\,\Theta, cr\,\Phi) \approx \mathrm{Tor}^\Gamma(\Theta, \Phi)$. On the other hand, $\Phi \approx \Gamma \otimes_\Omega (cr\,\Phi)$ by our formula above for the Morita equivalence, and this represents the total derived functor $\Gamma \overset{L}{\otimes}_\Omega (cr\,\Phi)$ because $\Lambda\Gamma$ is right $\Lambda\Omega$-free. Therefore*

$$\Theta \overset{L}{\otimes}_\Gamma \Phi \quad \approx \quad \Theta \overset{L}{\otimes}_\Gamma \Gamma \overset{L}{\otimes}_\Omega (cr\,\Phi) \quad \approx \quad \Theta \overset{L}{\otimes}_\Omega (cr\,\Phi),$$

whence by taking homology $\mathrm{Tor}^\Gamma(\Theta, \Phi) \approx \mathrm{Tor}^\Omega(\Theta, cr\,\Phi)$, where Θ is a right Ω-module by restriction. Similarly the freeness of $(\Lambda\Gamma)^*$ as a left $\Lambda\Omega$-module yields

$$\Theta \overset{L}{\otimes}_\Gamma \Phi \quad \approx \quad (cr\,\Theta) \overset{L}{\otimes}_\Omega (\Lambda\Gamma)^* \overset{L}{\otimes}_\Gamma \Phi$$

$$\approx \quad (cr\,\Theta) \overset{L}{\otimes}_\Omega \mathrm{Hom}_\Gamma(\Lambda\Gamma, \Phi)$$

$$\approx \quad (cr\,\Theta) \overset{L}{\otimes}_\Omega \Phi$$

and hence $\mathrm{Tor}^\Gamma(\Theta, \Phi) \approx \mathrm{Tor}^\Omega(cr\,\Theta, \Phi)$ where Φ is a left Ω-module by restriction.

In particular we may take Θ to be the cofunctor t of 4.3. Then $cr\,t$ is the Ω-module such that $(cr\,t)(\underline{1}) = \Lambda$ and $(cr\,t)(\underline{n}) = 0$ for $n \neq 1$.

These Tor-groups can in turn be calculated from a certain bicomplex [17] called the Ξ-complex. It is based upon a projective resolution of the module t, constructed from the representations Lie_n^*.

Theorem 4.5. *Let Φ be any Γ-module. There is a natural bicomplex $\Xi(\Phi)$ in which the $(q-1)$st row is the two-sided bar construction $\mathcal{B}(\mathrm{Lie}_q^*, \Sigma_q, \Phi[q])$, the vertical differential is induced by the Leibniz differential of [12] and the homology is*

$$H\Xi(\Phi) \approx \mathrm{Tor}_*^\Gamma(t, \Phi) .$$

The Morita equivalence converts the projective resolution of t into a projective resolution of the right Ω-module $cr\, t$ described above. We therefore have:

4.6.

$$H\Xi(\Phi) \;\approx\; \pi_*\Phi \;\approx\; \mathrm{Tor}_*^\Gamma(t, \Phi) \;\approx\; \mathrm{Tor}_*^\Omega(cr\, t, cr\, \Phi).$$

The Loday functor and the Γ-homology of commutative graded algebras 4.7. Let $R = \{R_n\}_{n\in\mathbb{Z}}$ be an associative graded ring with unit which is commutative in the usual graded sense: that is, $yx = (-1)^{mn}xy$ when $x \in R_m$ and $y \in R_n$. Let Λ be an R-algebra, and G a Λ-module. (We usually omit the word "graded", but it is to be understood.) Unmarked tensor products are over the ground ring R.

We denote by $(\Lambda|R)^\otimes$ the tensor algebra of Λ over R. Then $(\Lambda|R)^\otimes \otimes_R G$ has a natural $\Lambda\Gamma$-module structure: if $\varphi : [n] \to [m]$ is any morphism in Γ, we set

$$\varphi_*(\lambda_1 \otimes \cdots \otimes \lambda_n \otimes g) \;=\; \varepsilon\, \gamma_1 \otimes \cdots \otimes \gamma_m \otimes h$$

in which

$$\gamma_i = \lambda_{i_1}\ldots\lambda_{i_r} \quad \text{if } \varphi^{-1}(i) = \{i_1,\ldots,i_r\} \quad \text{where} \quad i_1 < i_2 < \cdots < i_r$$

$$h = \lambda_{j_1}\ldots\lambda_{j_s}g \quad \text{if } \varphi^{-1}(0) = \{0, j_1,\ldots,j_s\} \quad \text{where} \quad j_1 < j_2 < \cdots < j_s$$

and in which ε is the sign of the permutation that rearranges $\{1, 2, \ldots, n\}$ in the order in which $\lambda_1,\ldots,\lambda_n$ appear in the expansion of the product $\gamma_1 \ldots \gamma_m$. When φ is a permutation $\sigma : [n] \to [n]$ this means that φ_* rearranges the factors and multiplies by the sign (compare [13, p.158])

$$\sigma_*(\lambda_1 \otimes \cdots \otimes \lambda_n \otimes g) \;=\; \varepsilon(\sigma)\, \lambda_{\sigma^{-1}1} \otimes \cdots \otimes \lambda_{\sigma^{-1}n} \otimes g .$$

Definition 4.8. The above $\Lambda\Gamma$-module $(\Lambda|R)^\otimes \otimes_R G$ is called the *Loday functor* $\mathcal{L}(\Lambda|R; G)$ since it was first defined in the ungraded case by Loday [11]. The functor $\mathcal{L}(\Lambda|R; \Lambda)$ is also denoted $\mathcal{L}(\Lambda|R)$ and is called the *Γ-cotangent complex* of Λ over R.

The *Γ-homology* and *Γ-cohomology* of Λ relative to R, with coefficients in the Λ-module G, are defined as the homotopy and cohotopy of the Loday functor:

$$H\Gamma_*(\Lambda|R;\,G) = \pi_*(\mathcal{L}(\Lambda|R \otimes_\Lambda G))$$
$$H\Gamma^*(\Lambda|R;\,G) = \pi^* \mathrm{Hom}_\Lambda(\mathcal{L}(\Lambda|R),\,G) \ .$$

Since the Γ-modules here are graded, all these constructs have a further internal grading.

By Pirashvili's theorem and Theorem 4.5 above, we can write Γ-homology as a Tor-group, and therefore as the homology of a Ξ-complex:

$$H\Gamma_*(\Lambda|R;\,G) \ \approx \ \mathrm{Tor}^\Gamma_*(t,\,\mathcal{L}(\Lambda|R;\,G)) \ \approx \ H\Xi_*(\mathcal{L}(\Lambda|R;\,G)).$$

As the Ξ-complex is based upon a projective resolution of the right Γ-module t, this can be dualized to write Γ-cohomology in terms of Ext and the dual Ξ-cohomology complex:

$$H\Gamma^*(\Lambda|R;\,G) \ \approx \ \mathrm{Ext}^*_\Gamma(t,\,\mathrm{Hom}_\Lambda(\mathcal{L}(\Lambda|R),\,G)) \ \approx \ H\Xi^*(\mathcal{L}(\Lambda|R);\,G).$$

Our claim is that gamma homology of commutative algebras is a precise analogue of Hochschild homology of associative algebras; and the Ξ-complex of the Loday functor is the corresponding analogue of the standard Hochschild chain complex. As evidence for this, we show that gamma homology, like Hochschild homology, satisfies a normalization theorem.

Proposition 4.9. *The cross-effect functor* $\tilde{\mathcal{L}}(\Lambda|R;\,G)$ *of* $\mathcal{L}(\Lambda|R;\,G)$ *satisfies*

$$\tilde{\mathcal{L}}(\Lambda|R;\,G)[n] = cr\,\mathcal{L}(\Lambda|R;\,G)[n] = \tilde{\Lambda}^{\otimes n} \otimes G$$

where $\tilde{\Lambda}$ *is the quotient R-module Λ/R.*

Proof. See [13, 1.10] where an explicit formula is given for the action of morphisms of Ω on these tensor products. □

Corollary 4.10. *(Normalization Theorem for Γ-homology) The Ξ-complex for this reduced Loday functor* $\tilde{\mathcal{L}}(\Lambda|R;\,G)$ *also has homology $H\Gamma_*(\Lambda|R;\,G)$. The analogous result holds in cohomology.*

Proof. We use the results 4.4 and 4.6, which show that the homology of $\Xi(\Phi)$ is isomorphic both to $\mathrm{Tor}^\Omega(cr\,t,\,\Phi)$ and to $\mathrm{Tor}^\Omega(cr\,t,\,cr\,\Phi)$. For the cohomology case, we use the Ext-interpretation of 4.8. □

In the next section, we shall further justify the analogy with Hochschild theory, by showing that the Ξ-complex arises in the commutative obstruction theory exactly as the Hochschild complex arose in the associative case.

5. OBSTRUCTION THEORY IN THE COMMUTATIVE CASE

Resolving the commutative operad 5.1. In the theory of ring spectra we used the Stasheff operad as a convenient resolution of the associative operad $\mathcal{M}$. In the commutative case $\mathcal{M}$ is replaced by $\mathcal{C}$, where each space $\mathcal{C}_n$ is a single point upon which the symmetric group Σ_n acts. To build our obstruction theory, we need a suitable resolution $\mathcal{B} \to \mathcal{C}$. It must satisfy two properties: each space $\mathcal{B}_n$ should be contractible (in order for $\mathcal{B}$ to be a resolution) and Σ_n-free (in order for $\mathcal{B}$ to be cofibrant). That is, $\mathcal{B}$ must be an E_∞ operad. However, being E_∞ is not sufficient. The Barratt-Eccles E_∞ operad $\mathcal{D}$, in which $\mathcal{D}_n$ is the standard Eilenberg-Mac Lane model for $E\Sigma_n$, is not cofibrant because it fails the test that the faces (the images of compositions $\mathcal{D}_i \times \mathcal{D}_j \longrightarrow \mathcal{D}_n$) should intersect one another only in faces of faces. Better in this respect is the tree operad $\mathcal{T}$, (which differs from that discussed in [18] only in that stumps are permitted). Thus $\mathcal{T}_n$ is the Σ_n-space of trees having leaves labelled by $\{1, 2, \ldots, n\}$, and stumps. Though $\mathcal{T}_n$ is contractible and its faces intersect correctly, it is not Σ_n-free, and so $\mathcal{T}$ is not an E_∞ operad.

Definition 5.2. Our *standard resolution* is the product operad $\mathcal{B} = \mathcal{D} \times \mathcal{T}$. Since its factors are augmented over the commutative operad $\mathcal{C}$, this is augmented over $\mathcal{C} \times \mathcal{C} = \mathcal{C}$; furthermore, it inherits the facing properties of $\mathcal{T}$ and the Σ-freeness of $\mathcal{D}$. It follows that $\mathcal{B}$ is E_∞ and that $\mathcal{B} \to \mathcal{C}$ is a cofibrant resolution of $\mathcal{C}$.

In analogy with 2.1 we have the following. The operads now have permutations, but the proof is otherwise exactly as before.

Proposition 5.3. *Let E be an S-module which is also a ring spectrum. Suppose that there is a morphism of operads $\mathcal{B} \to \mathrm{End}(E)$ under which one point of $\mathcal{B}_2$ is mapped to the the given multiplication on E. Then E is weakly equivalent to a commutative S-algebra.*

The geometry of the operad $\mathcal{B}$ 5.4. The problem of replacing E by a commutative S-algebra is reduced by 5.3 to the problem of constructing a morphism of operads $\mathcal{B} \to \mathrm{End}(E)$. This in turn we shall tackle by using obstruction theory. As before, we impose a strict unitality condition: we only look for actions of $\mathcal{B}$ in which stumps are ignorable. In 3.7, this was equivalent to requiring the map of operads to commute with degeneracies. The operad $\mathcal{B}$ inherits degeneracies from its factors $\mathcal{D}$ and $\mathcal{T}$, and the ignorability of stumps (or the condition that $\eta : S \to E$ be a strict unit) is interpreted just as before. It means that at the nth inductive step, the map is already defined on the large wedge subspectrum $\bigvee^n E$ of $E^{(n)}$.

It is natural to try using induction on n to construct a sequence of Σ_n-equivariant maps $\mathcal{B}_n \to \mathrm{End}(E)_n = \mathrm{Map}(E^{(n)}, E)$ which satisfy the conditions, as far as these are defined, for a morphism of operads. This would be a direct analogue of our procedure in §3, but it turns out to be too naive. It leads to intractable obstructions, and we need a better way.

We recall that $\mathcal{B}_i = \mathcal{T}_i \times \mathcal{D}_i = \mathcal{T}_i \times E\Sigma_i$. The bar construction $E\Sigma_i$ has a well-known filtration: in the best-known model, $E\Sigma_{ij}$ is the join of $j + 1$ copies of the group Σ_i. Therefore $\mathcal{B}_i$ is also filtered by setting

$$\mathcal{B}_{ij} = \mathcal{T}_i \times E\Sigma_{ij} \, ,$$

and the composition in the operad $\mathcal{B}$ respects the filtration.

Definition 5.5. The *diagonal filtration* ∇ on the operad $\mathcal{B}$ is defined using the bar filtration described above: we set $\nabla^n \mathcal{B}_i = \mathcal{B}_{i,n-i}$. An *n-stage* for an E_∞ structure on E is a family of Σ_i-equivariant maps $\mu_i : \nabla^n \mathcal{B}_i \longrightarrow \mathrm{End}(E)_i$ preserving composition wherever defined, and such that a point of $\nabla^n \mathcal{B}_2$ represents the given multiplication in E.

We remark that the 2-stage representing the given multiplication upon E can be extended to a 3-stage. In fact, this is exactly equivalent to stating that the multiplication is homotopy associative and homotopy commutative, by homotopies strictly preserving the unit.

Extending an n-stage in the commutative case 5.6. In the associative case, the problem of extending an $\mathcal{A}_n$-structure to an $\mathcal{A}_{n+1}$-structure leads one to consider the Stasheff polyhedron formed of all coherent bracketings of $n + 1$ factors in fixed order. This polyhedron is a single cell of dimension $n - 1$.

The commutative case is more complex. First, the maps are required to be equivariant with respect to permutation of factors. Second, the lattice of coherent bracketings of $n + 1$ ordered factors is replaced by the lattice of all partitions of the set $\{1, 2, \ldots, n + 1\}$. The geometric realization of this lattice is a wedge of $n!$ spheres of dimension $n - 1$, and the action of the group Σ_{n+1} upon its homology is the twisted dual Lie_{n+1}^* of the Lie representation [21] described in 4.1 above. The connection between E_∞-structures and Lie representations, discovered by F. R. Cohen [7], underlies the appearance of the integral representations Lie_n^* in the bicomplex for Γ-homology. The next theorem gives the connection: it is a direct analogue of Theorem 3.9, with Hochschild cohomology replaced by Γ-cohomology. As before, R is the graded coefficient ring $\pi_* E$, and Λ the Hopf algebroid $E_* E$. A version of this theorem, with homotopy units in place of strict units, was published in [17].

Theorem 5.7. *Let E be a commutative ring spectrum which satisfies the perfect universal coefficient condition of 3.1. Then given an n-stage μ for an E_∞ structure on E, there is a natural $(n, 2 - n)$-cocycle $\theta(\mu)$ of the total complex* $\mathrm{Tot}\, \Xi(\Lambda|R; R)$ *which vanishes if and only if there exists an $(n + 1)$-stage extending μ. The cohomology class $[\theta(\mu)] \in H\Gamma^{n,2-n}(\Lambda|R; R)$ is zero if and only if there exists an $(n + 1)$-stage which has the same underlying $(n - 1)$-stage as μ.* □

Corollary 5.8. *If the groups $H\Gamma^{n,2-n}(\Lambda|R; R)$ are zero for all $n \geq 3$, then the commutative ring spectrum E has an E_∞ structure, and by 5.3 is therefore weakly equivalent to a commutative S-algebra.* □

The difference cochains belong to $\Xi^{n,1-n}(\Lambda|R; R)$. Therefore if the groups $H\Gamma^{n,1-n}(\Lambda|R; R)$ are zero for all $n \geq 2$, then E has at most one E_∞ structure. (The indexing of Γ-homology, like that of André-Quillen homology, differs by one from that of Hochschild homology, which is why the cohomological indices in 5.7 differ from those in 3.5)

The above results can be applied to a spectrum representing the Lubin-Tate theory corresponding to a Honda formal group law. Here $H\Gamma^{**}(\Lambda|R;R)$ ≈ 0 [15], so 5.7 and 5.8 imply that these spectra have one and only one E_∞ structure. This reproves theorems of Goerss, Hopkins and Miller.

Baker and Richter [1] have further applications of the results. They prove that the Adams summand $E(1)$ of the complex K-theory spectrum KU has one and only one E_∞ structure. Using a continuous Γ-cohomology, they prove that the completions of all the Johnson-Wilson spectra have unique E_∞ structures. (For the standard non-completed Johnson-Wilson spectra the question is still open.)

The analogue of Theorem 3.12 is also true in the E_∞ situation. The space of E_∞ structures on E (or of $\nabla^n \mathcal{B}$-structures for any n) is unchanged up to weak homotopy type if one neglects the strict unit condition and relies upon the homotopy unit.

REFERENCES

[1] A. Baker and B. Richter, *Γ-cohomology of rings of numerical polynomials and E_∞ structures on K-theory*, (to appear).

[2] M. Basterra, *André-Quillen cohomology of commutative S-algebras*, J. Pure Appl. Algebra **144** (1999), 111–144.

[3] M. Basterra and B. Richter, *Cohomology theories for commutative S-algebras*, this volume.

[4] J. M. Boardman, *Homotopy structures and the language of trees*, Proc. Symp. Pure Math., American Mathematical Society, Providence, RI **22** (1971), 37–58.

[5] J. M. Boardman and R. M. Vogt, *Homotopy invariant algebraic structures on topological spaces*, Lect. Notes in Math. **347**, Springer-Verlag, New York–Heidelberg–Berlin, (1973).

[6] A. K. Bousfield and E. M. Friedlander, *Homotopy theory of Γ-spaces, spectra, and bisimplicial sets*, Lect. Notes in Math. **658**, Springer-Verlag, New York–Heidelberg–Berlin, (1978) 80–130.

[7] F. R. Cohen, *The homology of $\mathcal{C}_{n+1}$-spaces, $n \geq 0$*, in: *The homology of iterated loop spaces*, Lect. Notes in Math. **533** Springer-Verlag, New York–Heidelberg–Berlin, (1976) 207–351.

[8] A. D. Elmendorf, I. Kriz, M. A. Mandell, and J. P. May, *Rings, modules and algebras in stable homotopy theory*, A.M.S. Mathematical Surveys and Monographs **47** (1996).

[9] V. Hinich, *Homological algebra of homotopy algebras*, Comm. Algebra **25**, (1997), 3291–3323.

[10] J. Hollender, *Hindernistheorie für A_n-Ringspektren*, Dissertation, Universität Osnabrück, (1995).

[11] J.-L. Loday, *Opérations sur l'homologie cyclique des algèbres commutatives*, Invent. Math. **96**, (1989) 205–230.

[12] J.-L. Loday, *Cyclic homology*, Grundlehren der math. Wissenschaften **301** (1992).

[13] T. Pirashvili, *Hodge decomposition for higher order Hochschild homology*, Ann. Scient. Éc. Norm. Sup **33** (2000) 151–179.

[14] C. Rezk, *Notes on the Hopkins-Miller theorem*, in: Homotopy Theory via Algebraic Geometry and Group Representations, eds. M. Mahowald and S. Priddy, Contemp. Math. **220** (1998) 313–366.

[15] B. Richter and A. Robinson, *Gamma homology of group algebras and of polynomial algebras*, to appear in Proc. Northwestern Univ. Conf. on Algebraic Topology 2002.

[16] A. Robinson, *Obstruction theory and the strict associativity of Morava K-theories*, in: Advances in homotopy theory, London Math. Soc. Lecture Notes **139**, (1989) 143–152.

[17] A. Robinson, *Gamma homology, Lie representations and E_∞ multiplications*, Invent. Math., **152** (2003) 331–348.

[18] A. Robinson and S. Whitehouse, *Operads and Γ-homology of commutative rings*, Math. Proc. Cambridge Philos. Soc. **132** (2002), 197–234.

[19] S. Schwede, *Stable homotopy of algebraic theories*, Topology **40**, (2001), 1–41.

[20] J. D. Stasheff, *Homotopy associativity of H-spaces I, II*, Trans. Amer. Math. Soc. **108** (1963) 275–292 and 293–312.

[21] S. Whitehouse, *The integral tree representation of the symmetric group*, J. Algebraic Combinatorics **13**, (2001) 317–326.

MATHEMATICS INSTITUTE, UNIVERSITY OF WARWICK, COVENTRY, ENGLAND CV4 7AL

email-address: car@maths.warwick.ac.uk

MODULI SPACES OF COMMUTATIVE RING SPECTRA

P. G. GOERSS AND M. J. HOPKINS

ABSTRACT. Let E be a homotopy commutative ring spectrum, and suppose the ring of cooperations E_*E is flat over E_*. We wish to address the following question: given a commutative E_*-algebra A in E_*E-comodules, is there an E_∞-ring spectrum X with $E_*X \cong A$ as comodule algebras? We will formulate this as a moduli problem, and give a way – suggested by work of Dwyer, Kan, and Stover – of dissecting the resulting moduli space as a tower with layers governed by appropriate André-Quillen cohomology groups. A special case is $A = E_*E$ itself. The final section applies this to discuss the Lubin-Tate or Morava spectra E_n.

Some years ago, Alan Robinson developed an obstruction theory based on Hochschild cohomology to decide whether or not a homotopy associative ring spectrum actually has the homotopy type of an A_∞-ring spectrum. In his original paper on the subject [35] he used this technique to show that the Morava K-theory spectra $K(n)$ can be realized as an A_∞-ring spectrum; subsequently, in [3], Andrew Baker used these techniques to show that a completed version of the Johnson-Wilson spectrum $E(n)$ can also be given such a structure. Then, in the mid-90s, the second author and Haynes Miller showed that the entire theory of universal deformations of finite height formal group laws over fields of non-zero characteristic can be lifted to A_∞-ring spectra in an essentially unique way. This implied, in particular, that the Morava E-theory (or Lubin-Tate) spectra E_n were A_∞ (which could have been deduced from Baker's work), but it also showed much more. Indeed, the theory of Lubin and Tate [25] gives a functor from a category of finite height formal group laws to the category of complete local rings, and one way to state the results of [34] is that this functor factors in an essentially unique way through A_∞-ring spectra. It was the solution of the diagram lifting problem that gave this result its additional heft; for example, it implied that the Morava stabilizer group acted on E_n – simply because Lubin-Tate theory implied that this group acted on $(E_n)_*$.

In this paper, we would like to carry this program several steps further. One step forward would be to address E_∞-ring spectra rather than A_∞-ring spectra. There is an existing literature on this topic developed by Robinson and others, some based on Γ-homology. See [36], [37], and [4]. This can be used, to prove, among other things, that the spectra E_n are E_∞, and we guess that the obstruction theory we uncover here reduces to that theory.

The authors were partially supported by the National Science Foundation.

Another step forward would be to write down and try to solve the realization problem as a moduli problem: what is the *space* of all possible realizations of a spectrum as an A_∞ or E_∞-ring spectrum, and how can one calculate the homotopy type of this space? Robinson's original work on Morava K-theory implied that this space would often have many components or, put another way, that there could be many A_∞-realizations of a fixed homotopy associative spectrum. A third step forward would be to build a theory that easily globalizes; that is, we might try to realize a diagram of commutative rings by a diagram of E_∞-ring spectra, or we might try to come to terms with some sheaf of commutative rings. One particular such sheaf we have in mind is the structure sheaf of a moduli stack of elliptic curves, but one could also consider the structure sheaf of the moduli stack of formal group laws. In fact, many of our examples arise by examining pieces of this latter stack. A final step forward would be to build a theory that passes directly from algebra to E_∞ or A_∞-ring spectra, rather than by an intermediate pass through the stable homotopy category. This would be in line with the Lubin-Tate lifting of the previous paragraph.

Let us expand on some of these points.

The E_∞-realization problem is more subtle than the A_∞-problem. If X is a spectrum, the free A_∞-ring spectrum $\mathcal{A}(X)$ on X has the homotopy type of $\bigvee_{n \geq 0} X^{\wedge n}$, so that if E_* is a homology theory with a Künneth spectral sequence and E_*X is flat over E_*, then $E_*\mathcal{A}(X)$ is isomorphic to the tensor algebra over E_* on $E_*(X)$. This basic computation underlies much of the rest of the theory. However, the free E_∞-ring spectrum $\mathcal{E}(X)$ has the homotopy type of

$$\bigvee_{n \geq 0} (E\Sigma_n)_+ \wedge_{\Sigma_n} X^{\wedge n}$$

where $E\Sigma_n$ is a free contractible Σ_n-space. To compute $E_*\mathcal{E}(X)$ would require, at the very least, knowledge of $E_*B\Sigma_n$ and, practically, one would need define and understand a great deal about the E_* Dyer-Lashof algebra. Even if this calculation could be made, one would be left with a another problem. In trying to realize some commutative E_*-algebra A in E_*E comodules as an E_∞ ring spectrum, one might not be able or might not want to stipulate a Dyer-Lashof algebra structure on A. Indeed, our problem is simply to realize A as a commutative algebra – *not* to realize A with some stipulated Dyer-Lashof algebra structure. Thus, any theory we build must allow for this flexibility.

Our solution is to resolve an E_∞ operad by a simplicial operad which at once yields this desired flexibility and the possibility of computing the E_*-homology of a free object. This has the drawback, of course, of getting us involved with the cohomology of simplicial objects over simplicial operads. Part of the point of this paper is to demonstrate that this is workable and, in fact, leads into familiar territory. See section 6 and, more generally, [18].

It is here that the Dyer-Lashof operations – which have to arise somewhere – reappear in an explicit manner.

The moduli space of all possible realizations of a commutative E_*-algebra A in E_*E-comodules is a Dwyer-Kan classification space in the sense of [12]. Let $\mathcal{E}(A)$ be the category whose objects are E_∞-spectra X with $E_*X \cong A$ as commutative E_*-algebras in E_*E-comodules. The morphisms are morphisms of E_∞-ring spectra which are E_*-isomorphisms. The moduli space $\mathcal{TM}(A)$ of all realizations of A is the nerve of this category. According to Dwyer and Kan, there is a weak equivalence

$$\mathcal{TM}(A) \simeq \coprod_X B\operatorname{Aut}(X)$$

where X runs over E_*-isomorphism classes of objects in $\mathcal{E}(A)$ and $\operatorname{Aut}(X)$ is the monoid of self equivalences of a cofibrant-fibrant model for X. Pleasant as this result is, it is not really a computation in this setting; for example, we cannot immediately tell if this space is non-empty or not. Thus, we need some sort of decomposition of $\mathcal{TM}(A)$ with computable and, ideally, algebraic input. This is accomplished in Section 5; the algebraic input is an André-Quillen cohomology of A with coefficients in shifted versions of A. The basic theory for this kind of construction is spelled out in [6]; the exact result we obtain is gotten by combining Proposition 5.2, Proposition 5.5, and Theorem 5.8. Keeping track of basepoints in the resulting tower decomposition of $\mathcal{TM}(A)$ yields an obstruction theory for realizing A. The details are in 5.9.

This material works equally well for A_∞-structures. In this case the André-Quillen cohomology we obtain is exactly the Quillen cohomology of associative algebras; see [30]. Except possibly in degree zero, this is a shift of Hochschild homology, as one might expect from Robinson's work.

One detail about this theory is worth examining here: the moduli space $\mathcal{TM}(A)$ and its decomposition do not require the existence of a homotopy associative or commutative ring spectrum X with $E_*X \cong A$. Of course, in practice, such an X might be required for another reason. For example, in the basic case where $A = E_*E$, then we need $X = E$ to exist and be a homotopy commutative ring spectrum.

Here is an outline of the paper. In the first section, we confront the foundations. There are many competing, but Quillen equivalent, models for spectra in the literature. We write down exactly what we need from any given model, and point out that there exist models which have the requisite properties. The next two sections are about resolutions, first of operads, and then of spectra and algebras in spectra over operads. Here is where the resolution (or "E_2") model category structures of Dwyer, Kan, and Stover ([14],[15]) come in. We use an elegant formulation of this theory due to Bousfield [9]. Section 4 is devoted to a definition of the requisite André-Quillen cohomology groups and to a spectral sequence for computing the homotopy type of the mapping space of E_∞-maps between two E_∞-ring spectra. Again we empha-size that the E_2-term of this spectral sequence requires no knowledge of a Dyer-Lashof structure. Section 5 introduces the decomposition of the moduli space. Section 6 talks about methods of calculation, and Section 7 applies these techniques to the example of the diagram of Lubin-Tate spectra – the Hopkins-Miller theorem in E_∞-ring spectra. The result is the same as for A_∞-case.

Two notation conventions: First, for two objects in a model category, the space of maps $\mathrm{map}(X, Y)$ will always mean the *derived* simplicial mapping set of maps. All our model categories will be simplicial model categories; hence $\mathrm{map}(X, Y)$ is weakly equivalent to the simplicial mapping set out of cofibrant model for X into a fibrant model for Y. Alternatively, one can write down $\mathrm{map}(X, Y)$ as the nerve of an appropriate diagram category, such as the Dwyer-Kan hammock localization [13].

Second, if X is a simplicial object in some category $\mathcal{C}$, then we will say X is s-free if the underlying degeneracy diagram is free. This means there are objects $Z_n \in \mathcal{C}$ and isomorphisms

$$X_n \cong \coprod_{\phi:[n]\to[m]} Z_m$$

where ϕ runs over the surjections in the ordinal number category. Further-more, these isomorphisms respect degeneracy maps of X in the obvious way.

CONTENTS

1. THE GROUND CATEGORY: WHICH CATEGORY OF SPECTRA TO USE?

In the original drafts of these notes, and in other papers on this subject, we used the category of spectra developed by Lewis, May, and Steinberger in [26]. This had a number advantages for us; in particular, every object is fibrant in this category, and the role of the operads is explicit, even elegant. We needed every object to be fibrant so that we could apply the theory of Stover resolutions and the E_2-model categories of [14] to build our resolutions.

However, since that time Bousfield [9] and Jardine [24] have both shown that it is possible to remove the condition that every object is fibrant and still have a theory of E_2-model categories – or, as we (and Bousfield) prefer to call them, *resolution* model categories. This opened up the possibility of using any one of a number of other models for spectra – in fact almost any will now do. We get a nice synergy with operads if the underlying category has a closed symmetric monoidal smash product, so we will choose one of the current such models with this property. The point of this section is to produce an exact statement of what we need, along with some examples. This statement is broken into two parts: see Axioms 1.1 and Axioms 1.4 below.

We note that we are surrendering one facet of the previous discussion by this move away from LMS spectra. It turns out the homotopy category of C-algebras in spectra, where C is some operad, depends only on the weak equivalence type of C in the naïvest possible sense, which is in sharp distinction to the usual results about, say, spaces. (The exact result is below, in Theorem 1.6.) For the LMS spectra this fact comes down to the fact that one must use operads over the linear isometries operad, and such operads always have free actions by the symmetric groups. For the categories under discussion here, however, the reasons are less transparent, because they are buried in the definition of the smash product – and only an avatar of this freeness appears in the last of our axioms (in 1.4) for spectra.

To begin, here is exactly we will need about the symmetric monoidal structure. For the language of model categories, see [21]. In particular, the concepts of a monoidal model category and of a module over a monoidal category is discussed in Chapter 4.2 of that work. Specifically, simplicial sets are a monoidal model category and a simplicial model category is a module category over simplicial sets. For any category of spectra, the action of a simplicial set K on a spectrum X should be, up to weak equivalence, given by the formula

$$X \otimes K = X \wedge K_+$$

whenever this makes homotopical sense. Here the functor $(-)_+$ means adjoin a disjoint basepoint. This is the point of axiom 3 below.

1.1. Axioms for Spectra. We will assume that we have some category $\mathcal{S}$ of spectra which satisfy the following conditions:

1.) The category $\mathcal{S}$ is a cofibrantly generated proper stable simplicial model category Quillen equivalent to the Bousfield-Friedlander [10] category of simplicial spectra; furthermore, $\mathcal{S}$ has a generating set of cofibrations and a generating set of acyclic cofibrations with cofibrant source.

2.) The category $\mathcal{S}$ has a closed symmetric monoidal smash product which descends to the usual smash product on the homotopy category; furthermore, with that monoidal structure, $\mathcal{S}$ is a monoidal model category.

3.) The smash product behaves well with respect to the simplicial structure; specifically, if S is the unit object of the smash product, then there is a natural monoidal isomorphism

$$X \otimes K \xrightarrow{\cong} X \wedge (S \otimes K).$$

Note that Axiom 1 guarantees, among other things, that the homotopy category is the usual stable category.

We immediately point out that at least three of the favorite candidates for such a category satisfy these axioms. Symmetric spectra built from simplicial sets are discussed in [22]; symmetric spectra and orthogonal spectra built using topological spaces are defined and discussed in [27]. The spectra known as S-modules are built from topological spaces and are discussed in [16]. It is worth pointing out that S-modules are built on and depend on LMS spectra [26]. The categories of symmetric spectra and of orthogonal spectra have at least two Quillen equivalent model category structures on them. For the next result either would do; later results will require the "positive" model category structure of [27], §14.

1.2. Theorem. *The category of symmetric spectra (in spaces or simplicial sets), the category of orthogonal spectra, and the category of S-modules satisfy the axioms 1.1.*

Proof. Axioms 1, 2, and 3 are explicit in [22] for symmetric spectra in simplicial sets. For othogonal spectra, symmetric spectra in spaces, and S-modules, we note that these categories are not immediately simplicial model categories, but topological model categories. But any topological model category is automatically a simplicial model category via the realization functor. Then Axioms 1, 2, and 3 are in [27] for symmetric and orthogonal spectra and in [16] for S-modules. □

As with all categories modeling the stable homotopy category one has to explicitly spell out what one means by some familiar terms.

1.3. Notation for Spectra. The following remarks and notation will be used throughout this paper.

1.) We will use the words cofibrant and *cellular* interchangeably. The generating cofibrations of $\mathcal{S}$ are usually some sort of inclusion of spheres into cells.

2.) We will write $[X, Y]$ for the morphisms in the homotopy category $\mathbf{Ho}(\mathcal{S})$. As usual, this is π_0 for some derived space of maps. See point (5) below.

3.) In the category $\mathcal{S}$ it is possible (indeed usual) that the unit object S for the smash product ("the zero-sphere") is not cofibrant. We will write S^k, $-\infty < k < \infty$ for a cofibrant model for the k-sphere unless we explicitly state otherwise. In this language the suspension functor on the homotopy category is induced by

$$X \mapsto X \wedge S^1.$$

Also the suspension spectrum functor from pointed simplicial sets to spectra is, by axiom 3, modeled by

$$K \mapsto S^0 \wedge K \stackrel{\text{def}}{=} \frac{S^0 \otimes K}{S^0 \otimes *}$$

Note that if the unit object S is not cofibrant, the functor $S \otimes (-)$ is not part of a Quillen pair.

4.) Let K be a simplicial set and $X \in \mathcal{S}$. We may write $X \wedge K_+$ for the tensor object $X \otimes K$. This is permissible by axiom 3 and in line with the geometry. The exponential object in $\mathcal{S}$ will be written X^K.

5.) We will write $\mathrm{map}(X, Y)$ or $\mathrm{map}_{\mathcal{S}}(X, Y)$ for the *derived* simplicial set of maps between two objects of $\mathcal{S}$. Thus, $\mathrm{map}(X, Y)$ is the simplicial mapping space between some fibrant-cofibrant models ("bifibrant") models for X and Y. This can be done functorially if necessary, as the category $\mathcal{S}$ is cofibrantly generated. Alternatively, we could use some categorical construction, such as the Dwyer-Kan hammock localization. Note that with this convention

$$\pi_0 \mathrm{map}(X, Y) = [X, Y].$$

6.) We will write $F(X, Y)$ for the function spectrum of two objects $X, Y \in \mathcal{S}$. The closure statement in Axiom 2 of 1.1 amounts to the statement that

$$\mathrm{Hom}_{\mathcal{S}}(X, F(Y, Z)) \cong \mathrm{Hom}_{\mathcal{S}}(X \wedge Y, Z).$$

This can be derived:

$$\mathrm{map}(X, RF(Y, Z)) \simeq \mathrm{map}(X \wedge^L Y, Z)$$

where the R and L refer to the total derived functors and $\mathrm{map}(-, -)$ is the derived mapping space. In particular

$$\pi_k RF(Y, Z) \cong [\Sigma^k Y, Z].$$

7.) If X is cofibrant and Y is fibrant, then there is a natural weak equivalence

$$\mathrm{map}(X, Y) \simeq \mathrm{map}(S^0, F(X, Y))$$

and the functor $\mathrm{map}(S^0, -)$ is the total right derived functor of the suspension spectrum functor from pointed simplicial sets to $\mathcal{S}$. Thus we could write

$$\mathrm{map}(X, Y) \simeq \Omega^\infty F(X, Y).$$

In particular, $\mathrm{map}(X, Y)$ is canonically weakly equivalent to an infinite loop space.

We need a notation for iterated smash products. So, define

$$X^{(n)} \overset{\text{def}}{=} \underset{\xleftarrow{\hspace{1em}} n \xrightarrow{\hspace{1em}}}{X \wedge \cdots \wedge X} .$$

This paper is particularly concerned with the existence of A_∞ and E_∞-ring spectrum structures. Thus we we must introduce the study of operads acting on spectra.

Let $\mathcal{O}$ denote the category of operads in simplicial sets. Our major source of results for this category is [33]. The category $\mathcal{O}$ is a cofibrantly generated simplicial model category where $C \to D$ is a weak equivalence or fibration if each of the maps $C(n) \to D(n)$ is a weak equivalence or fibration of Σ_n-spaces in the sense of equivariant homotopy theory. Thus, for each subgroup $H \subseteq \Sigma_n$, the induced map $C(n)^H \to D(n)^H$ is a weak equivalence or fibration. The existence of the model category structure follows from the fact that the forgetful functor from operads to the category with objects $X = \{X(n)\}_{n \geq 0}$ with each $X(n)$ a Σ_n-space has a left adjoint with enough good properties that the usual lifting lemmas apply.

If C is an operad in simplicial sets, then we have a category of Alg_C of algebras over C in spectra. These are exactly the algebras over the triple

$$X \mapsto C(X) \overset{\text{def}}{=} \vee_{n \geq 0} C(n) \otimes_{\Sigma_n} X^{(n)}.$$

Note that we should really write $X^{(n)} \otimes_{\Sigma_n} C(n)$, but we don't.

The object $C(*) \cong S \otimes C(0)$ is the initial object of Alg_C. If the operad is *reduced* – that is, $C(0)$ is a point – then this is simply S itself.

If $f : C \to D$ is morphism of operads, then there is a restriction of structure functor $f_* : \mathrm{Alg}_D \to \mathrm{Alg}_C$, and this has a left adjoint

$$f^* \overset{\text{def}}{=} D \otimes_C (-) : \mathrm{Alg}_C \to \mathrm{Alg}_D$$

The categories Alg_C are simplicial categories in the sense of Quillen and both the restriction of structure functor and its adjoint are continuous. Indeed, if $X \in \mathrm{Alg}_C$ and K is a simplicial set, and if X^K is the exponential object of K in $\mathcal{S}$, then X^K is naturally an object in Alg_C and with this structure, it is the exponential object in $\mathcal{S}^C$. Succinctly, we say the forgetful functor makes exponential objects. It also makes limits and reflexive coequalizers, filtered colimits, and geometric realization of simplicial objects.

Here is our second set of axioms. The numbering continues that of Axioms 1.1.

1.4. Axioms for Spectra. Suppose we are given some category $\mathcal{S}$ of spectra satisfying the axioms of 1.1. Then we further require that

4.) For a fixed operad $C \in \mathcal{O}$, define a morphism of $X \to Y$ of C-algebras in spectra to be a weak equivalence or fibration if it is so in spectra. Then the category Alg_C becomes a cofibrantly generated simplicial model category.

5.) Let $n \geq 1$ and let $K \to L$ be a morphism of Σ_n spaces which is a weak equivalence on the underlying spaces. Then for all cofibrant spectra X, the induced map on orbit spectra

$$K \otimes_{\Sigma_n} X^{(n)} \to L \otimes_{\Sigma_n} X^{(n)}$$

is a weak equivalence of spectra.

We immediately note that we have examples.

1.5. Proposition. *Let S be any of the categories of symmetric spectra in topological spaces, orthogonal spectra, or S-modules. Then S satisfies the axioms of 1.4.*

Proof. First, axiom 4. For S-modules, this is nearly obvious, from a standard argument that goes back to Quillen, but see also [33] or [5] for the argument in the context of operads. In brief, since Alg_C has a functorial path object and the forgetful functor to S creates filtered colimits in Alg_C, we need only supply a fibrant replacement functor for Alg_C. But every object is fibrant.

For symmetric or orthogonal spectra, the argument goes exactly as in §15 of [27]. The argument there is only for the commutative algebra operad, but it goes through with no changes for the geometric realization of an arbitrary simplicial operad.

Axiom 5 in all these cases follows from the observation that for cofibrant X (in the positive model category structure where required), the smash product $X^{(n)}$ is actually a free Σ_n-spectrum. For symmetric and orthogonal spectra, see Lemma 15.5 of [27]; for S-modules see Theorem III.5.1 of [16]. $\qquad\square$

We would guess this result is also true for symmetric spectra in simplicial sets, but this is not immediately obvious: the case of symmetric spectra in spaces uses that the inclusion of a sphere into a disk is an NDR-pair.

The following result is why we put the final axiom into our list 1.4.

1.6. Theorem. *Let $C \to D$ be a morphism of operads in simplicial sets. Then the adjoint pair*

$$f^* : \mathrm{Alg}_C \rightleftarrows \mathrm{Alg}_D : f_*$$

is a Quillen pair. If, in addition, the morphism of operads has the property that $C(n) \to D(n)$ is a weak equivalence of spaces for all $n \geq 0$, this Quillen pair is a Quillen equivalence.

Proof. The fact that we have a Quillen pair follows from the fact that the restriction of structure functor (the right adjoint) $f_* : \mathrm{Alg}_D \to \mathrm{Alg}_C$ certainly preserves weak equivalences and fibrations.

For the second assertion, first note that since f_* reflects weak equivalences, we need only show that for all cofibrant $X \in \mathrm{Alg}_C$, the unit of the adjunction

$$X \to f_* f^* X = D \otimes_C X$$

is a weak equivalence. If $X = C(X_0)$ is actually a free algebra on a cofibrant spectrum, then this map is exactly the map induced by f:

$$C(X_0) = \bigvee_n C(n)_+ \wedge_{\Sigma_n} X_0^{(n)} \to \bigvee_n D(n)_+ \wedge_{\Sigma_n} X_0^{(n)} = D(X_0).$$

For this case, Axiom 5 of 1.4 supplies the result. We now reduce to this case.

Let $X \in \mathcal{S}^C$ be cofibrant. We will make use of an augmented simplicial resolution

$$P_\bullet \longrightarrow X$$

with the following properties:

i.) the induced map $|P_\bullet| \to X$ from the geometric realization of $P_\bullet$ to X is a weak equivalence;

ii.) the simplicial C-algebra $P_\bullet$ is s-free on a set of C-algebras $\{C(Z_n)\}$ where each Z_n is a cofibrant spectrum. (The notion of s-free was defined at the end of the introduction.)

There are many ways to produce such a $P_\bullet$. For example, we could take an appropriate subdivision of a cofibrant model for X in the resolution model category for simplicial C-algebras based on the homotopy cogroup objects $C(S^n)$, $-\infty < n < \infty$. [1]

Given $P_\bullet$, consider the diagram

(1.1)
$$\begin{array}{ccc} |P_\bullet| & \longrightarrow & |f_* f^* P_\bullet| \\ \downarrow & & \downarrow \\ X & \longrightarrow & f_* f^* X \end{array}$$

For all n, we have an isomorphism

$$P_n \cong C\left(\bigvee_{\phi:[n]\to[k]} Z_k \right)$$

where ϕ runs over the surjections in the ordinal number category. Thus we can conclude that $P_n \to f_* f^* P_n$ is a weak equivalence and that both $P_\bullet$ and $f_* f^* P_\bullet$ are Reedy cofibrant. The result now follows from the diagram 1.1. $\quad\square$

We now make precise the observation that Theorem 1.6 implies that the notion of, for example, an E_∞ ring spectrum is independent of which E_∞ operad we choose.

First we recall the Dwyer-Kan [12] classification space in a model category. Let $\mathcal{M}$ be a model category and let $\mathbf{E}$ be a subcategory of $\mathcal{M}$ which has the twin properties that

1.) if X is an object in $\mathbf{E}$ and Y is weakly equivalent to X, then $Y \in \mathbf{E}$;

2.) the morphisms in $\mathbf{E}$ are weak equivalences and if $f : X \to Y$ is a weak equivalence in $\mathcal{M}$ between objects of $\mathbf{E}$, then $f \in \mathbf{E}$.

---

[1] Resolution model categories are reviewed in section 3.

For example, $\mathbf{E}$ might have the same objects as $\mathcal{M}$ and all weak equivalences, in which case we will write $\mathbf{E}(\mathcal{M})$.

Let $B\mathbf{E}$ denote the nerve of the category $\mathbf{E}$. While the category $\mathbf{E}$ might not be small, one of the theorems of Dwyer and Kan is that it is homotopically small in the sense that each component has the weak homotopy type of a simplicial set; thus, by limiting the objects of interest in some way we obtain a useful weak homotopy type. In fact, there is a formula for this weak homotopy type:

$$B\mathbf{E} \simeq \coprod_{[X]} B\mathrm{Aut}_{\mathcal{M}}(X)$$

where $[X]$ runs over the weak homotopy types in $\mathbf{E}$ and $\mathrm{Aut}_{\mathcal{M}}(X)$ is the (derived) monoid of self-weak equivalences of X.

To this can be added the following result, immediate from Theorem 1.6.

1.7. Corollary. *Let $C \to D$ be any morphism of simplicial operads so that $C(n) \to D(n)$ is a weak equivalence of spaces for all $n \geq 0$. Then the natural map induced by restriction of structure*

$$B\mathbf{E}(\mathcal{S}^D) \to B\mathbf{E}(\mathcal{S}^C)$$

is a weak equivalence.

1.8. Remark. 1.) Note that Theorem 1.6 and Corollary 1.7 do not require that the operads be cofibrant. Thus, if we define an E_∞-operad C to be an operad so that each $C(n)$ is a free and contractible Σ_n-space, then C is weakly equivalent to the commutative monoid operad **Comm** which is simply a point in each degree. These results then say that the category of E_∞-ring spectra (algebras over C) is Quillen equivalent to the category of commutative S-algebras (algebras over **Comm**).

2.) Let $C \in \mathcal{O}$ be an operad in simplicial sets and let X be a spectrum. We now *define* the moduli space $C[X]$ of C-algebra structures on X by the homotopy pull-back diagram

$$
\begin{array}{ccc}
C[X] & \longrightarrow & B\mathbf{E}(\mathcal{S}^D) \\
\downarrow & & \downarrow \\
\{X\} & \longrightarrow & B\mathbf{E}(\mathcal{S}).
\end{array}
$$

Thus, for example, $C[X]$ is not empty if and only if X has a C-algebra structure. Note that Corollary 1.7 implies that $C[X]$ is independent of C up to the naïvest sort of weak equivalence.

In the case of spaces (rather than spectra) Charles Rezk has shown in [33] that $C[X]$ is equivalent to the (derived) space of operad maps from C to the endomorphism operad of X. Furthermore, he gives a way of approaching the homotopy type of $C[X]$ using a type of Hochschild cohomology. A similar result is surely true here; see the last sections of [34] for more details on this approach.

Our project is slightly different. Rather than beginning with a spectrum X we will begin with an algebraic model and try to construct E_∞-ring spectrum from this data. In effect, we deal directly with $B\mathbf{E}(\mathcal{S}^D)$.

2. Simplicial spectra over simplicial operads

Simplicial objects have often been used to build resolutions and that is our main point here, also. However, given an algebra X in spectra over some operad, we will resolve not only X, but the operad as well. The main results of this section are that if X is a simplicial algebra over a simplicial operad T then the geometric realization $|X|$ is an algebra over the geometric realization $|T|$ and, furthermore, that geometric realization preserves level-wise weak equivalences between Reedy cofibrant objects, appropriately defined.

Let's begin by talking about simplicial operads. As mentioned in the previous section, the category of operads $\mathcal{O}$ is a simplicial model category. From this one gets the Reedy model category structure on simplicial operads $s\mathcal{O}$ ([32]), which are the simplicial objects in $\mathcal{O}$.[2] Weak equivalences are level-wise and cofibrations are defined using the latching objects. The Reedy model category structure has the property that geometric realization preserves weak equivalences between cofibrant objects. It also has a structure as a simplicial model category; for example if T is a simplicial operad and K is simplicial set, then

$$T^K = \{T_n^K\}.$$

However, note that this module structure over simplicial sets is inherited from $\mathcal{O}$ and is not the simplicial structure arising externally, as in [31], §II.2.

Let us next spell out the kind of simplicial operads we want. If E_* is the homology theory of a homotopy commutative ring spectrum and C is an operad in $\mathcal{O}$, one might like to compute $E_*C(X)$. As mentioned in the introduction, this is usually quite difficult, unless E_*X is projective as an E_* module and $\pi_0 C(q)$ is a free Σ_q-set for all q. Thus we'd like to resolve C using operads of this sort.

If T is a simplicial operad and E is a commutative ring spectrum in the homotopy category of spectra, then E_*T is a simplicial operad in the category of E_* modules. The category of simplicial operads in E_* modules has a simplicial model category structure in the sense of §II.4 of [31], precisely because there is a free operad functor. Cofibrant objects are retracts of diagrams which are "free" in the sense of [31]; meaning the underlying degeneracy diagram is a free diagram of free operads.

Given an operad $C \in \mathcal{O}$, we'd like to consider simplicial operads T of the following sort:

---

[2]These are bisimplicial operads, but when we say simplicial operad, we will mean a simplicial object in $\mathcal{O}$, emphasizing the second (external) simplicial variable as the resolution variable. The first (internal) simplicial variable will be regarded as the geometric variable.

2.1. Theorem. *Let $C \in \mathcal{O}$ be an operad. Then there exists an augmented simplicial operad*

$$T \longrightarrow C$$

so that

1.) *T is Reedy cofibrant as a simplicial operad;*
2.) *For each $n \geq 0$ and each $q \geq 0$, $\pi_0 T_n(q)$ is a free Σ_q-set;*
3.) *The map of operads $|T| \to C$ induced by the augmentation is a weak equivalence;*
4.) *If $E_* C(q)$ is projective as an E_* module for all q, then $E_* T$ is cofibrant as a simplicial operad in E_* modules and $E_* T \to E_* C$ is a weak equivalance of operads in that category.*

This theorem is not hard to prove, once one has the explicit construction of the free operad; for example, see the appendix to [33]. Indeed, here is a construction: first take a cofibrant model C' for C. Then, if $F_{\mathcal{O}}$ is the free operad functor on graded spaces, one may take T to be the standard cotriple resolution of C'. What this theorem does not supply is some sort of uniqueness result for T; nonetheless, what we have here is sufficient for our purposes.

Note that if C is the commutative monoid operad, then we can simply take T to be a cofibrant model for C in the category of simplicial operads and run it out in the simplicial (i.e., external) direction. Then T is, of course, an example of an E_∞-operad; and $E_* T$ will be an E_∞-operad in E_*-modules in the sense of Definition 6.1.

Now fix a simplicial operad $T = \{T_n\}$. Since the free algebra functor $X \mapsto C(X)$ is natural in X and the operad C, we see that if X is a simplicial spectrum, so is $T(X)$. Hence a simplicial algebra in spectra over T is a simplicial spectrum X equipped with a multiplication map

$$T(X) \longrightarrow X$$

so that the usual associativity and unit diagrams commute. In particular, if $X = \{X_n\}$, then each X_n is a T_n-algebra. Let $s \operatorname{Alg}_T$ be the category of simplicial T-algebras.

The category $s \operatorname{Alg}_T$ is a simplicial model category. Recall that given a morphism of operads $C \to D$, then the restriction of structure functor $\operatorname{Alg}_D \to \operatorname{Alg}_C$ is continuous. This implies that if K is a simplicial set and $X \in s \operatorname{Alg}_T$, we may define $X \otimes K$ and X^K level-wise; for example,

$$X \otimes K = \{X_n \otimes K\}.$$

We could use this structure to define a geometric realization functor; however, we prefer to proceed as follows.

If $\mathcal{M}$ is a module category over simplicial sets, then the geometric realization functor $|\cdot| : s\mathcal{M} \to \mathcal{M}$ has a right adjoint

$$Y \mapsto Y^\Delta = \{Y^{\Delta^n}\}.$$

where Δ^n is the standard n-simplex. In particular, this applies to simplical operads, and we are interested in the unit of the adjunction $T \to |T|^\Delta$. If C is any operad and Y is a C-algebra, then for all simplicial sets K, the spectrum Y^K is a C^K algebra. From this it follows that Y^Δ is a simplicial C^Δ algebra. Setting $C = |T|$ and restricting structure defines a functor

$$Y \mapsto Y^\Delta : \mathrm{Alg}_{|T|} \longrightarrow s\,\mathrm{Alg}_T .$$

The result we want is the following.

2.2. Theorem. *Let T be a simplicial operad and $X \in s\,\mathrm{Alg}_T$ a simplicial T-algebra. Then the geometric realization $|X|$ of X as a spectrum has a natural structure as a $|T|$ algebra and, with this structure, the functor*

$$X \mapsto |X|$$

is right adjoint to $Y \mapsto Y^\Delta$.

Proof. We know that for an operad $C \in \mathcal{O}$ the forgetful functor from Alg_C to spectra makes geometric realization. Actually, what one proves is that if X is a simplicial spectrum and $C(X)$ is the simplicial C-algebra on X, then there is a natural (in C and X) isomorphism

$$C(|X|) \longrightarrow |C(X)|.$$

Now use a diagonal argument. If T is a simplicial operad and X is a simplicial spectrum, then

$$T(X) = \mathrm{diag}\{T_p(X_q)\}.$$

Since the functor $D \mapsto D(Y)$ is a continuous left adjoint, taking the realization in the p-variable yields a simplicial object

$$\{|\{T_\bullet(X_q)\}|\} \cong \{|T|(X_q)\}.$$

Now take the realization in the q variable and get

$$|T(X)| \cong |T|(|X|)$$

using the fact about the constant case sited above. The result now follows. $\square$

In light of Theorem 2.1 and Theorem 1.6, this theorem gives a tool for creating homotopy types of algebras over operads.

The next item to study is the homotopy invariance of the geometric realization functor, in this setting. The usual result has been cited above: realization preserves level-wise weak equivalences between Reedy cofibrant objects. The same result holds in this case, but one must take some care when defining "Reedy cofibrant". The difficulty is this: the definition of Reedy cofibrant involves the latching object, which is the colimit

$$L_n X = \operatorname*{colim}_{\phi:[n]\to[m]} X_m$$

where ϕ runs over the non-identity surjections in the ordinal number category. We must define this colimit if each of the X_m is an algbera over a different operad. The observation needed is the following. Let $S : I \to \mathcal{O}$ be a diagram of operads. Then an I-diagram of S-algebras is an I-diagram $X : I \to \mathcal{S}$ of spectra equipped with a natural transformation of I-diagrams

$$S(X) \to X$$

satisfying the usual associativity and unit conditions. For example if $I = \Delta^{op}$ one recovers simplicial S-algebras. Call the category of such Alg_S.[3] Then one can form the colimit operad $\text{colim } S = \text{colim}_I S$ and there is a constant diagram functor

$$\text{Alg}_{\text{colim } S} \longrightarrow \text{Alg}_S$$

sending X to the constant I-diagram $i \mapsto X$ where X gets an S_i structure via restriction of structure along

$$S_i \longrightarrow \operatorname*{colim}_I S.$$

2.3. Lemma. *This constant diagram functor has a left adjoint*

$$X \to \text{colim}_I X.$$

Despite the notation, $\text{colim}_I X$ is not the colimit of X as an I diagram of spectra; indeed, if $X = S(Y)$ where Y is an I-diagram of spectra

$$\text{colim}_I X \cong (\text{colim}_I S)(\text{colim}_I Y).$$

If T is a simplicial operad we can form the latching object

$$L_n T = \operatorname*{colim}_{\phi:[n] \to [m]} T_m.$$

There are natural maps $L_n T \to T_n$ of operads. If X is a simplicial T-algebra we extend this definition slightly and define

$$L_n X = T_n \otimes_{L_n T} \operatorname*{colim}_{\phi:[n] \to [m]} X_m$$

where, again, ϕ runs over the non-identity surjections in Δ. In short we extend the operad structure to make $L_n X$ a T_n-algebra and the natural map $L_n X \to X_n$ a morphism of T_n-algebras.

With this construction on hand one can make the following definition. Let T be a simplicial operad and $f : X \to Y$ a morphism of simplicial T-algebras. Then f is a level-wise weak equivalence (or *Reedy weak equivalence*) if each of the maps $X_n \to Y_n$ is a weak equivalence of T_n-algebras – or, by definition, a weak equivalence as spectra. The morphism f is a Reedy cofibration if the morphism of $L_n T$-algebras

$$L_n Y \sqcup_{L_n X} Y_n \longrightarrow Y_n$$

is a cofibration of T_n-algebras. The coproduct here occurs in the category of T_n-algebras. The main result is then:

---

[3]This is a slight variation on the notation $s\,\text{Alg}_T$. If T is a simplicial operad, this new notation would simply have us write Alg_T for $s\,\text{Alg}_T$. No confusion should arise.

2.4. Theorem. *With these definitions, and the level-wise simplicial structure defined above, the category* $s\operatorname{Alg}_T$ *becomes a simplicial model category. Furthermore*

1.) *The geometric realization functor* $|-|: s\operatorname{Alg}_T \to \operatorname{Alg}_{|T|}$ *sends level-wise weak equivalences between Reedy cofibrant objects to weak equivalences; and*

2.) *if T is Reedy cofibrant as a simplicial operad, then any Reedy cofibration in $s\operatorname{Alg}_T$ is a Reedy cofibration of simplicial spectra.*

The importance of the second item in this result is that, in light of Theorem 2.2, one can calculate the homotopy type of the geometric realization of a T-algebra entirely in spectra.

3. RESOLUTIONS

Building on the results of the last section, we'd like to assert the following. Let X be a simplicial algebra over a simplicial operad T, and suppose T satisfies all the conclusions of Theorem 2.1. Then there is a simplicial T-algebra Y and a morphism of T-algebras $Y \to X$ so that a.) $|Y| \to |X|$ is a weak equivalence and b.) E_*Y is cofibrant as an E_*T algebra. The device for this construction is an appropriate Stover resolution ([38],[14],[15]) and, particularly, the concise and elegant paper of Bousfield [9].[4] We explain some of the details in this section.

We begin by specifying the building blocks of our resolutions. We fix a spectrum E which is a commutative ring object in the homotopy category of spectra. Let $D(\cdot)$ denote the Spanier-Whitehead duality functor.

3.1. Definition. A homotopy commutative and associative ring spectrum E satisfies *Adams's condition* if E can be written, up to weak equivalence, as a homotopy colimit of finite cellular spectra E_α with the properties that

1.) E_*DE_α is projective as an E_*-module; and

2.) for every module spectrum M over E the Künneth map

$$[DE_\alpha, M] \longrightarrow \operatorname{Hom}_{E_*-\mathrm{mod}}(E_*DE_\alpha, M_*)$$

is an isomorphism.

This is the condition Adams (following Atiyah) wrote down in [1] to guarantee that the (co-)homology theory over E has Künneth spectral sequences. If M is a module spectrum over E, then so is every suspension or desuspension of M; therefore, one could replace the source and target of the map in part 2.) of this definition by the corresponding graded objects.

Many spectra of interest satisfy this condition; for example, if E is the spectrum for a Landweber exact homology theory, it holds. (This is implicit in [1], and made explicit in [34].) In fact, the result for Landweber exact

---

[4]Bousfield's paper is written cosimplicially, but the arguments are so categorical and so clean that they easily produce the simplicial objects we require.

theories follows easily from the example of MU, which, in turn, was Atiyah's original example. See [2]. Some spectra do not satisfy this condition, however – the easiest example is $H\mathbb{Z}$.

We want to use the spectra DE_α as detecting objects for a homotopy theory, but first we enlarge the scope a bit.

3.2. Definition. Define $\mathcal{P}(E) = \mathcal{P}$ to be a set of finite cellular spectra so that

(1) the spectrum $S^0 \in \mathcal{P}$ and E_*X is projective as an E_*-module for all $X \in \mathcal{P}$;
(2) for each α there is finite cellular spectrum homotopy equivalent to DE_α in $\mathcal{P}$;
(3) $\mathcal{P}$ is closed under suspension and desuspension;
(4) $\mathcal{P}$ is closed under finite wedges; and
(5) for all $X \in \mathcal{P}$ and all E-module spectra M the Künneth map

$$[X, M] \longrightarrow \mathrm{Hom}_{E_*-\mathrm{mod}}(E_*X, M_*)$$

is an isomorphism.

The E_2 or *resolution* model category which we now describe uses the set $\mathcal{P}$ to build cofibrations in simplicial spectra and, hence, some sort of projective resolutions.

Because the category of spectra has all limits and colimits, the category of simplicial spectra is a simplicial category in the sense of Quillen using external constructions as in §II.4 of [31]. However, the Reedy model category structure on simplicial spectra is not a simplicial model category using the external simplicial structure; for example, if $i : X \to Y$ is a Reedy cofibration and $j : K \to L$ is a cofibration of simplicial sets, then

$$i \otimes j : X \otimes L \sqcup_{X \otimes K} Y \otimes K \to Y \otimes L$$

is a Reedy cofibration, it is a level-wise weak equivalence if i is, but it is not necessarily a level-wise weak equivalence if j is.

The following ideas are straight out of Bousfield's paper [9].

3.3. Definition. Let $\mathbf{Ho}(\mathcal{S})$ denote the stable homotopy category.

1.) A morphism $p : X \to Y$ in $\mathbf{Ho}(\mathcal{S})$ is $\mathcal{P}$-*epi* if $p_* : [P, X] \to [P, Y]$ is onto for each $P \in \mathcal{P}$.
2.) An object $A \in \mathbf{Ho}(\mathcal{S})$ is $\mathcal{P}$-*projective* if

$$p_* : [A, X] \longrightarrow [A, Y]$$

is onto for all $\mathcal{P}$-epi maps.
3.) A morphism $A \to B$ of spectra is called $\mathcal{P}$-*projective cofibration* if it has the left lifting property for all $\mathcal{P}$-epi fibrations in $\mathcal{S}$.

The classes of $\mathcal{P}$-epi maps and of $\mathcal{P}$-projective objects determine each other; furthermore, every object in $\mathcal{P}$ is $\mathcal{P}$-projective. Note however, that the class of $\mathcal{P}$-projectives is closed under arbitrary wedges. The class of $\mathcal{P}$-projective cofibrations will be characterizèd below; see Lemma 3.7.

3.4. Lemma. *1.) The category* $\mathbf{Ho}(\mathcal{S})$ *has enough* $\mathcal{P}$-*projectives; that is, for every object* $X \in \mathbf{Ho}(\mathcal{S})$ *there is a* $\mathcal{P}$-*epi* $Y \to X$ *with* Y $\mathcal{P}$-*projective.*

2.) Let X *be a* $\mathcal{P}$-*projective object. Then* $E_* X$ *is a projective* E_*-*module, and the Künneth map*

$$[X, M] \longrightarrow \mathrm{Hom}_{E_*-mod}(E_* X, M_*)$$

is an isomorphism for all E-*module spectra* M.

Proof. For part 1.) we can simply take

$$Y = \bigvee_{P \in \mathcal{P}} \bigvee_{f:P \to X} P$$

where f ranges over all maps $P \to X$ in $\mathbf{Ho}(\mathcal{S})$. Then, for part 2.), we note that the evaluation map

$$Y = \bigvee_{P \in \mathcal{P}} \bigvee_{f:P \to X} P \longrightarrow X$$

has a homotopy section. Then the result follows from the properties of the elements of $\mathcal{P}$. $\square$

We now come to the $\mathcal{P}$-resolution model category structure. Recall that a morphism $f : A \to B$ of simplicial abelian groups is a weak equivalence if $f_* : \pi_* A \to \pi_* B$ is an isomorphism. Also $f : A \to B$ is a fibration if the induced map of normalized chain complexes $Nf : NA \to NB$ is surjective in positive degrees. The same definitions apply to simplicial R-modules or even graded simplicial R-modules over a graded ring R. A morphism is a cofibration if it is injective with level-wise projective cokernel.

3.5. Definition. Let $f : X \to Y$ be a morphism of simplicial spectra. Then

1.) the map f is a $\mathcal{P}$-*equivalence* if the induced morphism

$$f_* : [P, X] \longrightarrow [P, Y]$$

is a weak equivalence of simplicial abelian groups for all $P \in \mathcal{P}$;

2.) the map f is a $\mathcal{P}$-*fibration* if it is a Reedy fibration and $f_* : [P, X] \to [P, Y]$ is a fibration of simplicial abelian groups for all $P \in \mathcal{P}$;

3.) the map f is a $\mathcal{P}$-*cofibration* if the induced maps

$$X_n \sqcup_{L_n X} L_n Y \longrightarrow Y_n, \qquad n \geq 0,$$

are $\mathcal{P}$-projective cofibrations.

Then, of course, the theorem is as follows. Let $s\mathcal{S}_\mathcal{P}$ denote the category of simplicial spectra with these notions of $\mathcal{P}$-equivalence, fibration, and cofibration.

3.6. Theorem. *With these definitions, the category $sS_{\mathcal{P}}$ becomes a simplicial model category.*

The proof is given in [9]. We call this the $\mathcal{P}$-resolution model category structure. It is cofibrantly generated; furthermore there are sets of generating cofibrations and generating acyclic cofibrations with cofibrant source. An object is $\mathcal{P}$-fibrant if and only if it is Reedy fibrant. The next result gives a characterization of $\mathcal{P}$-cofibrations.

Call a morphism $X \to Y$ of spectra $\mathcal{P}$-free if it can be written as a composition

$$X \xrightarrow{\;i\;} X \vee F \xrightarrow{\;q\;} Y$$

where i is the inclusion of the summand, F is cofibrant and $\mathcal{P}$-projective, and q is an acyclic cofibration. The following is also in [9].

3.7. Lemma. *A morphism $X \to Y$ of spectra is a $\mathcal{P}$-projective cofibration if and only if it is a retract of a $\mathcal{P}$-free map.*

There are two ways to characterize $\mathcal{P}$-equivalences. The first comes directly from the definition of $\mathcal{P}$-equivalences. If $X \in sS$ and $P \in \mathcal{P}$, then

$$[P, X] = \{[P, X_n]\}$$

is a simplicial abelian group, and we may define

$$\pi_i(X; P) = \pi_i[P, X].$$

Then, a morphism is a $\mathcal{P}$-equivalence if and only if it induces an isomorphism on $\pi_*(-; P)$ for all $P \in \mathcal{P}$.

There are other homotopy groups. Define "sphere objects" in sS as follows: let $P \in \mathcal{P}$, $n \geq 0$, and let $\Delta^n/\partial\Delta^n$ be the standard simplicial n-sphere. As always, $\Delta^0/\partial\Delta^0 = (\Delta^0)_+$ is the two-point simplicial set. Then the nth P-sphere $P \wedge \Delta^n/\partial\Delta^n$ is defined by the push-out diagram

$$
\begin{array}{ccc}
P \otimes * = P & \longrightarrow & P \otimes \Delta^n/\partial\Delta^n \\
\downarrow & & \downarrow \\
* & \longrightarrow & P \wedge \Delta^n/\partial\Delta^n.
\end{array}
$$

If $X \in sS$ is a simplicial spectrum, then the mapping space $\mathrm{map}(P, X)$ is a loop space – in fact, an infinite loop space. Now define the "natural" homotopy groups of a simplicial spectrum X by the formula

$$\pi_n^{\natural}(X; P) = [P \wedge \Delta^n/\partial\Delta^n, X]_{\mathcal{P}} \cong \pi_n \,\mathrm{map}(P, X)$$

where we take the constant map as the basepoint of the mapping space. The symbol $[\ ,\ ]_{\mathcal{P}}$ refers to morphisms in the homotopy category obtained from the $\mathcal{P}$-resolution model category structure.

The two notions of homotopy groups are related by the spiral exact sequence. Let $\Sigma : \mathbf{Ho}(S) \to \mathbf{Ho}(S)$ be the suspension operator on the homotopy category of spectra.

3.8. Proposition. *For all $P \in \mathcal{P}$ and all simplicial spectra X, there is a natural isomorphism*

$$\pi_0^{\natural}(X; P) \cong \pi_0(X; P)$$

and a natural long exact sequence

$$\cdots \to \pi_{n-1}^{\natural}(X; \Sigma P) \to \pi_n^{\natural}(X; P) \to \pi_n(X; P)$$
$$\to \pi_{n-2}^{\natural}(X; \Sigma P) \to \cdots \to \pi_1(X; P) \to 0.$$

See [15]. Note that this implies that a morphism of simplicial spectra is a $\mathcal{P}$-equivalence if and only if it induces an isomorphism on $\pi_*^{\natural}(-, P)$ for all $P \in \mathcal{P}$.

The long exact sequences of Proposition 3.8 can be spliced together to give a spectral sequence

$$(3.1) \qquad \pi_p(X; \Sigma^q P) \Longrightarrow \operatorname{colim}_k \pi_k^{\natural}(X; \Sigma^{p+q-k} P).$$

using the triangles

$$(3.2) \qquad \pi_{p-1}^{\natural}(X; \Sigma^{q+1} P) \longrightarrow \pi_p^{\natural}(X; \Sigma^q P)$$
$$\pi_p(X; \Sigma^q P)$$

as the basis for an exact couple. Here and below the dotted arrow means a morphism of degree -1. This is actually a very familiar spectral sequence in disguise.

We may assume that X is Reedy cofibrant, and let $\operatorname{sk}_n X$ denote the nth skeleton of X as a simplicial spectrum. Then geometric realization makes $\{|\operatorname{sk}_n X|\}$ into a filtration of $|X|$ and the standard spectral sequence of the geometric realization of a simplicial spectrum is gotten by splicing together the long exact sequences obtained by applying the functor $[\Sigma^{p+q} P, -]$ to the cofibration sequence

$$|\operatorname{sk}_{p-1} X| \longrightarrow |\operatorname{sk}_p X| \longrightarrow \Sigma^p (X_p / L_p X).$$

If we let

$$[\Sigma^{p+q} P, |\operatorname{sk}_p X|]^{(1)} = \operatorname{Im}\{[\Sigma^{p+q} P, |\operatorname{sk}_p X|] \longrightarrow [\Sigma^{p+q} P, |\operatorname{sk}_{p+1} X|]\}$$

then the first derived long exact sequence of this exact couple is

$$(3.3) \qquad [\Sigma^{p+q} P, |\operatorname{sk}_{p-1} X|]^{(1)} \longrightarrow [\Sigma^{p+q} P, |\operatorname{sk}_p X|]^{(1)}$$
$$\pi_p[\Sigma^q P, X]$$

and we obtain the usual spectral sequences

$$(3.4) \qquad \pi_p(X; \Sigma^q P) = \pi_p[\Sigma^q P, X] \Longrightarrow [\Sigma^{p+q} P, |X|].$$

Thus the two spectral sequences have isomorphic E^2-terms. More is true. The next result says that the two exact couples obtained from the triangles of 3.2 and 3.3 are isomorphic; hence, we have isomorphic spectral sequences and we can assert that geometric realization induces an isomorphism

$$\mathrm{colim}_k\, \pi_k^\natural(X; \Sigma^{p+q-k}P) \xrightarrow{\cong} [\Sigma^{p+q}P, |X|].$$

3.9. Lemma. *Geometric realization induces an isomorphism between the spiral exact sequence*

$$\cdots \to \pi_{p-1}^\natural(X; \Sigma^{q+1}P) \to \pi_p^\natural(X; \Sigma^q P) \to \pi_p(X; \Sigma^q P) \to \cdots$$

and the derived exact sequence

$$\cdots \to [\Sigma^{p+q}P, |sk_{p-1}X|]^{(1)} \longrightarrow [\Sigma^{p+q}P, |sk_p X|]^{(1)} \longrightarrow \pi_p[\Sigma^q P, X] \to \cdots$$

Proof. The difficulty is to construct the map of exact sequences inducing an isomorphism $\pi_p(X; \Sigma^q P) \cong \pi_p[\Sigma^q P, X]$. Once that is in place, the five lemma and an induction argument show that we must have an isomorphism.

In [15] the spiral exact sequence is obtained by deriving another exact sequence. If K is a finite pointed simplicial set and X is a simplicial spectrum, there is a spectrum $C_K X$ characterized by the natural isomorphism

$$\mathrm{Hom}_{sS}(Z \wedge K, X) \cong \mathrm{Hom}_S(Z, C_K X)$$

for all spectra Z. In particular, we write

$$Z_p X = C_{\Delta^p/\partial\Delta^p} X \qquad \text{and} \qquad C_p X = C_{\Delta^p/\Delta_0^p} X$$

where Δ_0^p is the union of all but the 0th face. If X is Reedy fibrant, there is a fibration sequence

$$Z_p X \longrightarrow C_p X \longrightarrow Z_{p-1} X$$

with maps induced by the cofibration sequence of simplicial sets

$$\Delta^{p-1}/\partial\Delta^{p-1} \xrightarrow{d^0} \Delta^p/\Delta_0^p \longrightarrow \Delta^p/\partial\Delta^p.$$

Then the spiral exact sequence is the first derived sequence of the triangle

(3.5)
$$[\Sigma^{q+1}P, Z_{p-1}X] \dashrightarrow [\Sigma^{q+1}P, Z_p X]^{(1)}$$
$$[\Sigma^{q+1}P, C_p X]$$

A key calculation is that $[\Sigma^q P, C_p X] \cong N_p[\Sigma^p, X]$ where $N_p(-)$ is the pth group in the normalized chain complex.

Geometric realization induces a function

$$\mathrm{Hom}_S(Z, C_K X) \cong \mathrm{Hom}_{sS}(Z \wedge K, X) \longrightarrow \mathrm{Hom}_S(Z \wedge |K|, X).$$

This does not induce a map out of the triangle of 3.5; however, after taking first derived triangles, we get a morphism from the triangle of 3.2 to the triangle 3.3, as required. □

3.10. Remark. At this point we can explain one of the reasons for using the models $\mathcal{P}$ to define the resolution model category. Suppose $X \to Y$ is a weak equivalence between cofibrant objects in the $\mathcal{P}$-resolution model category. Then for each of the spectra DE_α we have an isomorphism

$$f_* : \pi_p(X; \Sigma^q DE_\alpha) \xrightarrow{\cong} \pi_p(Y, \Sigma^q DE_\alpha).$$

However, if $E_*(-)$ is our chosen homology theory

$$\begin{aligned}
\pi_p E_q X &\cong \mathrm{colim}_\alpha \, \pi_p(E_\alpha)_q X \\
&\cong \mathrm{colim}_\alpha \, \pi_p[\Sigma^q DE_\alpha, X] \\
&= \mathrm{colim}_\alpha \, \pi_p(X; \Sigma^q DE_\alpha).
\end{aligned}$$

We note that the spectral sequence of Equation 3.4 is natural in P; thus, taking the colimit as this equation, we obtain a spectral sequence

(3.6) $$\pi_p E_q X \implies E_{p+q}|X|.$$

This is, of course, the standard homology spectral sequence of a simplicial spectrum. In any case, if $X \to Y$ is a $\mathcal{P}$-equivalence, then we get isomorphic E_* homology spectral sequences.

Finally, also note that Lemma 3.9 yields an isomorphism

(3.7) $$\mathrm{colim}_\alpha \, \pi_p^\natural(X, \Sigma^q DE_\alpha) \cong \mathrm{Im}\{E_q|\mathrm{sk}_p X| \to E_q|\mathrm{sk}_{p+1} X|\}.$$

3.11. Remark. The category $s\mathcal{S}$ of simplicial spectra, and the more structured simplicial spectra defined below have Postnikov sections. That is, for any X in $s\mathcal{S}$ we can produce a morphism of simplicial spectra $X \to P_n X$ so that $\pi_k^\natural(X; P) \cong \pi_k^\natural(P_n X; P)$ for $P \in \mathcal{P}$ and $k \leq n$, and, in addition, $\pi_k^\natural(P_n X; P) = 0$ for $k > n$. One way to construct $P_n X$ is to define $P_n X$ to the colimit of simplicial spectra $P_n^i X$ where $P_n^0 X$ to be a fibrant model for X and defining $P_n^i X$ to be a fibrant model for the spectrum simplicial spectrum Y obtained as a push-out

$$
\begin{array}{ccc}
\coprod_{k>n} \coprod_f P \wedge \Delta^k/\partial\Delta^k & \longrightarrow & P_n^{i-1} X \\
\downarrow & & \downarrow \\
\coprod_{k>n} \coprod_f P \wedge \Delta^k/\Delta_0^k & \longrightarrow & Y.
\end{array}
$$

where f runs over all morphisms

$$f : P \wedge \Delta^k/\partial\Delta^k \to P_n^{i-1} X.$$

Note that since $s\mathcal{S}_\mathcal{P}$ is cofibrantly generated, this can be made natural in X. If we are working with algebras in $s\,\mathrm{Alg}_T$ for some simplicial operad T, we would simply replace $P \wedge \Delta^k/\partial\Delta^k$ by $T(P \wedge \Delta^k/\partial\Delta^k)$, and so on.

It is worth remarking that one can now recover the universal coefficient theorem of Adams-Atiyah ([1] §III.13) from these constructions. If X is a spectrum, we regard it as constant simplicial spectrum and choose a $\mathcal{P}$-cofibrant replacement Y for X. Then the spectral sequence of Equation 3.4, with $P = S^0$, implies that $|Y| \simeq X$. The universal coefficient spectral sequence is the Bousfield-Kan spectral sequence of the cosimplicial spectrum $F(Y, M)$ for an E-module spectrum M:

$$\mathrm{Ext}^s_{E_*}(E_*X, M_{*+t}) \cong \pi^s \pi_{-t} F(Y, M) \Longrightarrow \pi_{-t-s} F(|Y|, M) \cong M^{s+t}X.$$

The E_2-term is identified using Definition 3.2.5. Here the symbol M_{*+t} means the graded group with $(M_{*+t})_n = M_{n+t}$.

The $\mathcal{P}$-resolution model category structure can be promoted to a model category for simplicial algebras over a simplicial operad. Fix a simplicial operad T and let $s\,\mathrm{Alg}_T$ be the category of algebras over T. This category has an external simplicial structure; indeed, if K is a simplicial set and $X \in s\,\mathrm{Alg}_T$, one has

$$(3.8) \qquad\qquad (X \otimes K)_n = \coprod_{K_n}{}^{T_n} X_n.$$

The superscript T_n is indicates that the coproduct is taken in the category of T_n algebras. The simplicial set of maps is defined again by

$$[n] \mapsto \mathrm{Hom}_{s\,\mathrm{Alg}_T}(X \otimes \Delta^n, Y).$$

We say that a morphism $X \to Y$ of simplicial T-algebras is a $\mathcal{P}$-fibration or $\mathcal{P}$-equivalence if the underlying morphism of simplicial spectra is. Then we have the *$\mathcal{P}$-resolution model category structure* on $s\,\mathrm{Alg}_T$:

3.12. Theorem. *With these definitions, the category $s\,\mathrm{Alg}_T$ becomes a simplicial model category. Furthermore, for each $X \in s\,\mathrm{Alg}_T$ there is a natural $\mathcal{P}$-equivalence*

$$P_T(X) \to X$$

so that

1.) *$P_T(X)$ is cofibrant in the $\mathcal{P}$-resolution model category structure on $s\mathcal{S}^T$;*

2.) *the underlying degeneracy diagram of $P_T(X)$ is of the form $T(Z)$ where Z is free as a degeneracy diagram and each Z_n is a wedge of elements of $\mathcal{P}$.*

Proof. The existence of the model category structure is the standard lifting argument. In fact, since $s\,\mathrm{Alg}_T$ has a functorial path object and the forgetful functor to $s\mathcal{S}$ creates filtered colimits in $s\,\mathrm{Alg}_T$, we need only supply a $\mathcal{P}$-fibrant replacement functor for $s\,\mathrm{Alg}_T$. However, every Reedy fibrant object in $s\,\mathrm{Alg}_T$ (as in the previous section) will be $\mathcal{P}$-fibrant, and the $s\,\mathrm{Alg}_T$ in its Reedy model category structure is cofibrantly generated, so we can choose a Reedy fibrant replacement functor. This will do the job.

The object $P_T(X)$ is produced by taking an appropriate subdivision (for example the big subdivision of [8] §XII.3, Example 3.4) of a cofibrant model for X.   □

Here is how one might use this model category structure. We fix an operad $C \in \mathcal{O}$ and a simplicial resolution $T \to C$ of C as in Theorem 2.1. If X is an C-algebra, then X can be regarded as a constant object in $s\,\mathrm{Alg}_T$ and, hence, we have the resolution $P_T(X) \to X$ of the previous result. Then $P_T(X)$ is Reedy cofibrant in $s\,\mathrm{Alg}_T$ and, by Theorem 2.4.2, also Reedy cofibrant as a simplicial spectrum. Thus we can use the spectral sequence of Equation 3.4 with $P = S^0$ to show that the natural map

$$|P_T(X)| \to X$$

is a weak equivalence. But also, arguing as in Remark 3.10 we have that the augmentation $P_T(X)$ induces an isomorphism

$$\pi_* E_* P_T(X) \cong E_* X.$$

Finally, if $E_* C(n)$ is projective as an E_* module for all n, then $E_* P_T(X)$ is a cofibrant $E_* T$ algebra. In fact, since each of the spectra $P \in \mathcal{P}$ has the property that $E_* X$ is projective as an E_*-module, Theorems 2.1 and 3.12 imply that there is an isomorphism of underlying degeneracy objects:

$$(3.9) \qquad E_*(P_T(X)) \cong (E_* T)(E_* Z).$$

The fact that $E_* P_T(X)$ is cofibrant can be read off of this equation.

As this discussion indicates, and as the reader may have already suspected, we are not really interested in the $\mathcal{P}$-equivalence classes of objects in simplicial spectra or simplicial T-algebras, but certain types of E_*-equivalences. There is an appropriate model category, and it is a localization of the one supplied in Theorem 3.12.

3.13. Theorem. *The category* $s\,\mathrm{Alg}_T$ *supports the structure of a cofibrantly generated simplicial model category with*

1.) *a morphism* $f : X \to Y$ *is an* E_*-*equivalence if*

$$\pi_* E_*(f) : \pi_* E_* X \longrightarrow \pi_* E_* Y$$

is an isomorphism;

2.) *a morphism is an* E_*-*cofibration if it is a* $\mathcal{P}$-*cofibration; and*

3.) *a morphism is an* E_*-*fibration if it has the right lifting property with respect to all morphisms which are at once an* E_*-*equivalence and an* E_*-*cofibration.*

Since every $\mathcal{P}$-equivalence in $s\,\mathrm{Alg}_T$, is an E_*-equivalence, this model category structure can be produced using the localization technology of Bousfield, et al. The are many minute details, but the technology is now available in [20].

4. ANDRÉ-QUILLEN COHOMOLOGY

If A is a commutative algebra over a commutative ring k, M an A-module and $X \to A$ a morphism of k-algebras, then the André-Quillen cohomology of X with coefficients in M is the non-abelian right derived functors of the functor

$$X \mapsto \mathrm{Der}_k(X, M)$$

which assigns to X the A-module of k-derivations from X to M. This cohomology has natural generalization to algebras over operads and their modules; indeed, much of the formalism of Quillen's paper [30] goes through without difficulty. This section outlines the details and explains the application to the computation of the homotopy type of the space of maps between structured ring spectra.

It should be said that for someone interested primarily in some A_∞ operad – that is, in producing associative ring spectra – then the André-Quillen cohomology produced and discussed here is exactly that of the associative algebra case in [30]. It is, except possibly in degree zero, a shift of Hochschild cohomology.

The first part of this section is written algebraically. We fix a commutative ring k, possibly graded, and we consider k-modules (again possibly graded), operads in k-modules, and so on. All tensor products will be over k. In our applications k will be E_* for some homotopy commutative ring spectrum E. Any omitted details can be found in [18].

Let C be an operad in k-modules and suppose A is a C algebra. We define what it means for M to be an A-module. Let $\Phi(A, M)$ to be the graded k-module with

$$\Phi(A, M)_n = \bigoplus_i A \otimes \cdots \otimes A \otimes \underset{i}{M} \otimes A \otimes \cdots \otimes A$$

with M appearing once in each summand and then in the ith slot. Note that $\Phi(A, M)_n$ has an obvious action of the symmetric group Σ_n. Define

$$C(A, M) = \bigoplus_n C(n) \otimes_{k\Sigma_n} \Phi(A, M)_n = \bigoplus_n C(n) \otimes_{k\Sigma_{n-1}} A^{\otimes(n-1)} \otimes M.$$

It is an exercise to show that there is a natural ismorphism of bifunctors

$$C(C(A), C(A, M)) \cong (C \circ C)(A, M))$$

where $(\cdot) \circ (\cdot)$ is the composition of operads. The k-module M is an A-*module over C* (or simply an A-module) if there is a morphism of k-modules $\eta : C(A, M) \to M$ which fits into a coequalizer diagram

$$C(C(A), C(A, M)) \cong (C \circ C)(A, M)) \overset{d_0}{\underset{d_1}{\rightrightarrows}} C(A, M) \overset{\eta}{\longrightarrow} M$$

where the maps d_0 and d_1 are induced by the operad multiplication of C, and by η and the algebra structure on A respectively. Furthermore, the unit

$1 \to \mathcal{C}$ defines a morphism of R-modules $M = \mathbf{1}(A, M) \to C(A, M)$ which is required to be a section of η.

We now come to derivations. If A is a commutative k-algebra, and M is an A-module, we can can form a new commutative algebra over A called $M \rtimes A$, which as an k-module is simply $M \oplus A$, but with algebra multiplication

$$(x, a)(y, b) = (xb + ay, ab).$$

The algebra $M \rtimes A$ is a *square-zero extension* and an abelian object in the category of algebras over A; all abelian group objects in this category have this form. It also represents the functor that assigns to an algebra over A the A-module of k-derivations from X to M:

$$\mathrm{Der}_R(X, M) \cong \mathrm{Alg}_{\mathbf{Comm}/A}(X, M \rtimes A)$$

where we write **Comm** for the commutative algebra operad in K-modules.

These concepts easily generalize. If C is an operad, A a C-algebra and M an A-module, define a new C-algebra over A called $M \rtimes A$ as follows: as an k-module $M \rtimes A$ is simply $M \oplus A$, but the C-algebra structure is defined by noting that there is a natural decomposition

$$C(M \oplus A) \cong E(A, M) \oplus C(A, M) \oplus C(A)$$

where $E(A, M)$ consists of those summands of $C(M \oplus A)$ with more than one M term. Since M is an A-module we get a composition

$$C(M \oplus A) \to C(A, M) \oplus C(A) \to M \oplus A$$

which defines the C-algebra structure on $M \rtimes A$. The algebra $M \rtimes A$ is again a square-zero extension and an abelian object in the category of algebras over A; again, all abelian objects have this form. This last observation makes it possible to *define* the category of A-modules over C to be the category of abelian C-algebras over A.

Note that if we are in a graded setting and M is an A-module, then the graded object $\Omega^t M$ with

$$(\Omega^t M)_k = M_{k+t}$$

is also an A-module. Also, as obvious example of an A-module is A itself. We will write

$$\Lambda_A(x_{-t}) = \Omega^t A \rtimes A$$

by analogy with the exterior algebra that arises in the commutative case.

The object $M \rtimes A$ in the category of C-algebras over A represents an abelian group valued functor which we might as well call *derivations*; in formulas we write

$$\mathrm{Der}_C(X, M) \overset{\mathrm{def}}{=} \mathrm{Alg}_{C/A}(X, M \rtimes A)$$

for all C-algebras over X. Such a derivation is determined by an k-module homomorphism $d : X \to M$ which fits into an appropriate diagram which reduces to the usual definition of derivation in the commutative or associative algebra case. We invite the reader to fill in the details.

Cohomology in this context should be derived functors of derivations; this immediately leads us to simplicial algebras. We also need simplicial operads.

Thus, we let $C = C_\bullet$ be simplicial operad in k-modules and $s\,\mathrm{Alg}_C$ the category of simplicial algebras over C. This is a simplicial category in the external simplicial structure; for example, if K is a simplicial set and $X \in s\,\mathrm{Alg}_C$ then

$$(A \otimes K)_n = \coprod_{K_n} A_n$$

with the coproduct in C_n-algebras. Also, among the morphisms of $s\,\mathrm{Alg}_C$ we single out the *free* maps: a morphism $X \to Y$ is free if the underlying morphism of degeneracy diagrams is isomorphic to a map of the form

$$X \to X \sqcup C(Z)$$

where Z is a free degeneracy diagram on a free R-module.

The main theorem of [31] §II.4 immediately implies the following:

4.1. Proposition. *The $s\,\mathrm{Alg}_C$ has the structure of a simplicial model category with a morphism $f : X \to Y$*

(1) *a weak equivalence if $\pi_* f : \pi_* X \to \pi_* Y$ is an isomorphism;*

(2) *a fibration if the induced map $Nf : NX \to NY$ of normalized chain complexes in k-modules is surjective in positive degrees;*

(3) *a cofibration if it is a retract of a free map.*

If $A \in s\,\mathrm{Alg}_C$ then $\pi_0 A$ is a $\pi_0 C$-algebra. If M is a $\pi_0 A$-module (over the operad $\pi_0 C$) then M is an A_n-module (over C_n) for all $n \geq 0$. Then we can form the simplicial module $K(M, n)$ over A whose normalization $NK(M, n) \cong M$ concentrated in degree n. From this object we can form the simplicial C-algebra $K_A(M, n) = K(M, n) \rtimes A$ over A and, for $X \in s\,\mathrm{Alg}_C /A$ an algebra over A we will define the André-Quillen cohomology of X with coefficients in M by the formula

$$(4.1) \quad D_C^n(X, M) \overset{\mathrm{def}}{=} [X, K_A(M, n)]_{s\,\mathrm{Alg}_C /A} \cong \pi_0 \,\mathrm{map}_{s\,\mathrm{Alg}_C /A}(X, K_A(M, n)).$$

We note immediately that there are natural isomorphisms

$$D_C^{n-i}(X, M) \cong \pi_i \,\mathrm{map}_{s\,\mathrm{Alg}_C /A}(X, K_A(M, n))$$

and that, in fact, the collection of spaces $\mathrm{map}_{s\,\mathrm{Alg}_C /A}(X, K_A(M, n))$, $n \geq 0$, assemble into a spectrum $\hom_{s\,\mathrm{Alg}_C /A}(X, K_A M)$ so that

$$D_C^n(X, M) \cong \pi_{-n} \hom_{s\,\mathrm{Alg}_C /A}(X, K_A M).$$

As usual, this cohomology can be written down as the cohomology of a chain complex. To be concrete about this, let us fix some notation. If C is our simplicial operad and Y is a simplicial C-algebra over a constant algebra A, and if M is an A-module, as above, then we have abelian groups

$$\mathrm{Der}_{C_n}(Y_n, M) = (\mathrm{Alg}_{C_n} /A)(Y_n, M \rtimes A).$$

Furthermore, if $\phi : [n] \to [m]$ is a morphism in the ordinal number category, the Y_n is a C_m-algebra by restriction of structure along $\phi^* : C_m \to C_n$ and then

$$\phi^* : Y_m \longrightarrow Y_n$$

is a morphism of C_m-algebras. Hence we get a map

$$\mathrm{Der}_{C_n}(Y_n, M) \longrightarrow \mathrm{Der}_{C_m}(Y_m, M)$$

and, in fact, $\mathrm{Der}_C(Y, M)$ becomes a cosimplicial abelian k-module. Then,

$$(4.2) \qquad D_C^n(X, M) = H^n N \, \mathrm{Der}_C(Y, M)$$

where Y is some cofibrant model for X and N is the normalization functor from cosimplicial k-modules to cochain complexes of k-modules. This concept is important enough that we will write

$$(4.3) \qquad \mathbb{D}_C(X, M) \in \mathbf{Ho}(C^* k)$$

for the well-defined object in the derived category of cochain complexes defined by $N \, \mathrm{Der}_C(Y, M)$, with Y a cofibrant model for X.

In our applications we will have a homology theory $E_*(\cdot)$ and $k = E_*$. We will also have a simplicial operad T – that is, a simplicial object in the category $\mathcal{O}$ of simplicial operads – so $C = E_* T$ and a typical C-algebra will be of the form $E_* X$ where $X \in s\,\mathrm{Alg}_T$. If $E_* E$ if flat over E_*, this will imply that we are actually working with operads, algebras and so forth in the category of $E_* E$-comodules, rather than simply in the more basic category of E_*-modules. Under appropriate hypotheses – for example, if E satisfies the Adams condition of Definition 3.1 – the $E_* E$-comodule version of Proposition 4.2 is true, and one can use this to define André-Quillen cohomology in the category of $E_* E$-comodules.

To do this requires a little care, as we are forced to resolve not only algebras, but also the modules; the short reason for this technical difficulty is that not every chain complex of comodules is fibrant. The same problem arose in [23] and our solution is not much different.

To get started, fix a simplicial operad C in $E_* E$-comodules and an $\pi_0 C$ algebra A, also all in $E_* E$-comodules.

To ease notation, let us abbreviate the extended comodule functor by

$$\Gamma(M) = E_* E \otimes_{E_*} M.$$

The functor Γ also induces a right adjoint to the forgetful functor from A-modules in $E_* E$-comodules to A-modules. Indeed, if M is an A-module, the module structure on $\Gamma(A)$ is determined by the top split row of the diagram

$$
\begin{array}{ccccc}
\Gamma(M) & \longrightarrow & \Gamma(M) \rtimes A & \rightleftarrows & A \\
{\scriptstyle =}\downarrow & & \downarrow & & \downarrow{\scriptstyle \psi_A} \\
\Gamma(M) & \longrightarrow & \Gamma(M \rtimes A) & \rightleftarrows & \Gamma(A)
\end{array}
$$

where the right square is a pull-back and where ψ_A is the comodule structure map, which, by assumption, is a morphism of algebras. The functor $\Gamma(-)$ thus becomes the functor of a triple on A-modules in E_*E-comodules and for a simplicial C-algebra Y in E_*E comodules we can form the bicosimplicial E_*-module

$$
\mathrm{Der}_C(Y, \Gamma^\bullet(M)) = \{\mathrm{Der}_{C_p}(Y_p, \Gamma^{q+1}(M))\}.
$$

We now write

(4.4) $$\mathbb{D}_{C/E_*E}(X, M) \in \mathbf{Ho}(C^* E_* E)$$

for the object in the derived category of comodules defined by taking Y to be some cofibrant model for X and then taking the total complex of the double normalization of the cosimplicial object $\mathrm{Der}_C(Y, \Gamma^\bullet(M))$. Then we have the André-Quillen cohomology

(4.5) $$D^n_{C/E_*E}(X, M) = H^n \mathbb{D}_{E_*E/C}(X, M).$$

However, with luck, one can reduce the calculation of this more complicated object to the first case. Here is the result we will use. The definitions should make the following results plausible; the proof is in [18].

4.2. Proposition. *Let C be a simplicial operad in E_*E comodules and A a $\pi_0 C$-algebra in E_*E-comodules. If M is a A-module in E_*-modules, then the extended comodule $\Gamma(M) = E_*E \otimes_{E_*} M$ is an A-module in E_*E-comodules and there is a natural isomorphism for simplicial C-algebras X over A in E_*E-comodules*

$$
D^*_{C/E_*E}(X, E_*E \otimes_{E_*} M) \cong D^*_C(X, M).
$$

A stronger assertion is true: there is an isomorphism

$$
\mathbb{D}_{C/E_*E}(X, E_*E \otimes_{E_*} M) \cong \mathbb{D}_C(X, M)
$$

in the derived category of E_*-modules.

With this technology at hand, we can now define a spectral sequence for computing the homotopy groups of the space of maps between two structured ring spectra. We fix a commutative S-algebra E; that is an algebra in spectra over the commutative algebra operad in $\mathcal{O}$. Suppose further that E satisfies the Adams condition of Definition 3.1. Let $\mathcal{F}$ be one of either the associative algebra operad or the commutative algebra operad in $\mathcal{O}$.[5] Now suppose X is an $\mathcal{F}$-algebra in spectra; thus, X is either an associative S-algebra or a commutative S-algebra. Then E_*X is an algebra over the operad $E_*\mathcal{F}$ in E_*E-comodules.

Now let $\phi : X \to Y$ be a morphism of $\mathcal{F}$-algebras in spectra. This amounts to choosing a basepoint $\phi \in \mathrm{map}_{\mathrm{Alg}_\mathcal{F}}(X, Y)$ for the space of $\mathcal{F}$-algebra maps from X to Y. The induced map

$$E_*\phi : E_*X \longrightarrow E_*Y$$

makes E_*Y into an E_*X module over the operad $E_*\mathcal{F}$, all in E_*E-comodules. Let $T \to \mathcal{F}$ be the resolution of operads supplied by Theorem 2.1. Also, for any spectrum Y, let Y_E denote the E-completion of Y, defined as the total space of the cobar complex:

$$Y_E \overset{\mathrm{def}}{=} \mathrm{Tot}(E^\bullet \wedge Y).$$

Since E is a commutative S-algebra, Y_E is an $\mathcal{F}$-algebra if Y is an $\mathcal{F}$-algebra.

4.3. Theorem. *Let $\phi : X \to Y$ be a morphism of $\mathcal{F}$-algebras and let E be a commutative S-algebra. Then there is a second quadrant spectral sequence abutting to*

$$\pi_{t-s}(\mathrm{map}_{\mathrm{Alg}_\mathcal{F}}(X, Y_E), \phi)$$

with E^2 term

$$E_2^{0,0} = \mathrm{Hom}_{E_*\mathcal{F}/E_*E}(E_*X, E_*Y)$$

and

$$E_2^{s,t} = D_{E_*T/E_*E}^s(E_*X, \Omega^t E_*Y) \qquad t > 0.$$

Of course, the $E_2^{0,0}$ term is either the associative or commutative algebra maps (in E_*E-comodules) from E_*X to E_*Y.

This is a Bousfield-Kan spectral sequence, as we will see in the next paragraph. The standard references are [8] and [7]. The latter work, for example, implies the following result:

4.4. Corollary. *There are succesively defined obstructions to realizing a map $f \in \mathrm{Hom}_{E_*\mathcal{F}/E_*E}(E_*X, E_*Y)$ in the groups*

$$D_{E_*T/E_*E}^{s+1}(E_*X, \Omega^s E_*Y) \qquad s \geq 1.$$

In particular, if these groups are all zero, then the Hurewicz map

$$(4.6) \qquad \pi_0(\mathrm{map}_{\mathrm{Alg}_\mathcal{F}}(X, Y_E)) \to \mathrm{Hom}_{E_*\mathcal{F}}(E_*X, E_*Y)$$

is surjective. If, in addition, the groups

$$D_{E_*T/E_*E}^s(E_*X, \Omega^s E_*Y) = 0$$

for $s \geq 1$, the Hurewicz map of Equation 4.6 is a bijection.

---

[5]The reader sensitive to generalization will note that this restriction is only aesthetics. A general operad in $\mathcal{O}$ will do.

The spectral sequence of Theorem 4.3 is constructed as follows. We may regard X as a constant simplicial T-algebra and take the simplicial resolution in T-algebras $P_T(X) \to X$ guaranteed by Theorem 3.12. In addition, since E is a commutative S-algebra, the cobar complex

$$Y \to E^\bullet \wedge Y = \{E^{q+1} \wedge Y\}$$

is a cosimplicial $\mathcal{F}$-algebra. We then obtain a cosimplicial space

$$\mathrm{map}_{s\,\mathrm{Alg}_T}(P_T(X), E^\bullet \wedge Y) = \{\mathrm{map}_{\mathrm{Alg}_{T_s}}(P_T(X)_s, E^{s+1} \wedge Y)\}^s$$
$$\cong \{\mathrm{map}_{\mathrm{Alg}_{\mathcal{F}}}(\mathcal{F} \otimes_{T_s} P_T(X)_s, E^{s+1} \wedge Y)\}^s$$

and the map $\phi : X \to Y$ supplies this with the basepoint. The Bousfield-Kan spectral sequence now reads

$$\pi^s \pi_t \, \mathrm{map}_{\mathrm{Alg}_T}(P_T(X), E^\bullet \wedge Y) \Longrightarrow \pi_{t-s}\mathrm{TOT} \, \mathrm{map}_{\mathrm{Alg}_T}(P_T(X), Y)^\bullet.$$

One uses standard adjunction arguments, Theorem 1.6, Theorem 2.1, and Theorem 2.2 to show

$$\mathrm{TOT}\,\mathrm{Alg}_T(P_T(X), E^\bullet \wedge Y) \cong \mathrm{Alg}_{|T|}(|P_T(X)|, \mathrm{TOT}(E^\bullet \wedge Y)) \simeq \mathrm{Alg}_{\mathcal{F}}(X, Y_E).$$

We then must identify the E_2-term. Let's abbreviate $E^{(q+1)} \wedge Y$ as $E^{(q)}Y$. Since the cosimplicial space in question is the diagonal of the bicosimplicial space

$$\{\mathrm{map}_{\mathrm{Alg}_{T_p}}(P_T(X)_p, E^{(q)}Y)\}$$

the E_2 term can be computed as the cohomology of the total complex of the double normalization of

$$\{\pi_t \, \mathrm{map}_{\mathrm{Alg}_{T_p}}(P_T(X)_p, E^{(q)}Y)\}.$$

If we let Y^{S^t} denote the spectrum of maps from the space S^t to the spectrum Y, then Y^{S^t} has a natural structure of an $\mathcal{F}$-algebra. Thus, because of Theorem 3.12, the choice of the spectra used to build $\mathcal{P}$-resolutions (Definition 3.2) and the conditions on the operads T_s from Theorem 2.1, we have, for $t > 0$:

$$\pi_t \, \mathrm{map}_{\mathrm{Alg}_{T_p}}(P_T(X)_p, E^{(q)}(Y)) = \pi_t \, \mathrm{map}_{\mathrm{Alg}_{T_p}}(T_p(Z_p), E^{(q)}(Y))$$
$$= \pi_t \, \mathrm{map}_S(Z_p, E^{(q)}(Y))$$
$$= \mathrm{Hom}_{E_*E}(E_*Z_p, E_*(E^{(q)}(Y)^{S^t}))$$
$$= \mathrm{Hom}_{E_*T_p/E_*E}(E_*T_p(E_*Z_p), \Lambda_{E_*E^{(q)}(Y)}(x_t))$$
$$= \mathrm{Der}_{E_*T_p/E_*E}(E_*T_p(E_*Z_p), \Omega^t E_*E^{(q)}Y).$$

In short, we obtain exactly the complex $\mathbb{D}_{E_*T/E_*E}(E_*X, \Omega^t E_*Y)$ used to define André-Quillen cohomology. See Equation 4.4.

As an amusing reduction of this theory, one can consider the case of the unit operad $\mathbf{1}$ in place of the commutative or associative algebra operad. An algebra over $\mathbf{1}$ is simply a spectrum, and an $E_*\mathbf{1}$-algebra is an E_*E-comodule. The formalism carries over and the spectral sequence of Theorem 4.3 becomes the Adams-Novikov spectral sequence

$$(4.7) \qquad \mathrm{Ext}^s_{E_*E}(E_*X, \Omega^t E_*Y) \Longrightarrow \pi_{t-s}\,\mathrm{map}_S(X, Y_E) \cong [\Sigma^{t-s}X, Y_E]$$

and the obstructions of Corollary 4.4 to realizing an E_*E-comodule map $E_*X \to E_*Y$ lie in

$$(4.8) \qquad\qquad \mathrm{Ext}^{s+1}_{E_*E}(E_*X, \Omega^s E_*Y) \qquad s \geq 1.$$

If Y is an E-module, then the cobar complex $E^\bullet \wedge Y$ has a contraction; in particular, $Y = Y_E$. We would expect a corresponding simplification of the spectral sequence of Theorem 4.3. Indeed, the André-Quillen cohomology groups simplify: we need only use the derived functors of derivations in E_*-modules. We can also weaken the assumption that E be a commutative S-algebra. The result then reads:

4.5. Theorem. *Suppose that E is a homotopy commutative ring spectrum, satisfying the Adams condition of Definition 3.1 and let $\phi : X \to Y$ be a morphism of $\mathcal{F}$-algebras in spectra. Then there is a second quadrant spectral sequence abutting to*

$$\pi_{t-s}(\mathrm{map}_{\mathrm{Alg}_{\mathcal{F}}}(X, Y), \phi)$$

with E_2 term

$$E_2^{0,0} = \mathrm{Hom}_{E_*\mathcal{F}}(E_*X, Y_*)$$

and

$$E_2^{s,t} = D^s_{E_*T}(E_*X, \Omega^t Y_*) \qquad t > 0.$$

Here the $E_2^{0,0}$-term is the set of $E_*\mathcal{F}$-algebra morphisms from E_*X to Y_*.
The proof is identical to the proof of Theorem 4.3; the relevant cosimplicial space is

$$\mathrm{map}_{\mathrm{Alg}_T}(P_T(X), Y) = \{\mathrm{map}_{\mathrm{Alg}_{T_s}}(P_T(X)_s, Y)\}^s.$$

Because we have a cosimplicial space, we again have obstructions to realizing maps. In fact, there are succesively defined obstructions in

$$(4.9) \qquad D^{s+1}_{E_*T}(E_*X, \Omega^s Y_*) \cong D^{s+1}_{E_*T/E_*E}(E_*X, \Omega^s E_*Y), \quad n \geq 1$$

to the realization of a map in $\mathrm{Hom}_{E_*\mathcal{F}/E_*E}(E_*X, E_*Y)$.

5. THE MODULI SPACE OF REALIZATIONS

We now fix a homotopy commutative ring spectrum E satisfying the Adams condition of Definition 3.1. Let $\mathcal{F}$ be an operad in $\mathcal{O}$ and suppose that A is an $E_*\mathcal{F}$-algebra in E_*E-comodules. The purpose of this section is to discuss the homotopy type of the space $\mathcal{TM}(A)$ of realizations of A in $\mathrm{Alg}_{\mathcal{F}}$. In practice, of course, $\mathcal{F}$ is either the associative or commutative monoid operad and, hence, A is an associative or commutative algebra in E_*E-comodules. The method here is exactly that of [6].

By definition, $\mathcal{TM}(A)$ is the nerve (or classifying space) of the category $\mathcal{E}(A)$ with objects the $\mathcal{F}$-algebra spectra X with $E_*X \cong A$ as $E_*\mathcal{F}$-algebras and morphisms are E_*-isomorphisms.[6] As in section 1, the Dwyer-Kan decomposition of $\mathcal{TM}(A)$ supplies a weak equivalence

$$\mathcal{TM}(A) \simeq \coprod_{[X]} B\operatorname{Aut}(X)$$

where X ranges over the E_*-equivalence classes of realizations of A and $\operatorname{Aut}(X)$ is the derived space of self-equivalences of X in the E_*-local model category structure on $\operatorname{Alg}_{\mathcal{F}}$. It is worth emphasizing that this result uses the identification

(5.1) $$B\operatorname{Aut}(X) \simeq \mathcal{M}(X)$$

where $\mathcal{M}(X)$ is the nerve of the category with objects $Y \in \operatorname{Alg}_{\mathcal{F}}$ so that there a chain of E_*-isomorphisms in $\operatorname{Alg}_{\mathcal{F}}$ between Y and X. The morphisms are E_*-isomorphisms in $\operatorname{Alg}_{\mathcal{F}}$.

The initial question, of course, is whether $\mathcal{TM}(A)$ is non-empty.

We now decompose $\mathcal{TM}(A)$. As always, we will let $T \to \mathcal{F}$ be a simplicial resolution of the sort supplied by Theorem 2.1.

5.1. Definition. Let $X \in s\operatorname{Alg}_T$ be a simplicial T-algebra. We say that X is a potential n-stage for the $E_*\mathcal{F}$-algebra A if

(1) $\pi_0 E_*X$ is isomorphic to A as an $E_*\mathcal{F}$-algebra in E_*E-comodules;
(2) $\pi_i E_*X = 0$ for $1 \leq i \leq n+1$; and
(3) For all $P \in \mathcal{P}$, the groups $\pi_i^\natural(X; P) = 0$ for $i > n$.

The partial moduli space $\mathcal{TM}_n(A)$ is defined to be the moduli space of all simplicial T-algebras which are potential n-stages for A. The weak equivalences are the simplicial E_*-equivalences of Section 4.

It follows from the spiral exact sequence of Proposition 3.8 that a potential n-stage X for A has

(5.2) $$\pi_i E_*X \cong \begin{cases} A & i = 0; \\ \Omega^{n+1}A & i = n+2; \\ 0 & i \neq 0, n+2. \end{cases}$$

Furthermore, by the spiral exact sequence or, more exactly, the isomorphism 3.7, the A-module structure on $\pi_{n+1}E_*X$ is the evident shifted A-module structure. The same calculation shows that for the natural homotopy groups

$$\pi_i^\natural E_*X \overset{\text{def}}{=} \operatorname{colim} \pi_i^\natural(X, \Sigma^* DE_\alpha) \cong \begin{cases} \Omega^i A & 0 \leq i \leq n; \\ 0 & i > n. \end{cases}$$

---

[6]The isomorphism $E_*X \cong A$ is *not* part of the data.

Again, the A-module structure on $\pi_i^\natural E_* X$ is the evident shifted module structure.

Definition 5.1 makes sense for $n = \infty$: a potential ∞-stage is simply an object $X \in s\,\mathrm{Alg}_T$ so that $\pi_0 E_* X \cong A$ and $\pi_i E_* X = 0$ for $i > 0$. Let $\mathcal{TM}_\infty(A)$ be the resulting moduli space.

Theorem 2.2 and the spectral sequence of 3.4 imply that geometric realization defines a map

$$| - | : \mathcal{TM}_\infty(A) \longrightarrow \mathcal{TM}(A)$$

and the Postnikov stage construction of Remark 3.11 implies that there are maps

$$\mathcal{TM}_n(A) \longrightarrow \mathcal{TM}_m(A); \qquad 0 \le m < n \le \infty.$$

Here is the first part of the decomposition result.

5.2. Proposition. *The map induced by geometric realization*

$$| - | : \mathcal{TM}_\infty(A) \longrightarrow \mathcal{TM}(A)$$

is a weak equivalence. Furthermore the map

$$\mathcal{TM}_\infty(A) \longrightarrow \operatorname*{holim}_{n < \infty} \mathcal{TM}_n(A)$$

is a weak equivalence.

Proof. The first assertion is formal. Compare Theorem 9.3 of [6]. The second assertion is not formal; however, it follows from the main theorem of [12]. Compare Theorem 9.4 of [6]. $\square$

The next step is to investigate the tower $\{\mathcal{TM}_n(A)\}$. To do this we will identify the bottom space as a $K(G,1)$, then tell how to pass from the $(n-1)$st stage to the nth stage using André-Quillen cohomology. We begin by constructing the 0-stage; in particular, we show $\mathcal{TM}_0(A) \ne \phi$.

5.3. Definition. A simplicial T-algebra X is said to be of *type B_A* if

  1.) $\pi_0 E_* X \cong A$ as an $E_* \mathcal{F}$-algebra in $E_* E$ comodules; and
  2.) for $Y \in s\,\mathrm{Alg}_T$, the natural map

$$[Y, X]_{s\,\mathrm{Alg}_T} \longrightarrow \mathrm{Hom}_{E_* \mathcal{F}/E_* E}(\pi_0 E_* Y, A)$$

  is an isomorphism. Here the homotopy classes of maps are in the E_*-local homotopy category of $s\,\mathrm{Alg}_T$.

We write B_A for any of the (essentially unique) objects of type B_A and, if we need to, will assume B_A is E_*-local without saying so.

Simplicial T-algebras of type B_A exist. This can be seen by a generators and relations argument or by some generalized Brown representability theorem. See [19].

5.4. Remark. We have that for any simplicial T-algebra X of type B_A that

$$\pi_i^\flat(X; P) = \begin{cases} \mathrm{Hom}_{E_*E}(E_*P, A) & i = 0; \\ 0 & i > 0. \end{cases}$$

Thus a simplicial spectrum of type B_A is potential 0-stage for A. Furthermore, the spiral exact sequence implies

$$\pi_i E_* X \cong \begin{cases} A & i = 0; \\ \Omega A & i = 2; \\ 0 & i \neq 0, 2. \end{cases}$$

The following result says, among other things, that there is a unique potential 0-stage of A up to E_*-equivalence and that it is of type B_A.

5.5. Proposition. *Let* $\mathrm{Aut}(A)$ *denote the discrete group of automorphisms of the* $E_*\mathcal{F}$*-algebra* A *in* E_*E*-comodules. Then there is a natural weak equivalence*

$$\mathcal{TM}_0(A) \cong B\,\mathrm{Aut}(A).$$

Proof. Fix a choice B_A of an E_*-local space of type B_A. Let X be a potential 0-stage for A. Then a choice of isomorphism $\pi_0 E_* X \cong A$ defines a morphism in $s\,\mathrm{Alg}_T$

$$X \longrightarrow B_A$$

which defines an isomorphism on $\pi_* E(-)$ by the spiral exact sequence. Thus $\mathcal{TM}_0(A)$ is connected and, by the Dwyer-Kan analysis (5.1)

$$\mathcal{TM}_0(A) = B\,\mathrm{Aut}(B_A).$$

But it is an easy calculation that

$$\pi_n B\,\mathrm{Aut}(B_A) \cong \begin{cases} \mathrm{Aut}(A) & n = 0; \\ 0 & n \neq 0. \end{cases}$$

$\square$

To pass between the various stages of the tower, we need to know that André-Quillen cohomology is representable in the homotopy category of $s\,\mathrm{Alg}_T$. Specifically, we have the following ideas.

5.6. Definition. Let A be an $E_*\mathcal{F}$-algebra in E_*E-comodules and let M be an A-module, also in E_*E-comodules. We say that a map $X \to Y$ in $s\,\mathrm{Alg}_T$ is of type $B_A(M, n)$, $n \geq 1$ if

   1.) X is of type B_A and the induced map

$$\pi_i E_* X \to \pi_i E_* Y$$

    is an isomorphism for $i < n$;

   2.) $\pi_n E_* Y \cong M$ as a $\pi_0 E_* Y \cong A$ module; and

   3.) $\pi_i^\flat(Y; P) = 0$ if $i > n$.

We may abuse notation and refer to the simplicial T-algebra Y as being of type $B_A(M, n)$. Again, it is possible to construct such objects by a generators and relations argument, or by Brown representability using the evident homotopy characterization supplied by Proposition 5.7 below.

We would like to give a homotopical interpretation of the simplicial T-algebras of type $B_A(M, n)$; in fact, such objects will – in some sense – represent the functor

$$Z \mapsto D^n_{E_*T/E_*E}(E_*Z, M).$$

The exact result is below in Proposition 5.7, but to get there requires some preliminaries.

If $X \to Y$ is of type $B_A(M, n)$, we may assume that X is E_*-fibrant and that the map from X to Y is a cofibration to an E_*-fibrant object – and we may make this assumption without repeating it and then we will write

$$B_A \longrightarrow B_A(M, n)$$

for such a map. If we suppose $n \geq 2$, then the spiral exact sequence implies that

$$(5.3) \qquad \pi_i E_* B_A(M, n) \cong \pi_i E_* B_A \times \begin{cases} M & i = n; \\ \Omega M & i = n + 2; \\ 0 & i \neq n, n + 2. \end{cases}$$

In particular, we get a natural isomorphism $\pi_0 E_* B_A(M, n) \cong A$ and then Remark 5.3 supplies a map $B_A(M, n) \to B_A$ so that the composite $X \to Y \to B_A$ is an E_*-equivalence. In this way we will regard $B_A(M, n)$ as an object over B_A.

Because of the isomorphism of Equation 5.3, the simplicial E_*T-algebra $E_* B_A(M, n)$ is not weakly equivalent to $K_A(M, n) = K(M, n) \rtimes A$ in the category of simplicial E_*T-algebras over E_*E-comodules. However, there is a natural map of E_*T algebras

$$(5.4) \qquad \epsilon : E_* B_A(M, n) \longrightarrow K_A(M, n)$$

over the constant simplicial E_*T-algebra A. This we now produce.

Let C be the push-out in $s\,\mathrm{Alg}_{E_*E/E_*T}$ of the two-source

$$E_* B_A(M, n) \longleftarrow E_* B_A \longrightarrow \pi_0 E_* B_A \cong A.$$

Then Equation 5.3 implies that the $(n + 1)$st Postnikov section $P_{n+1}C$ of C in $s\,\mathrm{Alg}_{E_*E/E_*T}$ has the property that

$$\pi_i P_{n+1}C \cong \begin{cases} A & i = 0; \\ M & i = n; \\ 0 & i \neq 0, n. \end{cases}$$

This alone is not enough to identify the homotopy type of $P_{n+1}C$. However the map

$$A \cong \pi_0 E_* B_A \to P_{n+1}C$$

is a section of the map $P_{n+1}C \to \pi_0 P_{n+1}C \cong A$; hence $P_{n+1}C$ is canonically weakly equivalent to $K_A(M, n)$, and the composition

$$E_* B_A(M, n) \to C \to P_{n+1}C$$

is a model for the morphism ϵ of Equation 5.4.

5.7. Proposition. *Let $B_A \to B_A(M, n)$ be of type $B_A(M, n)$ and suppose $n \geq 2$. Let $X \in s\,\mathrm{Alg}_T$ and suppose a morphism of $\mathcal{F}$-algebras in $E_* E$-comodules $\pi_0 E_* X \to A$ is represented by a map $f : X \to B_A$. Then the morphism of simplicial $E_* T$-algebras*

$$\epsilon : E_* B_A(M, n) \to K_A(M, n)$$

induces a natural weak equivalence

$$\mathrm{map}_{s\,\mathrm{Alg}_T / B_A}(X, B_A(M, n)) \xrightarrow{\simeq} \mathrm{map}_{s\,\mathrm{Alg}_{E_* T / E_* E} / A}(E_* X, K_A(M, n)).$$

In particular

$$\pi_i \,\mathrm{map}_{s\,\mathrm{Alg}_T / B_A}(X, B_A(M, n)) \cong D^{n-i}_{E_* T / E_* E}(E_* X, M).$$

Proof. We have a natural map induced by ϵ. Since both source and target take homotopy colimits to homotopy limits, it is sufficient to check the result for objects of the form $X = T(P \otimes K)$ where $P \in \mathcal{P}$ and K is a simplicial set. Inducting over the skeleta of K, we find it is sufficient to check the result for objects of the form $T(P \otimes \partial \Delta^n)$ equipped with some choice of map

$$E_* T(P \otimes \partial \Delta^n) \to \pi_0 E_* T(P \otimes \partial \Delta^n) \cong E_* \mathcal{F}(E_* P) \to A.$$

But the objects of type $B_A(M, n)$ are built exactly so the result holds in this case. For more details see Proposition 8.7 of [6]. $\qquad\square$

To shorten notation, let us write

$$\mathcal{H}^n(A, M) \overset{\mathrm{def}}{=} \mathrm{map}_{s\,\mathrm{Alg}_{E_* T / E_* E} / A}(E_* X, K_A(M, n)).$$

Let $\mathrm{Aut}(A, M)$ of be the group of automorphisms of the pair (A, M). Then $\mathrm{Aut}(A, M)$ acts in a natural way on the space $\mathcal{H}^n(A, M)$; let $\hat{\mathcal{H}}^n(A, M)$ denote the Borel construction. The space $\mathcal{H}^n(A, M)$ has a basepoint given by

$$0 \in \pi_0 \mathcal{H}^n(A, M) = D^n_{E_* T / E_* E}(A, M).$$

There is a choice of representative for 0 which is invariant under the action of $\mathrm{Aut}(A, M)$; therefore we get an induced map

$$B\,\mathrm{Aut}(A, M) \to \hat{\mathcal{H}}^n(A, M).$$

5.8. Theorem. *For all $n \geq 1$ there is a homotopy pull-back diagram*

$$
\begin{array}{ccc}
\mathcal{T}\mathcal{M}_n(A) & \longrightarrow & B\,\mathrm{Aut}(A, M) \\
\downarrow & & \downarrow \\
\mathcal{T}\mathcal{M}_{n-1}(A) & \longrightarrow & \hat{\mathcal{H}}^{n+2}(A, \Omega^n A).
\end{array}
$$

To interpret this result, let $Y \in \mathcal{TM}_{n-1}(A)$ be a basepoint – that is, a potential $(n-1)$st stage of of A. Then the homotopy fiber of $\mathcal{TM}_n(A) \to \mathcal{TM}_{n-1}(A)$ is non-empty if and only if Y is weakly equivalent to $P_{n-1}X$ for some potential nth stage of A. This, in turn, will occur if the image of Y in $\pi_0 \hat{\mathcal{H}}^{n+2}(A, \Omega^n A)$ is the zero element. Furthermore, if it is the zero element, then homotopy fiber at Y is weakly equivalent to $\mathcal{H}^{n+1}(A, \Omega^n A)$. Therefore, by trying to lift the basepoint of $\mathcal{TM}_0(A) = B\operatorname{Aut}(A)$ up the tower, we obtain the following corollary.

5.9. Corollary. *There are successively defined obstructions, well defined up to the action of* $\operatorname{Aut}(A, M)$,

$$\theta_n \in D^{n+2}_{E_*T/E_*E}(A, \Omega^n A), \qquad n \geq 1$$

to realizing the $E_*\mathcal{F}$-*algebra* A *by an* $\mathcal{F}$-*algebra* X. *Obstructions to uniqueness lie in*

$$D^{n+1}_{E_*T/E_*E}(A, \Omega^n A).$$

Theorem 5.8 is proved exactly in the same manner as the main theorem of [6]. If one is interested only in the obstructions to realization, one can proceed as follows. Let Y be a potential $(n-1)$st stage for A. We'd like to construct a potential nth stage X so that $P_{n-1}X \simeq Y$. We may assume that Y is a cofibrant simplicial T-algebra. By a Postnikov section argument, we see that it is necessary and sufficient to produce a map of simplicial T-algebras over B_A

$$Y \longrightarrow B_A(\Omega^n A, n+1)$$

which induces an isomorphism on $\pi_{n+1}E_*(-)$. Because the space $B_A(\Omega^n A, n+1)$ represents André-Quillen cohomology, this is equivalent to producing a map of simplicial E_*T-algebras over A

$$E_*Y \longrightarrow K_A(\Omega^n A, n+1)$$

which (by calculating with the spiral exact sequence) is a weak equivalence. Since, as a simplicial E_*T-algebra, E_*Y is a two-stage Postnikov tower, it is determined up to weak equivalence by a morphism in $s\operatorname{Alg}_{E_*T}$ over A

$$A \simeq P_0 E_*Y \longrightarrow K_A(\Omega^n A, n+2).$$

The class of this map in

$$\pi_0 \hat{\mathcal{H}}^{n+2}(A, \Omega^n A) \cong D^{n+2}_{E_*T/E_*E}(A, \Omega^n A)/\operatorname{Aut}(A, \Omega^n A)$$

is the obstruction. The Borel construction is necessary as we have not fixed our various isomorphisms to A and $\Omega^n A$.

The obstructions to uniqueness can found in Equation 4.9.

6. COMPUTING WITH E_∞ OPERADS.

If **Comm** is the commutative monoid operad, then Theorem 2.1 supplies an augmented simplicial operad $T \to$ **Comm** so that the augmented simplicial operad $E_* T \to E_*$ **Comm** is an algebraic E_∞ operad in a sense to be defined shortly. Since it is the simplicial operad T and the methods of the previous section that we will use to attempt to impose E_∞ structures on spectra, we need to be able to compute the André-Quillen cohomology functor $D^*_{E_* T}$. The purpose of this section is to reduce that computation, at least in some cases, to the calculation of ordinary André-Quillen homology or cohomology. The main result is the two spectral sequences supplied by the Propositions 6.4 and 6.5 below. Note that E_* **Comm** is the commutative algebra operad in E_* modules; hence ordinary André-Quillen cohomology is cohomology over the operad E_* **Comm**.

We will first say what we mean by an E_∞ operad. If k is a commutative ring, we will write **Comm** for the commutative monoid operad in k-modules—rather than, for example, $k[$**Comm**$]$.

6.1. Definition. For any commutative ring k (possibly graded) an E_∞-operad $\mathcal{E}$ is a simplicial operad in k-modules equipped with a weak equivalence $\mathcal{E} \to$ **Comm** and so that for each $q \geq 0$, $\mathcal{E}(q)$ is a cofibrant (i.e., level-wise projective) simplicial $k[\Sigma_q]$ module.

There is a canonical such operad—namely a cofibrant model for **Comm** in the category of simplicial k-operads —but we don't need that much structure in this discussion.

If V_* is a cofibrant simplicial k-module, the shuffle chain equivalence of normalized chain complexes

$$\underbrace{N(V) \otimes \cdots \otimes N(V)}_{n} \to \underbrace{N(V \otimes \cdots \otimes V)}_{n}$$

is Σ_n-equivariant; thus if C is any simplicial k-module operad, the normalized object $NC = \{NC(k)\}_{k \geq 0}$ is an operad of k-chain complexes. In particular, if $\mathcal{E}$ is an E_∞ operad in the sense of Definition 6.1, then $N\mathcal{E}$ is an E_∞ operad in the category of chain complexes over k. More is true. If V is a simplicial k-module, and C is a simplicial operad, then there is a natural map of chain complexes

$$(6.1) \qquad NC(q) \otimes_{k\Sigma_q} NV^{\otimes q} \to N(C(q) \otimes_{k\Sigma_q} V^{\otimes q})$$

and if $C(q)$ is cofibrant as a $k\Sigma_q$-module and V is cofibrant as a k-module, this is a quasi-isomorphism. In shorthand,

$$(6.2) \qquad NC(NV) \to NC(V)$$

is a quasi-isomorphism of NC algebras. Furthermore, if A is any simplicial C-algebra, NA is an NC algebra via

$$NC(NA) \to NC(A) \to NA.$$

From these considerations, and from [28], it immediately follows that if k is an algebra over a field $\mathbb{F}$ of characteristic $p > 0$, and $\mathcal{E}$ is a simplicial E_∞ operad, then the homotopy of any simplicial $\mathcal{E}$-algebra is an "unstable" algebra over the Dyer-Lashof algebra. That is, if $A \in s\,\mathrm{Alg}_\mathcal{E}$, then $\pi_* A$ is a graded commutative algebra equipped with operations

$$Q^i : \pi_n A \to \pi_{n+i} A, \qquad i \geq 0, p = 2$$

or

$$\beta^\epsilon Q^i : \pi_n A \to \pi_{n+2i(p-1)-\epsilon} A \qquad i \geq 0, \epsilon = 0, 1, p > 2$$

subject to the Adem and Cartan formulas of [11], §I.1. Unstable in this context means, at $p = 2$,

$$Q^i(x) = \begin{cases} 0 & i < \deg(x) \\ x^2 & i = \deg(x) \end{cases}$$

and at $p > 2$

$$\beta^\epsilon Q^i(x) = \begin{cases} 0 & 2i - \epsilon < \deg(x) \\ x^p & 2i = \deg(x) \text{ and } \epsilon = 0. \end{cases}$$

This condition arises because we are dealing with a normalized object, not an arbitrary algebra over an E_∞ operad in chain complexes.

Also if k is not the prime field $\mathbb{F}_p$, these operations are not k-linear; if $\phi : k \to k$ is the Frobenius, $a \in k$ and $x \in \pi_* A$, then

$$(6.3) \qquad \beta^\epsilon Q^i(ax) = \phi(a)\beta^\epsilon Q^i(x).$$

There is an obvious category $\mathcal{UR}$ of unstable algebras over the Dyer-Lashof algebra. The forgetful functor $\mathcal{UR} \to n\mathcal{M}_k$ to graded k-modules has a left adjoint $\mathcal{S}_\mathcal{R}$. It follows from the quasi-isomorphisms of 6.1 and 6.2 and the calculations of [28] that if $\mathcal{E}$ is an E_∞ operad and $V \in s\mathcal{M}_k$, then the natural map

$$(6.4) \qquad \mathcal{S}_\mathcal{R}(\pi_* V) \to \pi_* \mathcal{E}(V)$$

is an isomorphism in $\mathcal{UR}$ provided that $\pi_* V$ is a graded projective k-module. Note that this isomorphism does not depend on $\mathcal{E}$: if V a cofibrant simplicial k-module, $\pi_* \mathcal{E}(V)$ is independent of the E_∞ operad $\mathcal{E}$ and we have:

6.2. Proposition. *Let $f : \mathcal{E} \to \mathcal{E}'$ be a morphism of E_∞ operads over an $\mathbb{F}$ algebra k, where $\mathbb{F}$ is a field of positive characteristic. Then the restriction of structure functor and its left adjoint induce a Quillen equivalence*

$$\mathcal{E}' \otimes_\mathcal{E} (\cdot) = f^* : s\,\mathrm{Alg}_\mathcal{E} \rightleftarrows s\,\mathrm{Alg}_{\mathcal{E}'} : f_*.$$

This result is true over any ground ring k, although in general a less computational proof is required. Furthermore, any two E_∞-operads are connected by a chain of such weak equivalences.

The algebra $S_\mathcal{R}(W)$ has a simple description, at least when W is a graded projective k-module. See [11] §I.1. The operations $\beta^\epsilon Q^i$ can be assembled into an algebra $\mathcal{R}$ over $\mathbb{F}_p$ using the Adem relations (see [11], §I.2). This is the Dyer-Lashof algebra. If $W \in n\mathcal{M}_k$ is a graded k-module, let

(6.5) $$\mathcal{R}(W) = \mathcal{R} \otimes_{\mathbb{F}_p} W/U$$

where U is the sub-$\mathcal{R}$-module generated by elements of the form

$$Q^i \otimes x, \quad i < \deg(x) \quad (p = 2)$$
$$\beta^\epsilon Q^i \otimes x, \quad 2i - \epsilon < \deg(x) \quad p > 2.$$

Then, if W is a graded projective k-module,

(6.6) $$S_{\mathcal{R}}(W) = S(\mathcal{R}(W))/I$$

where S is the graded symmetric algebra functor over k and I is the ideal generated by the elements

$$Q^i(x) - x^2 \quad \deg(x) = i, \quad p = 2$$
$$Q^i(x) - x^p \quad \deg(x) = 2i, \quad p > 2$$

In particular $S_{\mathcal{R}}(W)$ is a free graded commutative k-algebra.

If $\Gamma \in \mathcal{U}\mathcal{R}$ then Γ is, among other things, a graded commutative algebra and, as such, we can form its André-Quillen homology $D_*\Gamma$ as a graded commutative algebra:

$$D_*\Gamma \overset{\text{def}}{=} \pi_* \mathbb{L}_{\Gamma/k}$$

where $\mathbb{L}_{\Gamma/k}$ is the cotangent complex as a graded commutative algebra. As usual (cf. [29],[17]), the André-Quillen homology inherits structure from the Dyer-Lashof operations. We next spell out exactly what this structure is.

Let $\mathcal{U}$ be the category of non-negatively-graded modules over the Dyer-Lashof algebras $\mathcal{R}$. These are graded k-modules and $\mathcal{R}$ acts with the Frobenius twist as in Equation 6.3, and unstable means that

$$Q^i(x) = 0 \quad \text{if} \quad i \leq \deg(x) \quad (p = 2)$$
$$\beta^\epsilon Q^i(x) = 0 \quad \text{if} \quad 2i - \epsilon < \deg(x) \text{ or } 2i = \deg(x) \quad (p > 2)$$

If $\Gamma \in \mathcal{U}\mathcal{R}$ then $\mathcal{U}(\Gamma)$ is the category of objects M which are at once in $\mathcal{U}$, and graded Γ-modules subject to the compatibility condition that the multiplication map

$$\Gamma \otimes_k M \to M$$

is a morphism in $\mathcal{U}$.

Such structures arise naturally as follows: If $M \in \mathcal{U}(\Gamma)$, let $\text{Der}_{\mathcal{R}}(\Gamma, M)$ be the module of commutative k-algebra derivations that commute with the elements of $\mathcal{R}$. The following is proved with a minor variation of the (entirely standard) techniques of [17] §1.

6.3. Lemma. *Let $\Gamma \in \mathcal{U}\mathcal{R}$. The graded module $\Omega_{\Gamma/k}$ of commutative algebra derivations is naturally an object in $\mathcal{U}(\Gamma)$ and there is a natural isomorphism*

$$\text{Der}_{\mathcal{R}}(\Gamma, M) \cong \text{Hom}_{\mathcal{U}(\Gamma)}(\Omega_{\Gamma/k}, M).$$

The functor which assigns to an algebra $\Gamma \in \mathcal{UR}$ the module of derivations $\mathrm{Der}_{\mathcal{R}}(\Gamma, M)$ has non-abelian right derived functors. Choose a cofibrant X model for $\Gamma \in \mathcal{UR}$ regarded as a constant object in $s\mathcal{UR}$. Then these derived functors are a kind of André-Quillen cohomology:

$$(6.7) \qquad D_{\mathcal{R}}^n(\Gamma, M) = \pi^n \, \mathrm{Der}_{\mathcal{R}}(X, M).$$

This cohomology can be dissected. We may assume $X_q = S_{\mathcal{R}}(V_q)$ for some graded, projective k-module V_q; hence, as a simplicial graded commutative algebra, $X \to \Gamma$ is still a cofibrant model for Γ. Thus

$$(6.8) \qquad \mathbb{L}_{\Gamma/k} \simeq \Gamma \otimes_X \Omega_{X/k}$$

acquires, by Lemma 6.3, the structure of a cofibrant simplicial $s\mathcal{U}(\Gamma)$ module. This implies that the ordinary André-Quillen homology

$$(6.9) \qquad D_*\Gamma \cong \pi_*\mathbb{L}\Omega_{\Gamma/k}$$

is a graded object in $\mathcal{U}(\Gamma)$, and this structure is independent of the choice of X.

This noted, it is not surprising that the natural isomorphism of Lemma 6.3 yields a composite functor spectral sequence:

6.4. Proposition. *Let $\Gamma \in \mathcal{UR}$. Then there is a spectral sequence*

$$\mathrm{Ext}_{\mathcal{U}(\Gamma)}^p(D_q(\Gamma), M) \Rightarrow D_{\mathcal{R}}^{p+q}(\Gamma, M).$$

This is important because of the following result. Let k be an algebra over a field of positive characteristic, and $\mathcal{E}$ a simplicial E_∞ operad over k. If A is a simplicial $\mathcal{E}$ algebra and M is a $\pi_0 A$ module (over the operad $\pi_0\mathcal{E} = \mathbf{Comm}$), then M is an object in $\mathcal{U}(\pi_* A)$.

6.5. Proposition. *Let $\mathcal{E}$ be an E_∞ operad over an $\mathbb{F}$-algebra k, where $\mathbb{F}$ is a field of characteristic $p > 0$. Let $A \in s\,\mathrm{Alg}_{\mathcal{E}}$. Then there is a spectral sequence*

$$D_{\mathcal{R}}^p(\pi_* A, M)^q \Rightarrow D_{\mathcal{E}}^{p+q}(A, M).$$

Proof. Here is an outline of the proof. We may assume A is cofibrant. Let $P_\bullet^{\mathcal{E}} A \in s(s\mathcal{A}^T)$ be a simplicial resolution of A by $\mathcal{E}$ algebras of the form $\mathcal{E}(W)$ where W is a cofibrant simplicial k-module with the property that $\pi_* W$ is a projective k-module. Here resolution means that

$$\pi_* \pi_* P_\bullet^{\mathcal{E}} A \cong \pi_* A$$

via the augmentation. It is possible to construct such by a Stover resolution argument. Compare section 3. Note that this and Equation 6.4 imply that

$$\pi_* P_\bullet^{\mathcal{E}} A \longrightarrow \pi_* A$$

is a cofibrant model for $\pi_* A$ as a simplicial object in $\mathcal{UR}$.

Taking the geometric realization, which is possible because $s\,\mathrm{Alg}_{\mathcal{E}}$ is a simplicial model category, we obtain a weak equivalence

$$|P_\bullet^{\mathcal{E}} A| \longrightarrow A$$

and hence a spectral sequence

$$\pi^p D_{\mathcal{E}}^q(P_\bullet^{\mathcal{E}}(A), M) \Longrightarrow D^{p+q}(|P_\bullet^{\mathcal{E}} A|, M) \cong D^{p+q}(A, M).$$

The last isomorphism follows because $|P_\bullet^{\mathcal{E}} A|$ is cofibrant. The claim is that

$$D_{\mathcal{E}}^*(\mathcal{E}(W), M) \cong \operatorname{Der}_{\mathcal{R}}(S_{\mathcal{R}}(\pi_* W), M).$$

This is easily verified, completing the proof. □

6.6. Remark. If k is an algebra over a field of characteristic 0 and $\mathcal{E}$ is an E_∞-operad for k-modules, then the weak equivalence of simplicial operads $\epsilon : \mathcal{E} \to \mathbf{Comm}$ induces a Quillen equivalence

$$\epsilon^* : s\operatorname{Alg}_{\mathcal{E}} \rightleftarrows s\operatorname{Alg}_{\mathbf{Comm}} : \epsilon_*.$$

Furthermore, André-Quillen cohomology over $\mathcal{E}$ reduces to André-Quillen cohomology for commutative k-algebras.

In our applications, we will encounter simplicial algebras of the form $E_* X$ where X is some simplicial algebra over some simplicial operad. In this case, the ground ring will be $k = E_*$ and very rarely will this be an algebra over a field of characteristic p. Therefore, we close this section with two results intended to reduce calculations to the case considered above.

The first is this. Suppose $\mathfrak{m} \subseteq k$ is an ideal with the property that $k/\mathfrak{m}$ is an algebra over a field of chacteristic p. Then if $\mathcal{E}$ is an E_∞ operad over k in the sense of Definition 6.1, then $k/\mathfrak{m} \otimes_k \mathcal{E}$ is an E_∞ operad over $k/\mathfrak{m}$. Furthermore, if A is a simplicial $\mathcal{E}$ algebra, then $k/\mathfrak{m} \otimes_k A$ is a simplicial $k/\mathfrak{m} \otimes_k \mathcal{E}$ algebra. If M is module over

$$\pi_0(k/\mathfrak{m} \otimes_k A) \cong k/\mathfrak{m} \otimes_k \pi_0 A$$

then M is a module over $\pi_0 A$ and we'd like to use these facts to compute $D_{\mathcal{E}}^*(A, M)$. If X is a cofibrant $\mathcal{E}$ algebra, then X is cofibrant as simplicial k module; hence if $X \to A$ is a weak equivalence of $\mathcal{E}$ algebras with X cofibrant, then $k/\mathfrak{m} \otimes_k X$ is a model for the derived tensor product $k/\mathfrak{m} \otimes_k^{\mathbb{L}} A$.

6.7. Proposition. *Let A be a simplicial $\mathcal{E}$ algebra over an E_∞ operad over k and let M be a $k/\mathfrak{m} \otimes_k \pi_0 A$ module. Then there is a natural isomorphism*

$$D_{\mathcal{E}}^*(A, M) \cong D_{k/\mathfrak{m} \otimes_k \mathcal{E}}^*(k/\mathfrak{m} \otimes_k^{\mathbb{L}} A, M).$$

If $\pi_* A$ is flat over k, then

$$\pi_*(k/\mathfrak{m} \otimes_k^{\mathbb{L}} A) \cong k/\mathfrak{m} \otimes_k \pi_* A.$$

Now suppose M is simply a module over $\pi_0 A$ and suppose that M is flat as a k-module. Then we can filter the module M by powers of the ideal $\mathfrak{m} \subseteq k$ to get a spectral sequence:

6.8. Proposition. *There is a spectral sequence*

$$E_1^{p,q} = D_{k/\mathfrak{m} \otimes_k \mathcal{E}}^p(k/\mathfrak{m} \otimes_k^{\mathbb{L}} A, \mathfrak{m}^q M / \mathfrak{m}^{q+1} M) \Longrightarrow \lim_q D_{\mathcal{E}}^p(A, M / \mathfrak{m}^q M).$$

If M is $\mathfrak{m}$-complete in the sense that $M \cong \lim_q M/\mathfrak{m}^q M$ and there is an r so that for all (p,q) we have $E_r^{p,q} = E_\infty^{p,q}$, then

$$\lim_q D_{\mathcal{E}}^p(A, M/\mathfrak{m}^q M) \cong D_{\mathcal{E}}^q(A, M).$$

7. THE LUBIN-TATE THEORIES

In the section we apply the technology developed in the previous sections to show that the techniques used by Haynes Miller and the second author (cf. [34]) to show that the algebraic theory of deformations of height n formal group laws actually lifts to E_∞-ring spectra.

The Lubin-Tate theory [25] of deformations of finite height formal group laws works over an arbitrary perfect field of characteristic p. However, we will specialize to algebraic extensions of the prime field $\mathbb{F}_p$ to keep the language simple.

Fix a such a field k and a formal group law Γ over k. A *deformation* of Γ to a complete local ring A (with maximal ideal $\mathfrak{m}$) is a pair (G, i) where G is a formal group law over A, $i : k \to A/\mathfrak{m}$ is a morphism of fields and one requires $i^*\Gamma = \pi^*G$, where $\pi : A \to A/\mathfrak{m}$ is the quotient map. Two such deformations (G, i) and (H, j) are $\star$-isomorphic if there is an isomorphism $f : G \to H$ of formal group laws which reduces to the identity modulo $\mathfrak{m}$. Write $\mathbf{Def}_\Gamma(A)$ for the set of $\star$-isomorphism classes of deformations of Γ over A.

A common abuse of notation is to write G for the deformation (G, i); i is to be understood from the context.

Now suppose the height of Γ is finite. Then the theorem of Lubin and Tate [25] says that the functor $A \mapsto \mathbf{Def}_\Gamma(A)$ is representable. Indeed let

$$(7.1) \qquad A(\Gamma, k) = W(k)[[u_1, \cdots, u_{n-1}]]$$

where $W(k)$ denotes the Witt vectors on k and n is the height of Γ. This is a complete local ring with maximal ideal $\mathfrak{m} = (p, u_1, \cdots, u_{n-1})$ and there is a canonical isomorphism $q : k \cong A(\Gamma, k)/\mathfrak{m}$. Then Lubin and Tate prove there is a deformation (G, q) of Γ over $A(\Gamma, k)$ so that the natural map

$$(7.2) \qquad \mathrm{Hom}_c(A(\Gamma, k), A) \to \mathbf{Def}_\Gamma(A)$$

sending a continuous map $f : A(\Gamma, k) \to A$ to $(f_*G, \bar{f}q)$ (where $\bar{f}$ is the map on residue fields induced by f) is an isomorphism. Continuous maps here are very simple: they are the *local* maps; that is, we need only require that $f(\mathfrak{m})$ be contained in the maximal ideal of A. Furthermore, if two deformations are $\star$-isomorphic, then the $\star$-isomorphism between them is unique.

We'd like to now turn the assignment $(\Gamma, k) \mapsto A(\Gamma, k)$ into a functor. For this we introduce the category $\mathcal{FGL}_n$ of height n formal group laws over fields which are algebraic extensions of $\mathbb{F}_p$. The objects are pairs (Γ, k) where Γ is of height n. A morphism

$$(f, j) : (\Gamma_1, k_1) \to (\Gamma_2, k_2)$$

is a homomorphism of fields $j : k_1 \to k_2$ and an isomorphism of formal group laws $f : j^*\Gamma_1 \to \Gamma_2$. This is the opposite of the category considered by Rezk in [34]. We make this choice so we get a covariant functor. As a result, some of our results below also have an opposite flavor – nonetheless, these are the same results.

Let (f, j) be such a morphism and let G_1 and G_2 be the fixed universal deformations over $A(\Gamma_1, k)$ and $A(\Gamma_2, k)$ respectively. If $\bar{f} \in A(\Gamma_2, k_2)[[x]]$ is any lift of $f \in k_2[[x]]$, then we can define a formal group law H over $A(\Gamma_2, k_2)$ by requiring that $\bar{f} : H \to G_2$ is an isomorphism. Then the pair (H, j) is a deformation of Γ_1, hence we get a homomorphism $A(\Gamma_1, k_1) \to A(\Gamma_2, k_2)$ classifying the $\star$-isomorphism class of H – which, one easily checks, is independent of the lift $\bar{f}$. Thus if $Rings_c$ is the category of complete local rings and local homomorphims, we get a functor

$$A(\cdot, \cdot) : \mathcal{FGL}_n \longrightarrow Rings_c.$$

In particular, note that any morphism in $\mathcal{FGL}_n$ from a pair (Γ, k) to itself is an isomorphism. Thus, these endomorphisms form the "big" Morava stabilizer group of the formal group law. It contains the usual Morava stabilizer group as the subgroup of elements of the form (f, id_k). The formal group law and hence also its automorphism group is determined up to isomorphism by the height of Γ if k is separably closed.

Next we put in the gradings. This requires a paragraph of introduction. For any commutative ring R, the morphism $R[[x]] \to R$ of rings sending x to 0 makes R into a $R[[x]]$-module. Let $\mathrm{Der}_R(R[[x]], R)$ denote the R-module of continuous R-derivations; that is, continuous R-module homomorphisms

$$\partial : R[[x]] \longrightarrow R$$

so that

$$\partial(f(x)g(x)) = \partial(f(x))g(0) + f(0)\partial(g(x)).$$

If ∂ is any derivation, write $\partial(x) = u$; then, if $f(x) = \sum a_i x^i$,

$$\partial(f(x)) = a_1 \partial(x) = a_1 u.$$

Thus ∂ is determined by u, and we write $\partial = \partial_u$. We then have that the module $\mathrm{Der}_R(R[[x]], R)$ is a free R-module of rank one, generated by any derivation ∂_u so that u is a unit in R. In the language of schemes, ∂_u is a generator for the tangent space at 0 of the formal scheme $\mathbb{A}^1_R$ over $\mathrm{Spec}(R)$.

Now consider pairs (F, u) where F is a formal group law over R and u is a unit in R. Thus F defines a smooth one dimensional commutative formal group scheme over $\mathrm{Spec}(R)$ and ∂_u is a chosen generator for the tangent space at 0. A morphism of pairs

$$f : (F, u) \longrightarrow (G, v)$$

is an isomorphism of formal group laws $f : F \to G$ so that

$$u = f'(0)v.$$

Note that if $f(x) \in R[[x]]$ is a homomorphism of formal group laws from F to G, and ∂ is a derivation at 0, then $(f^*\partial)(x) = f'(0)\partial(x)$. In the context of deformations, we may require that f be a $\star$-isomorphism.

This suggests the following definition: let Γ be a formal group law of height n over a field k which is an algebraic extension of $\mathbb{F}_p$ and let A be a complete local ring. Define $\mathbf{Def}_\Gamma(A)_*$ to be equivalence classes of pairs $((G, i), u)$ where (G, i) is a deformation of Γ to A and u is a unit in A. The equivalence relation is given by $\star$-isomorphisms transforming the unit as in the last paragraph. We now have that there is a natural isomorphism

$$\mathrm{Hom}_c(A(\Gamma, k)[u^{\pm 1}], A) \cong \mathbf{Def}_\Gamma(A)_*.$$

We impose a grading by giving an action of the multiplicative group scheme $\mathbb{G}_m$ on the scheme $\mathbf{Def}_\Gamma(\cdot)_*$ (on the right) and thus on $A(\Gamma, k)[u^{\pm 1}]$ (on the left): if $v \in A^\times$ is a unit and (G, u) represents an equivalence class in $\mathbf{Def}_\Gamma(A)_*$ define an new element in $\mathbf{Def}_\Gamma(A)_*$ by $(G, v^{-1}u)$. In the induced grading on $A(\Gamma, k)[u^{\pm 1}]$, one has $A(\Gamma, k)$ in degree 0 and u in degree -2.

This grading is essentially forced by topological considerations. See the remarks before Theorem 20 of [39] for an explanation.

We now collect a sequence of results, mostly from Rezk's paper [34], to develop the input to our machine.

7.1. Proposition. *For all pairs $(\Gamma, k) \in \mathcal{FGL}_n$, the universal deformation over $A(\Gamma, k)[u^{\pm 1}]$ is a Landweber exact formal group law. Furthermore, the resulting homology theory $E(\Gamma, k)_*$ is of Adams-type.*

Proof. See Propositions 6.5 and 15.3 of [34]. □

We will write $E(\Gamma, k)$ for the representing spectrum of this homology theory.

7.2. Remark. The importance of these homology theories – and of the whole moduli problem we are discussing here – was first recognized by Morava. Hence we might call these homology theories Morava E-theories. If we choose $k = \mathbb{F}_{p^n}$ and Γ to be the Honda formal group law of height n, the $E(\Gamma, k)_*$ is what is commonly written $(E_n)_*$. A mild variant of the resulting spectrum was shown to be an A_∞-ring spectrum by Baker [3]; his methods apply equally to all of the spectra $E(k, \Gamma)$.

Note that the ring $E(\Gamma, k)_0 \cong A(\Gamma, k)$ and, hence, it is a complete local ring. Fix two objects (Γ_1, k_1) and (Γ_2, k_2) is $\mathcal{FGL}_n$ and let $F = E(\Gamma_1, k_1)$, $E = E(\Gamma_2, k_2)$.

7.3. Proposition. *Let A_* be a graded commutative ring so that A_0 is a complete local ring with maximal ideal $\mathfrak{m}$. Suppose $i : E_* \to A_*$ is a morphism of graded commutative rings which is continuous in degree 0. Then the set*

$$\mathrm{Hom}_{E_* - alg}(E_*F, A_*)$$

is isomorphic to the set of morphisms in $\mathcal{FGL}_n$

$$(\Gamma_1, k_1) \to (i^*\Gamma_2, A_0/\mathfrak{m}).$$

Proof. This is a consequence of Landweber exactness and the groupoid point of view to deformations. See section §17 of [34]. □

For example, if we set $A_* = E_*$, the get that

(7.3) $\qquad \text{Hom}_{E_*-\text{alg}}(E_*F, E_*) = \text{Hom}_{\mathcal{FGL}_n}((\Gamma_1, k_1), (\Gamma_2, k_2)).$

If k is field of characteristic p and A a k-algebra, let $\sigma : A \to A$ denote the Frobenius. This is not a k-algebra homomorphism, but the commutative diagram

$$
\begin{array}{ccc}
k & \longrightarrow & A \\
\downarrow{\sigma} & & \downarrow{\sigma} \\
k & \longrightarrow & A
\end{array}
$$

yields a k-algebra homomorphism

$$\sigma : k \otimes_\sigma A \to A$$

called the relative Frobenius. Now let $\mathfrak{m}_E \subseteq E_0F$ be extension of the maximal ideal $\mathfrak{m} \subseteq A(\Gamma_1, k_1) = E_0$; that is $\mathfrak{m}_E = \mathfrak{m}E_0F$ and

$$E_0F/\mathfrak{m}_E = k_1 \otimes_{E_0} E_0F.$$

7.4. Proposition. *The relative Frobenius*

$$\sigma : k_1 \otimes_\sigma E_0F/\mathfrak{m}_E \to E_0F/\mathfrak{m}_E$$

is an isomorphism. As a consequence

$$\mathbb{L}_{k_1 \otimes_{E_0} E_0F/k_1} \simeq 0.$$

Proof. The first statement follows easily from Proposition 7.3 and facts about powers series. See [34], Proposition 21.5. The second statement follows from the fact that

$$\sigma_* : \mathbb{L}_{(k_1 \otimes_R E_0F)/k_1} \to \mathbb{L}_{E_0F/k_1}$$

is both an isomorphism and the zero map. See Proposition 21.2 of [34]. □

7.5. Corollary. *The graded cotangent complex is contractible:*

$$\mathbb{L}_{(k_1 \otimes_{E_0} E_*F)/(k_1[u^{\pm 1})]} \simeq 0.$$

Proof. This is a consequence of the previous result and flat base-change (see [30]) for the square

$$
\begin{array}{ccc}
k_1 & \longrightarrow & k_1[u^{\pm 1}] \\
\downarrow & & \downarrow \\
k_1 \otimes_{E_0} E_0F & \longrightarrow & k_1 \otimes_{E_0} E_*F
\end{array}
$$

□

7.6. Corollary. *The moduli space of a realizations of $E(\Gamma, k)_*E(\Gamma, k)$ as a commutative $E(\Gamma, k)_*$ algebra in $E(\Gamma, k)_*E(\Gamma, k)$-comodules has the homotopy type of*

$$B\operatorname{Aut}(\Gamma, k)$$

where the automorphism group is computed in $\mathcal{FGL}_n$. In particular, $E(\Gamma, k)$ has a unique E_∞-structure realizing $E(\Gamma, k)_*E(\Gamma, k)$ as a commutative $E(\Gamma, k)_*$-algebra.

Proof. Let's write E_* and E_*E for $E(\Gamma, k)_*$, etc. We first show $T\mathcal{M}(E_*E) \simeq B\operatorname{Aut}(E_*E)$. Putting together the decomposition of the moduli space given Proposition 5.2, Proposition 5.5, and Theorem 5.8, we see that it is sufficient to calculate that

$$D^*_{E_*T/E_*E}(E_*E, \Omega^t E_*E) = 0$$

for all t. By Proposition 4.2, these groups are isomorphic to

$$D^*_{E_*T}(E_*E, \Omega^t E_*).$$

Now Proposition 6.8, and the spectral sequences of Propositions 6.4 and 6.5, and the previous result imply that this latter cohomology group is zero.

To finish the result we see that Proposition 7.3 – or more exactly its consequence Equation 7.3 – implies that

$$Aut(E_*E) \cong \operatorname{Aut}(\Gamma, k).$$

$\square$

7.7. Corollary. *Let $E(\Gamma_i, k_i)$ be two of the Lubin-Tate E_∞ ring spectra. Then the space of E_∞-maps between these spectra has contractible components; furthemore the set of path components is isomorpic to the set of morphisms*

$$(\Gamma_1, k_1) \to (\Gamma_2, k_2)$$

is $\mathcal{FGL}_n$.

Proof. This is the same line of argument, where the mapping space is decomposed via the spectral sequence of Theorem 4.3 or 4.5. $\square$

REFERENCES

[1] Adams, J.F., *Stable homotopy and generalised cohomology*, University of Chicago Press, Chicago, 1974.

[2] Atiyah, M. F., "Vector bundles and the Künneth formula", *Topology*, 1 (1962), 245–248.

[3] Baker, A., "A_∞ structures on some spectra related to Morava K-theories", *Quart. J. Math. Oxford Ser. (2)*, 42 (1991), No. 168, 403–419.

[4] Basterra, M., "André-Quillen cohomology of commutative S-algebras", *J. Pure and Applied Algebra*, to appear.

[5] Berger, C. and Moerdijk, I,, "Axiomatic homotopy theory for operads", preprint 2002, http://front.math.ucdavis.edu/math.AT/0206094.

[6] Blanc, D., Dwyer, W., and Goerss, P., "The realization space of a Π-algebra: a moduli problem in algebraic topology", preprint 2002, available at the Hopf archive.

[7] Bousfield, A. K., "Homotopy spectral sequences and obstructions", *Isr. J. Math.* 66 (1989) 54-104.

[8] Bousfield, A. K. and Kan, D. M., *Homotopy limits, completions, and localizations*, Lecture Notes in Math. 304 (2nd corrected printing), Springer-Verlag, Berlin-Heidelberg-New York, 1987.

[9] Bousfield, A. K., "Cosimplicial resolutions and homotopy spectral sequences in model categories", manuscript, University of Illinois at Chicago, 2001.

[10] Bousfield, A. K. and Friedlander, E. M., "Homotopy theory of Γ-spaces, spectra, and bisimplicial sets", *Geometric applications of homotopy theory* (Proc. Conf., Evanston, Ill., 1977), II, Lecture Notes in Math, 658, 80–130, Springer-Verlag, Berlin 1978.

[11] Cohen, F. R., Lada, T., and May, J. P., *The Homology of Iterated Loop Spaces*, Lecture Notes in Mathematics, Vol. 533, Springer-Verlag, Berlin, 1976.

[12] Dwyer, W. G. and Kan, D. M., "A classification theorem for diagrams of simplicial sets", *Topology*, 23 (1984) No.2, 139–155.

[13] Dwyer, W. G. and Kan, D. M., "Function complexes in homotopical algebra", *Topology*, 18 (1980), No. 4, 427–440.

[14] W. G. Dwyer, D. M. Kan, C. R. Stover, E_2 model category structure for pointed simplicial spaces," *J. of Pure and Applied Algebra* 90 (1993), 137–152.

[15] W. G. Dwyer, D. M. Kan, C. R. Stover, "The bigraded homotopy groups $\pi_{i,j}X$ of a pointed simplicial space X", *J. of Pure and Applied Algebra* 103 (1995), 167–188.

[16] A. D. Elmendorff, I. Kriz, M. A. Mandell, J. P. May, "Rings, modules, and algebras in stable homotopy theory", *Mathematical Surveys and Monographs* 47, AMS, Providence, RI, 1996.

[17] Goerss, P. G., "André-Quillen cohomology and the homotopy groups of mapping spaces: understanding the E_2 term of the Bousfield-Kan spectral sequence," *J. of Pure and Applied Algebra* 63 (1990), pp. 113-153.

[18] Goerss, P. G. and Hopkins, M. J., "André-Quillen (co)-homology for simplicial algebras over simplicial operads", *Une dégustation topologique: homotopy theory in the Swiss Alps (Arolla, 1999)*, Contemp. Math., 265, 41–85, Amer. Math. Soc., Providence,RI, 2000.

[19] Heller, A., "On the representability of homotopy functors", *J. London Math. Soc.* 23 (1981), 551-562.

[20] Hirschhorn, P., *Model categories and their localizations*, Mathematical Surveys and Monographs, 99, American Mathematical Society, Providence, RI, 2002.

[21] Hovey, M., *Model categories*, Mathematical Surveys and Monographs, 63, American Mathematical Society, Providence, RI, 1999.

[22] Hovey, M. and Shipley, B. and Smith, J., "Symmetric spectra", *J. Amer. Math. Soc.*, 13 (2000) No. 1, 149–208.

[23] Illusie, L,, *Complexe cotangent et déformations. I*, Lecture Notes in Mathematics, Vol. 239, Springer-Verlag, Berlin, 1971.

[24] Jardine, J. F., "Bousfield's E_2 model theory for simplicial objects", preprint University of Western Ontario, 2002.

[25] Lubin, J. and Tate, J., "Formal moduli for one-parameter formal Lie groups", *Bull. Soc. Math. France* 94 (1966), 49-60.

[26] Lewis, L. G., Jr., May, J. P., Steinberger, M., *Equivariant Stable Homotopy Theory*, Lecture Notes in Mathematics 1213, Springer-Verlag, Berlin, 1986.

[27] Mandell, M. A. and May, J. P. and Schwede, S. and Shipley, B., "Model categories of diagram spectra", *Proc. London Math. Soc. (3)*, 82 (2001), No. 2, 441–512.

[28] May, J.P., "A general approach to Steenrod operations", *The Steenrod Algebra and its Applications (Proc. Conf. to Celebrate N.E. Steenrod's Sixtieth Birthday, Battelle*

Memorial Inst. Columbus, Ohio, 1970, pp. 153-231, Lecture Notes in Mathematics, Vol. 168, Springer-Verlag, Berlin.

[29] Miller, H., "The Sullivan conjecture on maps from classifying spaces," *Ann. of Math.* 120 (1984), pp. 39-87.

[30] Quillen, D.G., On the (co)-homology of commutative rings, *Proc. Symp. Pure Math.* 17 (1970), 65–87.

[31] Quillen D.G., *Homotopical Algebra*, Lecture Notes in Math. 43, Springer-Verlag, Berlin-Heidelberg-New York, 1967.

[32] Reedy, C. L.,"Homotopy theory of model categories", Preprint, 1973. Available from http://math.mit.edu/~psh.

[33] Rezk, C. W., "Spaces of algebra structures and cohomology of operads", Thesis, MIT, 1996.

[34] Rezk, C. W., "Notes on the Hopkins-Miller theorem", in *Homotopy Theory via Algebraic Geometry and Group Representations*, M. Mahowald and S. Priddy, eds., Contemporary Math. 220 (1998) 313-366.

[35] Robinson, A., "Obstruction theory and the strict associativity of Morava K-theory," *Advances in homotopy theory*, London Math. Soc. Lecture Notes 139 (1989), 143-152.

[36] Robinson, A., "Gamma homology, Lie representations and E_∞ multiplications", *Invent. Math.*, 152 (2003) No. 2, 331–348.

[37] Robinson, A. and Whitehouse, S., "Operads and Γ-homology of commutative rings", *Math. Proc. Cambridge Philos. Soc.*, 132 (2002), No. 2, 197–234.

[38] Stover, C. R., "A Van Kampen spectral sequence for higher homotopy groups," *Topology* 29 (1990), 9–26.

[39] Strickland, N. P., "Gross-Hopkins duality", *Topology*, 39 (2000) No. 5 (1021–1033).

DEPARTMENT OF MATHEMATICS, NORTHWESTERN UNIVERSITY, EVANSTON IL 60208

pgoerss@math.nwu.edu

DEPARTMENT OF MATHEMATICS, MIT, CAMBRIDGE MA, 02139

mjh@math.mit.edu

COHOMOLOGY THEORIES FOR HIGHLY STRUCTURED RING SPECTRA

A. LAZAREV

ABSTRACT. This is a survey paper on cohomology theories for A_∞ and E_∞ ring spectra. Different constructions and main properties of topological André-Quillen cohomology and of topological derivations are described. We give sample calculations of these cohomology theories and outline applications to the existence of A_∞ and E_∞ structures on various spectra. We also explain the relationship between topological derivations, spaces of multiplicative maps and moduli spaces of multiplicative structures.

1. INTRODUCTION

In recent years algebraic topology witnessed renewed interest to highly structured ring spectra first introduced in [23]. To a large extent it was caused by the discovery of a strictly associative and symmetric smash product in the category of spectra in [8], [15]. This allowed one to replace the former highly technical notions of A_∞ and E_∞ ring spectra by equivalent but conceptually much simpler notions of S-algebras and commutative S-algebras respectively.

An *S-algebra* is just a monoid in the category of spectra with strictly symmetric and associative smash product (hereafter referred to as the category of *S-modules*). Likewise, a *commutative S-algebra* is a commutative monoid in the category of S-modules.

The most important formal property of categories of S-algebras and commutative S-algebras is that both are *topological model categories* in the sense of Quillen, [27] as elaborated in [8]. That means that together with the usual structure of the closed model category (fibrations, cofibrations and weak equivalences subject to a set of axioms) these categories are topologically enriched, that is their *Hom* sets are topological spaces and the composition of morphisms is continuous. Moreover, there exist *tensors* and *cotensors* of objects with topological spaces that satisfy the usual adjunction isomorphisms which hold for cartesian products and mapping spaces in the category of topological spaces. In addition, this enrichment is supposed to be compatible with the closed model structure in the sense that an appropriate analogue of Quillen's corner axiom SM7 is satisfied.

This rich structure allows one to translate a lot of the notions of conventional homotopy theory (homotopy relation, cellular approximation, the

The author was partially supported by the EPSRC grant No. GR/R84276/01 .

formation of homotopy limits and colimits) in the abstract setting. The language of closed model categories will be freely used here and we refer the reader to the monograph [14] where the necessary details may be found.

The categories of S-algebras and commutative S-algebras are in some sense analogous to the unstable category of topological spaces. In these categories there are certain natural homotopy invariant theories which play the role of singular cohomology for topological spaces. In particular they are natural homes for various obstruction groups. The corresponding theory is called *topological André-Quillen cohomology* in the commutative case and *topological derivations* in the associative case. As the name suggests topological André-Quillen cohomology is an extension of the cohomology theory for commutative algebras introduced by Quillen and André in [28, 1]. As far as we know the analogue of André-Quillen cohomology for associative algebras has not been considered in the literature. The reason for this is perhaps that the resulting theory coincides with the Hochschild theory up to a shift. However in the topological context topological derivations cannot be reduced completely to topological Hochschild cohomology; rather they are related to it by a certain fibre sequence.

We will use the abbreviation TAQ for the topological André-Quillen cohomology and Der for topological derivations. The purpose of the present paper is to give an overview of recent results on these cohomology theories. The technical level will be kept to a minimum, however we will try to outline proofs of the results we formulate, especially when these proofs are easy and conceptual. Occasionally the theorems we obtain are 'new', that is, have not appeared in print before; but in each case they are either direct extensions of the known results or could be obtained following similar patterns. Still, we tried to put these results in different perspectives from those in the published sources; in particular the associative and commutative cases are treated as uniformly as possible.

The paper does not pretend to be comprehensive by any means. One notable omission is the work of Robinson and Whitehouse on Γ-homology of commutative rings [32] and the subsequent work of Robinson [34] where Γ-homology was used to show the existence of a commutative S-algebra structure on E_n, the Morava E-theory spectrum, also called the Lubin-Tate spectrum. One should also mention the important papers [30] and [26] containing further results on Γ-homology. Our limited expertise as well as the lack of space prevented us from including these results in the survey. Another theme only briefly mentioned here is the existence of the action of the Morava stabilizer group on the spectrum E_n (the Hopkins-Miller theorem). For this and related topics we refer the reader to [29], [18] and [11].

The paper is organized as follows. In sections 2 and 3 we outline the construction and basic properties of TAQ and Der. In section 4 we investigate the problem of lifting a map of S-algebras or commutative S-algebras and introduce the notion of a *topological singular extension*. In section 5 we discuss an alternative construction of TAQ and Der as *stabilizations* of appropriate forgetful functors. Section 6 presents calculations of TAQ and Der of the Eilenberg-MacLane spectrum $H\mathbb{F}_p$. In section 7 we show how the developed technology could be used to produce MU-algebra structures on many complex oriented spectra. Here MU is the spectrum of complex cobordisms. In section 8 we discuss the relationship between spaces of algebra maps and Der. In section 9 we make a link between Der and spaces of multiplicative structures on a given spectrum.

Notation and conventions. The paper is written in the language of S-modules of [8] and we adopt the terminology of the cited reference. There is one exception: we use the terms 'cofibrant' and 'cofibration' for what [8] calls 'q-cofibrant' and 'q-cofibration'. This is because we never have the chance to use cofibrations in the classical sense (maps that satisfy the homotopy extension property). Except for Section 6 we work over an arbitrary (but fixed) cofibrant commutative S-algebra R, smash products $\wedge$ and function objects $F(-,-)$ mean $\wedge_R$ and $F_R(-,-)$; when we use a different 'ground' S-algebra A this is indicated by a subscript such as $\wedge_A$ and $F_A(-,-)$. The topological space of maps in a topological category $\mathcal{C}$ is denoted by $\mathrm{Map}_{\mathcal{C}}$. The category of unbased topological spaces is denoted by $\mathcal{T}op$ and that of based spaces - by $\mathcal{T}op_*$. For an S-algebra A the category of (left) A-modules is denoted by $\mathcal{M}_A$. The category of A-bimodules (which is the same as the category of $A \wedge A^{op}$-modules where A^{op} is the R-algebra A with opposite multiplication) will be denoted by $\mathcal{M}_{A-A}$. The category of R-algebras is denoted by $\mathcal{C}_R$ and that of *commutative* R-algebras - by $\mathcal{C}_R^{comm}$ The homotopy category of a closed model category $\mathcal{C}$ is denoted by $h\mathcal{C}$ so, for example $h\mathcal{C}_R^{comm}$ is the homotopy category of commutative R-algebras. For an R-algebra A we will denote by $[-,-]_{A-mod}$ and $[-,-]_{A-bimod}$ the sets of homotopy classes of A-module or A-bimodule maps respectively. The field consisting of p elements is denoted by $\mathbb{F}_p$.

Acknowledgement. I would like to express my sincere gratitude to Andy Baker and Birgit Richter for doing such a wonderful job of organizing the Workshop on Structured Ring Spectra in Glasgow in January 2002. I am also thankful to Maria Basterra for explaining to me her joint work with Mike Mandell on TAQ and to Stefan Schwede for pointing out that Theorem 9.2 is a direct consequence of Theorem 8.1. Thanks are also due to Mike Mandell, Paul Goerss, Nick Kuhn and Bill Dwyer for useful discussions and comments made at various times.

2. Topological André-Quillen cohomology for commutative *S*-algebras

The exposition in this section is based on Basterra's paper [2]. Let A be a commutative R-algebra which without loss of generality will assumed to be cofibrant. Denote by $\mathcal{C}_{R/A}^{comm}$ the category of commutative R-algebras over A. An object of $\mathcal{C}_{R/A}^{comm}$ is a commutative R-algebra B supplied with an R-algebra map $B \longrightarrow A$ (an augmentation). A morphism between two such objects B and C is the following commutative diagram in $\mathcal{C}_R^{comm}$.

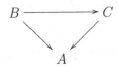

Then $\mathcal{C}_{R/A}^{comm}$ inherits a topological model category structure from $\mathcal{C}_R^{comm}$ so that a map in $\mathcal{C}_{R/A}^{comm}$ is a cofibration if it is so considered as a map in $\mathcal{C}_R^{comm}$. Note that in the case $R = A$ the category $\mathcal{C}_{A/A}^{comm}$ is pointed and therefore is enriched over the category of pointed topological spaces.

Let us denote by $\mathcal{N}_A$ the category of *nonunital* commutative A-algebras. An object in this category is an A-module M together with a strictly associative multiplication map $M \wedge_A M \longrightarrow M$. The morphisms in $\mathcal{N}_A$ are defined in the obvious fashion. Following [2] we will refer to an object of $\mathcal{N}_A$ as an A-NUCA. It is not hard to prove that $\mathcal{N}_A$ has a topological model structure where weak equivalences are those maps which are weak equivalences on underlying A-modules. The fibrations then are the maps that are fibrations of underlying A-modules and the cofibrations are the maps which have the left lifting property, LLP, with respect to the acyclic fibrations. Then we could form the homotopy category $h\mathcal{N}_A$ of $\mathcal{N}_A$.

We want to show that the categories $h\mathcal{C}_{A/A}^{comm}$ and $h\mathcal{N}_A$ are equivalent. In fact, more is true, see Proposition 2.1.

Let $K : \mathcal{N}_A \longrightarrow \mathcal{C}_A^{comm}$ denote the functor which assigns to an A-NUCA M the commutative A-algebra $A \vee M$ with multiplication

$$(A \vee M) \wedge_A (A \vee M) \cong A \vee M \vee M \vee M \wedge_A M \longrightarrow A \vee M$$

given by the obvious maps on the first three wedge summands and by the multiplication of M on the last one.

Clearly the A-algebra $A \vee M$ can be considered as an object in $\mathcal{C}_{A/A}^{comm}$ via the canonical projection map $A \vee M \longrightarrow A$. In other words the functor K lands in fact in the category $\mathcal{C}_{A/A}^{comm}$.

Now let B be an A-algebra over A and denote by I its 'augmentation ideal', i.e. the fibre of the augmentation map $B \longrightarrow A$. Then $I(B)$ is naturally an A-NUCA and we can consider I as a functor from $\mathcal{C}_{A/A}^{comm}$ to $\mathcal{N}_A$.

Then we have the following

Proposition 2.1. *The functor K is left adjoint to I. Moreover the functors K and I establish a Quillen equivalence between the categories $\mathcal{N}_A$ and $\mathcal{C}_{A/A}^{comm}$.*

To see that the functors K and I are adjoint notice that the category $\mathcal{N}_A$ is in fact a category of algebras over a certain monad $\mathbb{A}$ in A-modules. This monad is specified by $\mathbb{A}M := \bigvee_{i>0} M^{\wedge_A i}/\Sigma_i$ where Σ_i is the symmetric group on i symbols. Further note that $K(\mathbb{A}M)$ is a free commutative A-algebra on M. Therefore for an A-module M we have

$$\mathcal{C}_{A/A}^{comm}(K(\mathbb{A}M), B) \cong \mathcal{M}_A(M, I(B)) \cong \mathcal{N}_A(\mathbb{A}M, I(B)).$$

To get the adjointness isomorphism for a general A-NUCA X it suffices to notice that there is a canonical split coequalizer (the beginning of the monadic bar-construction for X):

$$\mathbb{A}\mathbb{A}X \rightrightarrows \mathbb{A}X \longrightarrow X.$$

Explicitly, if $g : X \longrightarrow I(B)$ is a morphism in $\mathcal{N}_A$ then its adjoint $\tilde{g} : A \vee X \longrightarrow B$ is the composite

$$\tilde{g} : A \vee X \xrightarrow{id \vee g} A \vee I(B) \xrightarrow{1 \vee i} B,$$

where i is the canonical map $I(B) \longrightarrow B$.

Further straightforward arguments show that the functor K preserves cofibrations and acyclic cofibrations and that the adjunction described above determines an equivalence on the level of homotopy categories.

We will denote the homotopy invariant extension (also called a *total derived functor*) of the functor I by $\mathbf{R}I$. For an object B in $\mathcal{C}_{A/A}^{comm}$ denote by $\tilde{B}$ its fibrant replacement. Then $\mathbf{R}I(B) = I(\tilde{B})$. Next we define the functor of 'taking the indecomposables' that assigns to an A-NUCA N the A-module $Q(N)$ given by the cofibre sequence

$$N \wedge_A N \xrightarrow{m} N \longrightarrow Q(N).$$

Here m stands for the multiplication map. The functor Q has a right adjoint functor $Z : \mathcal{M}_A \longrightarrow \mathcal{N}_A$ which is given by considering A-modules as objects in $\mathcal{N}_A$ with zero multiplication. It is easy to see that this adjunction passes to the homotopy categories. Let $\mathbf{L}Q$ denote the total derived functor of Q. Explicitly, $\mathbf{L}Q(N) = Q(\tilde{N})$ where $\tilde{N}$ is a cofibrant replacement of the A-NUCA N.

It is clear that $Q(\mathbb{A}N) \cong N$. However for a general cofibrant A-NUCA the functor $Q(N)$ is very hard to compute. One can approach its computation as follows. The functor $\mathbb{A} : \mathcal{M}_A \longrightarrow \mathcal{N}_A$ from A-modules to A-NUCA's (equivalently, to $\mathbb{A}$-algebras) is left adjoint to the forgetful functor $U : \mathcal{N}_A \longrightarrow \mathcal{M}_A$. There results a monad in $\mathcal{M}_A$. Given an A-NUCA N denote by $B_*(N) = B_*(\mathbb{A}, U\mathbb{A}, UN)$ its monadic bar-construction. Explicitly B_*N is the cosimplicial A-NUCA with $B_n(N) = \mathbb{A}^{n+1}N$ and the faces and codegeneracies are the standard ones, cf. [24]. The geometric realization $|B_*(N)|$ of $B_*(N)$ is weakly equivalent to N. Then one has the following

Proposition 2.2. *For a cofibrant A-NUCA N one has the following weak equivalence of A-modules: $Q(|B_*(N)|) \simeq Q(N)$.*

Despite the innocent appearance of this proposition it is not at all obvious. The problem is that the functor Q only preserves weak equivalences between cofibrant objects and $|B_*(N)|$ is not a cofibrant A-NUCA. However in the end it turns out that it is close enough to being cofibrant to give the result; much of the work [2] is devoted to overcoming this point.

Further associated to the simplicial A-NUCA $B_*(N)$ is the spectral sequence which computes $\pi_*|B_*(N)| = \pi_* N$:

$$(2.1) \qquad E^1_{i,j} = \pi_i Q(\mathbb{A}^{j+1} N) = \pi_i \mathbb{A}^j N \implies \pi_{i-j} Q(N).$$

We now have all the ingredients to define the *abelianization* functor:

$$\Omega_A^{comm} : h\mathcal{C}_{R/A}^{comm} \longrightarrow h\mathcal{M}_A.$$

Definition 2.3. Let C be a commutative R-algebra over A and M be a A-module. Then $\Omega_A^{comm}(C) := \mathbf{L}Q \circ \mathbf{R}I(C \wedge_R^{\mathbf{L}} A)$.

Note that $C \wedge_R^{\mathbf{L}} A$ is an object of $\mathcal{C}_{A/A}^{comm}$ in an obvious way. Of course we only defined the abelianization functor on objects, but its extension to morphisms is immediate. The main property of the abelianization functor is that it is left adjoint to the 'square-zero' extension functor: if M is a A-module then the square zero extension of A by M is the A-algebra over A whose underlying R-module is $A \vee M$ with the obvious multiplication. More precisely we have the following

Theorem 2.4. *There is the following natural isomorphism:*

$$h\mathcal{C}_{R/A}^{comm}(C, A \vee M) \cong h\mathcal{M}_A(\Omega_A^{comm}(C), M)$$

where C is a commutative R-algebra over A and M is a A-module.

The proof of the theorem is rather formal and relies on the already established properties of the functors I, K, Z and Q.

Remark 2.5. We would like to emphasize here that the isomorphism of Theorem 2.4 holds on the level of *homotopy* categories and is not a reflection of an adjunction between strict categories. Indeed, even though the abelianization functor could be considered as a point-set level functor its definition involves composition of right adjoint and left adjoint functors. Therefore one cannot expect any good formal properties of Ω_A^{comm} on the point-set level. The situation improves upon passing to homotopy categories because one of these functors, namely $\mathbf{R}I$, becomes an equivalence.

We will sometimes need the enriched version of Theorem 2.4. The category $\mathcal{C}_{R/A}^{comm}$ is enriched over unbased topological spaces. For two objects B, A of a topological category $\mathcal{C}$ the space of maps $\text{Map}_{\mathcal{C}/A}(B, C)$ from B to C *over* A is defined from the pullback diagram

$$
\begin{array}{ccc}
\text{Map}_{\mathcal{C}/A}(B, C) & \longrightarrow & pt \\
\downarrow & & \downarrow \\
\text{Map}_{\mathcal{C}}(B, C) & \longrightarrow & \text{Map}_{\mathcal{C}}(B, A)
\end{array}
$$

where the right downward arrow is just picking the structure map $B \longrightarrow A$ in $\text{Map}_{\mathcal{C}}(B, A)$. It is easy to see that the cotensor $(A \vee M)^X$ of X and $A \vee M$ in $\mathcal{C}_{R/A}^{comm}$ is weakly equivalent to $A \vee M^X$. Then for any CW-complex X we have the following isomorphisms:

$$
h\mathcal{T}op(X, \text{Map}_{\mathcal{M}_A}(\Omega_A^{comm}(C), M)) \cong h\mathcal{M}_A(\Omega_A^{comm}(C), M^X)
$$
$$
\cong h\mathcal{C}_{R/A}^{comm}(C, A \vee M^X)
$$
$$
\cong h\mathcal{T}op(X, \text{Map}_{\mathcal{C}_{R/A}}(C, A \vee M)).
$$

Therefore the topological spaces $\text{Map}_{\mathcal{C}_{R/A}}(C, A \vee M)$ and $\text{Map}_{\mathcal{M}_A}(\Omega_A^{comm}(C), M)$ are weakly equivalent.

We will denote the A-module $\Omega_A^{comm}(A) = \mathbf{L}Q \circ \mathbf{R}I(A \wedge_R^{\mathbf{L}} A)$ simply by Ω_A^{comm} so that

$$
h\mathcal{C}_{R/A}^{comm}(A, A \vee M) \cong h\mathcal{M}_A(\Omega_A^{comm}, M).
$$

The set $h\mathcal{C}_{R/A}^{comm}(A, A \vee M)$ could be interpreted as derivations of the commutative S-algebra A with coefficients in the A-module M. In particular we see, that the set of such derivations is an abelian group (note that a priori the set of homotopy classes of (commutative) S-algebra maps does not carry a structure of a group, let alone an abelian group).

We can now define the *topological André-Quillen cohomology* spectrum of A relative to R with coefficients in M by

$$
\mathbf{TAQ}_R(A, M) := F_A(\Omega_A^{comm}, M)
$$

and similarly the *topological André-Quillen homology* spectrum of A relative to R with coefficients in M:

$$
\mathbf{TAQ}^R(A, M) := \Omega_A^{comm} \wedge_A M.
$$

We will refer to the homotopy groups of these spectra as topological André-Quillen (co)homology of A relative to R:

$$
TAQ_R^*(A, M) := \pi_{-*}F_A(\Omega_A^{comm}, M);
$$

$$
TAQ_*^R(A, M) := \pi_*\Omega_A^{comm} \wedge_A M.
$$

Remark 2.6. The enriched version of Theorem 2.4 gives us the weak equivalence of spaces $\text{Map}_{\mathcal{C}_{R/A}^{comm}}(A, A \vee M) \simeq \text{Map}_{\mathcal{M}_A}(\Omega_A^{comm}, M)$. The latter mapping space is in turn weakly equivalent to the zeroth space of the spectrum

$F_A(\Omega_A^{comm}, M) = \mathbf{TAQ}_R(A, M)$. We obtain the weak equivalence

$$\mathrm{Map}_{\mathcal{C}_{R/A}^{comm}}(A, A \vee M) \simeq \Omega^{\infty} \mathbf{TAQ}_R(A, M).$$

We are most interested in topological André-Quillen *cohomology* groups since they are related to the obstruction theory. Repeat, that for a commutative R-algebra A and an A-module M the abelian group $TAQ_R^0(A, M)$ is identified with the set of *commutative derivations* of A with values in M that is the homotopy classes of maps in $\mathcal{C}_{R/A}^{comm}$ from A into $A \vee M$. We say 'commutative derivations' to distinguish them from just 'derivations' or 'topological derivations' which will be considered in the next section and refer to the *bimodule* derivations in the context of associative R-algebras.

Let us now discuss the functoriality of TAQ. Since by definition $\mathbf{TAQ}_R(A, M) = F_A(\Omega_A^{comm}, M)$ we see that it is *covariant* with respect to the variable M. On the level of commutative derivations it can be seen as follows. A map of A-modules $f : M \longrightarrow N$ determines a map in $\mathcal{C}_{R/A}^{comm} : A \vee M \longrightarrow A \vee N$. Then a topological derivation $d : A \longrightarrow A \vee M$ determines a topological derivation $f_* d : A \longrightarrow A \vee M \longrightarrow A \vee N$.

On the other hand, TAQ is *contravariant* in the algebra argument. If $g : X \longrightarrow A$ is a map of R-algebras then the composite map

$$X \xrightarrow{\ g\ } A \xrightarrow{\ d\ } A \vee M$$

is a map in $\mathcal{C}_{R/A}^{comm}$. Therefore by Theorem 2.4 it corresponds to a map of in $h\mathcal{M}_A : \Omega_A^{comm}(X) \longrightarrow M$ which is the same as a map in $h\mathcal{M}_X : \Omega_X^{comm} \longrightarrow M$. The latter corresponds, again by Theorem 2.4 to a commutative derivation of X with values in the X-module M. We will denote this derivation by $g^* d : X \longrightarrow X \vee M$.

We can relate the topological André-Quillen cohomology to ordinary R-module cohomology as follows. For a cofibrant R-algebra A consider the composite map of R-modules

$$l_R : A \longrightarrow A \vee \Omega_A^{comm} \longrightarrow \Omega_A^{comm}.$$

Here the first map is the R-algebra map adjoint to the identity on Ω_A^{comm} and the second map is the projection onto the wedge summand. The map l_R induces a forgetful map from TAQ to R-module cohomology:

$$l_R^* : TAQ_R^*(A, M) = [\Omega_A^{comm}, M]_{A-mod}^* \longrightarrow [\Omega_A^{comm}, M]_{R-mod}^* \longrightarrow [A, M]_{R-mod}^*.$$

To describe the image of l_R^* let us introduce the notion of a *primitive operation* $E \longrightarrow M$ for a R-ring spectrum E and an E-bimodule spectrum M. Namely, the map $p \in [E, M]_{R-mod}^*$ is primitive if the following diagram is commutative in $h\mathcal{M}_R$.

$$
\begin{array}{ccc}
E \wedge E & \xrightarrow{\ m\ } & E \\
{\scriptstyle 1 \wedge p \vee p \wedge 1} \downarrow & & \downarrow {\scriptstyle p} \\
E \wedge M \vee M \wedge E & \xrightarrow{m_l \vee m_r} & M
\end{array}
$$

where m_l and m_r denote the left and right actions of E on M. (In other words, p is a derivation up to homotopy but we refrain from using this term to avoid confusion.) Then it is easy to see that the image of l_R^* is contained in the subspace of primitive operations in $[A, M]_{R-mod}^*$. Of course the left and right A-module spectrum structures on M coincide in the commutative case.

3. TOPOLOGICAL DERIVATIONS

In this section we construct the analogue of TAQ for not necessarily commutative S-algebras following [17]. Traditionally the André-Quillen cohomology was considered for commutative algebras only. Therefore we will reserve this term for the commutative case and call its analogue for associative S-algebras *topological derivations*. The main result we are going to describe here is the analogue of Theorem 2.4 in the context of noncommutative S-algebras. Note, however, that the construction and the proof differ considerably from the commutative case.

Let A be a cofibrant R-algebra. We define the module of differentials Ω_A from the following homotopy fibre sequence

$$(3.1) \qquad \Omega_A \longrightarrow A \wedge A \longrightarrow A .$$

Here the second arrow is the multiplication map. Clearly Ω_A is an A-bimodule. Note that the sequence (3.1) splits in the homotopy category of left A-modules by the map

$$A \xrightarrow{\cong} A \wedge R \xrightarrow{id \wedge 1} A \wedge A$$

and therefore the A-bimodule Ω_A is equivalent as a left A-module to $A \wedge A/R$. Similarly Ω_A is equivalent as a *right* A-module to $A/R \wedge A$.

Denote the category of objects over A inside $\mathcal{C}_R$ by $\mathcal{C}_{R/A}$. The category $\mathcal{C}_{R/A}$ has a topological model category structure inherited from $\mathcal{C}_R$.

Let M be an A-bimodule. Then we could form the R-algebra with the underlying R-module $A \vee M$, the 'square-zero extension' of A by M. There are obvious product and augmentation maps making $A \vee M$ into an R-algebra over A.

We now introduce the analogue of the abelianization functor in the noncommutative context. Let B be a cofibrant R-algebra over A and denote by $\Omega_A(B)$ the A-bimodule

$$A \wedge_B \Omega_B \wedge_B A \cong A \wedge A^{op} \wedge_{B \wedge B^{op}} \Omega_B.$$

Theorem 3.1. *There is a natural equivalence*

$$h\mathcal{C}_{R/A}(B, A \vee M) \cong h\mathcal{M}_{A-A}(\Omega_A(B), M).$$

Let us show how to associate to any map $B \longrightarrow A \vee M$ in $h\mathcal{C}_{R/A}$ a map of A-bimodules $\Omega_A(B) \longrightarrow M$.

Denote by $\overline{A \vee M}$ the fibrant cofibrant approximation of $A \vee M$ in the category of R-algebras over A. It suffices to construct a 'universal' map of A-bimodules

$$(3.2) \qquad \Omega_A(\overline{A \vee M}) \longrightarrow M.$$

Indeed for a map $B \longrightarrow \overline{A \vee M}$ in $\mathcal{C}_{R/A}$ we have a composite map

$$\Omega_A(B) \longrightarrow \Omega_A(\overline{A \vee M}) \longrightarrow M$$

and therefore a correspondence

$$(3.3) \qquad h\mathcal{C}_{R/A}(B, \overline{A \vee M}) = h\mathcal{C}_{R/A}(B, A \vee M) \longrightarrow h\mathcal{M}_{A-A}(\Omega_A(B), M)$$

as desired.

Further, since

$$\Omega_A(\overline{A \vee M}) = A \wedge_{\overline{A \vee M}} \Omega_{\overline{A \vee M}} \wedge_{\overline{A \vee M}} A,$$

the map of (3.2) is the same as a map of $\overline{A \vee M}$-bimodules $\Omega_{\overline{A \vee M}} \longrightarrow M$. Instead of the latter map we construct a map $\Omega_{A \vee M} \longrightarrow M$. This would be good enough since even $A \vee M$ is not a cofibrant object in $\mathcal{C}_{R/A}$ the smash product $(A \vee M) \wedge (A \vee M)$ clearly represents the derived smash product and therefore $\Omega_{\overline{A \vee M}}$ is weakly equivalent to $\Omega_{A \vee M}$. Consider the following diagram in the homotopy category of $A \vee M$-bimodules.

$$
\begin{array}{ccc}
A \wedge A \vee A \wedge M \vee M \wedge A \vee M \wedge M & \longrightarrow & A \vee M \\
\downarrow & & \| \\
A \vee M \vee M & \longrightarrow & A \vee M
\end{array}
$$

Here the left vertical arrow is determined by the R-algebra structure on A, an A-bimodule structure on M and is zero on the last summand. The lower horizontal arrow is zero on A, identity on the first M-summand and minus identity on the last M-summand. Then the homotopy fibre of the upper row is equivalent to $\Omega_{A \vee M}$ by definition and the homotopy fibre of the lower row is equivalent to M. There results a map of $A \vee M$-bimodules $\Omega_{A \vee M} \longrightarrow M$ as desired.

The proof of the theorem is then completed by showing that the above correspondence (3.3) is in fact an equivalence if B is equal to the free algebra

$$B = TV = R \vee V \vee V^{\wedge 2} \vee \ldots$$

where V is a cofibrant R-module over A and then resolving a general B by a monadic bar-construction. Note that the homotopy fibre of the multiplication map $TV \wedge TV \longrightarrow TV$ is equivalent to the TV-bimodule $TV \wedge V \wedge TV$ and therefore

$$\Omega_{TV} \simeq TV \wedge V \wedge TV.$$

We have the 'universal derivation' map $A \longrightarrow A \vee \Omega_A$ which is adjoint to the identity map $\Omega_A(A) = \Omega_A \longrightarrow \Omega_A$. Recall that as an R-module (even as a left A-module) Ω_A is weakly equivalent to $A \wedge A/R$. Taking the composition of the universal derivation with the canonical projection $A \vee \Omega_A \longrightarrow \Omega_A$ we get a map $\phi : A \longrightarrow \Omega_A \simeq A \wedge A/R$. Then as a map of R-modules ϕ is homotopic to the map

$$A \cong R \wedge A \xrightarrow{1 \wedge d} A \wedge A/R ,$$

where d is is defined as the second arrow in the homotopy cofibre sequence

$$R \xrightarrow{1} A \xrightarrow{d} A/R.$$

This could be seen by first taking A to be the free algebra TV and then resolving a general R-algebra A by a simplicial construction consisting of free R-algebras.

Let us denote by $Der_R^0(A, M)$ the set of homotopy classes of maps $A \longrightarrow A \vee M$ in $h\mathcal{C}_{A/R}$. We will call elements of $Der_R^0(A, M)$ *topological derivations* of A with values in the A-bimodule M. Then Theorem 3.1 shows that $Der_R^0(A, M)$ is in fact an abelian group. Moreover, $Der_R^0(A, M)$ is the zeroth homotopy group of the function spectrum $F_{A \wedge A^{op}}(\Omega_A, M)$. We will call it the spectrum of topological derivations of A with values in M and denote it by $\mathbf{Der}_R(A, M)$. The homotopy groups of $\mathbf{Der}_R(A, M)$ will be denoted by $Der_R^i(A, M)$, so $Der_R^i(A, M) = \pi_{-i}F_{A \wedge A^{op}}(\Omega_A, M)$.

Remark 3.2. Arguing in the same way as for commutative derivations we see that there is a weak equivalence of topological spaces

$$\mathrm{Map}_{\mathcal{C}_{R/A}}(A, A \vee M) \simeq \Omega^\infty \mathbf{Der}_R(A, M).$$

In particular the space $\mathrm{Map}_{\mathcal{C}_{R/A}}(A, A \vee M)$ is an infinite loop space.

Let $\tilde{A}$ be a cofibrant replacement of A as an A-bimodule. Replacing A with $\tilde{A}$ in the homotopy fibre sequence (3.1) and applying the functor $F_{A \wedge A^{op}}(?, M)$ to it we obtain the homotopy fibre sequence:

$$F_{A \wedge A^{op}}(\tilde{A}, M) \longrightarrow M \longrightarrow \mathbf{Der}_R(A, M).$$

The function spectrum $F_{A \wedge A^{op}}(\tilde{A}, M)$ is called *topological Hochschild cohomology* spectrum of A with values in M and has a special notation $\mathbf{THH}_R(A, M)$. So we obtain the following homotopy fibre sequence relating topological Hochschild cohomology and topological derivations:

$$(3.4) \qquad \mathbf{THH}_R(A, M) \longrightarrow M \longrightarrow \mathbf{Der}_R(A, M).$$

This fibre sequence to a large extent reduces the calculation of $Der_R^*(A, M)$ to that of topological Hochschild cohomology groups

$$THH_R^*(A, M) := \pi_{-*}\mathbf{THH}_R(A, M).$$

The latter groups are relatively computable due to the hypercohomology spectral sequence of [8]:

$$\mathrm{Ext}_{(A \wedge A^{op})_*}^{**}(A_*, M_*) \Longrightarrow THH_R^*(A, M).$$

Similarly to the commutative case there is a forgetful map

$$l_R^* : Der_R^*(A, M) \longrightarrow [A, M]_{R-mod}^*$$

whose image is contained in the subspace of primitive operations in $[A, M]_{R-mod}^*$.

Remark 3.3. There is also the notion of topological Hochschild *homology* spectrum

$$\mathbf{THH}^R(A, M) := \tilde{A} \wedge_{A \wedge A^{op}} M.$$

Topological Hochschild homology plays an important role in computations of algebraic K-theory of rings, cf. [21]. If the R-algebra A is commutative then both $\mathbf{THH}_R$ and $\mathbf{THH}^R$ are R-modules and there is a duality between them in $h\mathcal{M}_A$:

$$\mathbf{THH}_R(A, M) \simeq F_A(\mathbf{THH}^R(A, M), A).$$

Remark 3.4. Just as topological André-Quillen cohomology the spectrum $\mathbf{Der}_R(A, M)$ is covariant with respect to M and contravariant with respect to A. The arguments are precisely the same as in the commutative context except that we refer to Theorem 3.1 instead of Theorem 2.4 and A-modules are replaced with A-bimodules. For a map of A-bimodules $f : M \longrightarrow N$ and a topological derivation $d : A \longrightarrow A \vee M$ we have a topological derivation f_*d of A with values in N. For a map of R-algebras $g : X \longrightarrow A$ we obtain a topological derivation g^*d of X with values in M.

Sometimes in the noncommutative context it is useful to consider the so-called *relative topological derivations* which we will now define. Let R' be a not necessarily commutative R-algebra. Denote by $\mathcal{C}_{R'}^{ass}$ the category of R-algebras A supplied with a fixed R-algebra map $R' \longrightarrow A$, not necessarily central. In other words $\mathcal{C}_{R'}^{ass}$ is the undercategory of R' in $\mathcal{C}_R$. The objects of $\mathcal{C}_{R'}^{ass}$ will be called R'-*algebras* by a slight abuse of language. Let $m : A \wedge A \longrightarrow A$ be the multiplication map. Note that there exists a unique map $m' : A \wedge_{R'} A \longrightarrow A$ such that the diagram

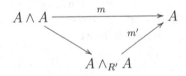

is commutative.

Now suppose without loss of generality that the R-algebra R' is cofibrant and that the structure map $R' \longrightarrow A$ is a cofibration of R-algebras. We define the R'-relative module of differentials $\Omega_{A|R'}^{ass}$ from the following homotopy fibre sequence:

$$\Omega_{A|R'}^{ass} \longrightarrow A \wedge_{R'} A \xrightarrow{m'} A .$$

Let $\mathcal{C}^{ass}_{R'/A}$ denote the category whose objects are R'-algebras B supplied with an R'-algebra map $B \longrightarrow A$. Note that if M is an A-bimodule then $A \vee M$ is an object of $\mathcal{C}^{ass}_{R'/A}$. This gives a functor $\mathcal{M}_{A-A} \longrightarrow \mathcal{C}^{ass}_{R'/A}$. It turns out that on the level of homotopy categories this functor admits a left adjoint

$$B \longmapsto \Omega^{ass}_{R'|A}(B) := A \wedge_B \Omega^{ass}_{R'|B} \wedge_B A.$$

More precisely there is a natural equivalence

$$h\mathcal{C}^{ass}_{R'/A}(B, A \vee M) \cong h\mathcal{M}_{A-A}(\Omega^{ass}_{A|R'}(B), M).$$

The proof is similar to the one in the absolute case. One makes use of the 'free' R'-algebra functor $T_{R'}$ given for an R'-bimodule V as

$$T_{R'}V = R' \vee V \vee (V \wedge_{R'} V) \vee \ldots$$

instead of the usual free R-algebra on an R-module V.

The homotopy classes of maps in $h\mathcal{C}^{ass}_{R'/A}(A, A \vee M)$ are called topological derivations of A with values in M *relative to R'*. An algebraic analogue of relative topological derivations are the derivations of an algebra A with coefficients in an A-bimodule M vanishing on a subalgebra R' of A. Relative modules of differentials were studied in the context of noncommutative differential forms in [6]. Replacing $\Omega^{ass}_{A|R'}$ with $\tilde{\Omega}^{ass}_{A|R'}$, its cofibrant approximation as an A-bimodule we define the R-module of R'-relative derivations of A with values in an A-bimodule M to be

$$\mathbf{Der}_{R/R'}(A, M) := F_{A \wedge A^{op}}(\tilde{\Omega}^{ass}_{A|R'}, M).$$

It is easy to see that there is the following homotopy fibre sequence of R-modules:

$$\mathbf{THH}_R(A, M) \longrightarrow \mathbf{THH}_R(R', M) \longrightarrow \mathbf{Der}_{R/R'}(A, M)$$

which specializes to (3.4) when $R = R'$.

4. Obstruction theory

In this section we show that topological Andre-Quillen cohomology in the commutative case and topological derivations in the associative case are a natural home for lifting algebra maps. For definiteness we work in the context of associative algebras, however the results have obvious analogues, with the same proofs, in the commutative case. Topological derivations are replaced with commutative derivations, spectra $\mathbf{Der}_R(-, -)$ with spectra $\mathbf{TAQ}_R(-, -)$ and the spaces $\mathrm{Map}_{\mathcal{C}_R}(-, -)$ with spaces $\mathrm{Map}_{\mathcal{C}^{comm}_R}(-, -)$.

Let A be a cofibrant R-algebra, I be an A-bimodule and $d : A \longrightarrow A \vee \Sigma I$ be a derivation of A with coefficients in ΣI, the suspension of I. Let $\epsilon : A \longrightarrow A \vee \Sigma A$ be the inclusion of a retract. Define the R-algebra B from the following homotopy pullback square of R-algebras:

$$
\begin{array}{ccc}
B & \longrightarrow & A \\
\downarrow & & \downarrow{\scriptstyle \epsilon} \\
A & \xrightarrow{\;d\;} & A \vee \Sigma I
\end{array}
$$

It is clear that the homotopy fibre of the map to A is weakly equivalent to the A-bimodule I.

Definition 4.1. The homotopy fibre sequence

$$
I \longrightarrow B \longrightarrow A
$$

is called the topological singular extension of A by I associated with the derivation $d : A \longrightarrow A \vee \Sigma I$.

Theorem 4.2. *Let $I \to B \to A$ be a singular extension of R-algebras associated with a derivation $d : A \to A \vee \Sigma I$ and $f : X \to A$ a map of R-algebras where the R-algebra X is cofibrant. Then we have the following.*

(1) *The map f lifts to an R-algebra map $X \to B$ if and only if the induced derivation $f^*d \in \mathrm{Der}^1_R(X, I) = \mathrm{Der}^0_R(X, \Sigma I)$ is homotopic to zero.*

(2) *Assuming that a lifting exists in the fibration*

(4.1) $$ \mathrm{Map}_{\mathcal{C}_R}(X, B) \to \mathrm{Map}_{\mathcal{C}_R}(X, A) $$

the homotopy fibre over the point $f \in \mathrm{Map}_{\mathcal{C}_R}(X, A)$ is weakly equivalent to $\Omega^\infty \mathbf{Der}_R(X, I)$ (the 0th space of the spectrum $\mathbf{Der}_R(X, I)$).

The proof is rather formal. Assume first that a lifting of f exists. Then $d \circ f$ factors as

$$
X \xrightarrow{\;f\;} B \longrightarrow A \xrightarrow{\;\epsilon\;} A \vee \Sigma I
$$

which means that the derivation f^*d is trivial, Conversely, if $d \circ f$ is trivial then the diagram of R-algebras

$$
\begin{array}{ccc}
X & \xrightarrow{\;f\;} & A \\
\downarrow{\scriptstyle f} & & \downarrow{\scriptstyle \epsilon} \\
A & \xrightarrow{\;d\;} & A \vee \Sigma I
\end{array}
$$

commutes up to homotopy and by the universal property of the homotopy pullback there is a map $\tilde{f} : X \to B$ lifting f.

For the second part suppose that the lifting of f exists, so the (homotopy) fibre of the map (4.1) is nonempty. We have the following diagram of R-algebras.

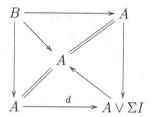

Changing B in its homotopy class if necessary we can arrange that this diagram be strictly commutative. Notice that the outer square is a homotopy pullback of R-algebras. Applying the functor $\mathrm{Map}_{\mathcal{C}_R}(X, -)$ to this diagram we get the following diagram of spaces.

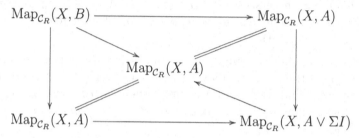

Again the outer square is a homotopy pullback (of spaces). Taking the homotopy fibres of the maps from the outer square to the center (over $f \in \mathrm{Map}_{\mathcal{C}_R}$) we get the following homotopy pullback of spaces.

(4.2)
$$\begin{array}{ccc} \mathrm{hofib}\,\mathrm{Map}_{\mathcal{C}_R}(X, B) \to \mathrm{Map}_{\mathcal{C}_R}(X, A) & \longrightarrow & pt \\ \downarrow & & \downarrow \\ pt & \longrightarrow \mathrm{hofib}\,\mathrm{Map}_{\mathcal{C}_R}(X, A \vee \Sigma I) \to \mathrm{Map}_{\mathcal{C}_R}(X, A) \end{array}$$

Notice that the space in the right lower corner of (4.2) is canonically weakly equivalent to $\Omega^\infty \mathbf{Der}_R(X, \Sigma I)$, see Remark 3.2. Since according to our assumption the space

$$\mathrm{hofib}\,\mathrm{Map}_{\mathcal{C}_R}(X, B) \longrightarrow \mathrm{Map}_{\mathcal{C}_R}(X, A)$$

is nonempty the images of the lower and right arrows in (4.2) coincide and therefore

$$\mathrm{hofib}\,\mathrm{Map}_{\mathcal{C}_R}(X, B) \to \mathrm{Map}_{\mathcal{C}_R}(X, A) \simeq \Omega(\Omega^\infty \mathbf{Der}_R(X, \Sigma I))$$
$$\simeq \Omega^\infty \mathbf{Der}_R(X, I).$$

This finishes the proof of Theorem 4.2.

A large supply of topological singular extensions can be obtain by taking Postnikov stages of connective R-algebras or those of connective commutative R-algebras. We have the following result, first proved in [16], cf. also [2].

Theorem 4.3. *Let A be a connective R-algebra where R is a connective commutative S-algebra. Then there exists a tower of R-algebras under A*

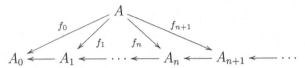

such that

- $\pi_k A_n = 0$ *for $k > n$;*
- *the map $f_n : A \longrightarrow A_n$ induces an isomorphism $\pi_k A \longrightarrow \pi_k A_n$ for $0 \leq k \leq n$;*
- *the homotopy fibre sequences $H\pi_{n+1}A \longrightarrow A_{n+1} \longrightarrow A_n$ are topological singular extensions associated with topological derivations $k_n \in Der_R^1(A_n, H\pi_{n+1}A)$.*

In the case when A is a connective commutative R-algebra we have the precise analogue of the theorem above: there is a tower $\{A_n\}$ of *commutative* R-algebras under A and the homotopy fibre sequences $H\pi_{n+1}A \longrightarrow A_{n+1} \longrightarrow A_n$ are topological singular extensions associated with *commutative* derivations $k_n \in TAQ_R^1(A_n, H\pi_{n+1}A)$.

Theorem 4.3 is proved by induction on n. By glueing cells in the category of R-algebras or commutative R-algebras we construct a map $A \longrightarrow A_0 = H\pi_0 A$. Suppose that the nth stage A_n for a commutative R-algebra A_n is already constructed. At this point let us introduce the notation $\Omega_{A_n/A}^{comm}$ to denote the abelianization of A_n *as a commutative A-algebra* (as opposed to a commutative R-algebra). This notation is only needed here and will not be used later on. Then it is not hard to see using the spectral sequence (2.1) that the lowest nonzero homotopy group of $\Omega_{A_n/A}^{comm}$ is in dimension $n + 1$ and is equal to $\pi_{n+1}A$. We then construct a map $k_n : \Omega_{A_n/A}^{comm} \longrightarrow H\pi_{n+1}A$ by attaching cells to $\Omega_{A_n/A}^{comm}$ to kill its higher homotopy groups. The map k_n is the required commutative derivation. In the associative context we use the A-bimodule of relative differentials $\Omega_{A_n|A}^{ass}$ and proceed similarly.

Theorem 4.3 provides a large supply of R-algebras or commutative R-algebras. For example Postnikov stages of MU or kU are commutative S-algebras. Other examples of commutative S-algebras include the Morava E-theory spectra E_n and Eilenberg-MacLane spectra Hk for a commutative ring k. It is generally very hard to prove that a spectrum possesses a commutative S-algebra structure unless there is a geometric reason for it. We will see later on that the situation with associative structures is better; a large class of complex-oriented spectra could be given structures of MU-algebras and, therefore of S-algebras.

5. STABILIZATION

In this section we give an interpretation of topological derivation and topological André-Quillen cohomology as 'stabilizations' of a certain forgetful functors. This interpretation in the commutative case is due to Basterra-McCarthy [3] and Basterra-Mandell (private communication).

Let $F : \mathcal{C} \longrightarrow \mathcal{D}$ be a continuous functor between pointed topological model categories. We assume that F is a homotopy functor which means that it preserves homotopy equivalences. We also assume that F is reduced, that is it takes the initial object of $\mathcal{C}$ into a contractible object in $\mathcal{D}$ (i.e. homotopy equivalent to the initial object in $\mathcal{D}$). For an object X in $\mathcal{D}$ we will denote by ΩX the cotensor of X and the pointed circle S^1. In addition the tensor of ? with a pointed topological space X will be denoted by $X \hat{\otimes} ?$.

We have the following pushout diagram of pointed spaces.

(5.1)
$$\begin{array}{ccc} S^0 & \longrightarrow & I \\ \downarrow & & \downarrow \\ I & \longrightarrow & S^1 \end{array}$$

where I is the unit interval. Tensoring it with X and applying the functor F we get the diagram

$$\begin{array}{ccc} F(X) = F(S^0 \hat{\otimes} X) & \longrightarrow & F(I \hat{\otimes} X) \\ \downarrow & & \downarrow \\ F(I \hat{\otimes} X) & \longrightarrow & F(S^1 \hat{\otimes} X) \end{array}$$

Define the functor $X \longrightarrow TF(X)$ by the requirement that the diagram

$$\begin{array}{ccc} TF(X) & \longrightarrow & F(I \hat{\otimes} X) \\ \downarrow & & \downarrow \\ F(I \hat{\otimes} X) & \longrightarrow & F(S^1 \hat{\otimes} X) \end{array}$$

be a homotopy pullback in $\mathcal{D}$. There results a natural transformation $\xi : F(X) \longmapsto TF(X)$.

Definition 5.1. The stabilization of F is the functor F^{st} defined as

$$F^{st}(X) = \text{hocolim}(F(X) \longrightarrow TF(X) \longrightarrow TTF(X) \longrightarrow \ldots)$$

where the colimit is taken over the maps $T^i(\xi) : T^i(F(X)) \longrightarrow T^i(TF(X))$.

Note that since $I \hat{\otimes} X$ is contractible the object $TF(X)$ is homotopy equivalent to $\Omega F(S^1 \hat{\otimes} X)$. Furthermore, it is easy to see that there is a homotopy equivalence

$$T^n FX \simeq \Omega^n F(S^n \hat{\otimes} X)$$

where Ω^n is the nth iterate of Ω.

Remark 5.2. The stabilization (or linearization) of a homotopy functor was introduced by Goodwillie in [12] and constitutes the first layer of its *Taylor tower*, [13]. It would be interesting to investigate the higher layers of this tower in our context.

We are now ready to give the interpretation of Der and TAQ as stabilizations of certain forgetful functors. We start with the associative case. Let A be a cofibrant R-algebra and consider the category $\mathcal{C}_{A/A}$ of A-algebras over A. Recall that $\mathcal{C}_{A/A}$ consists of R-algebras B supplied with two maps $\eta_B : B \longrightarrow A$ and $\zeta_B : A \longrightarrow B$ such that $\eta_A \circ \zeta_A = id_A$. The category $\mathcal{C}_{A/A}$ is a topological model category. Let us first describe its cotensors. Let $\tilde{B}$ be a fibrant approximation of B as an object in $\mathcal{C}_{A/A}$. Then $\eta_{\tilde{B}} : \tilde{B} \longrightarrow A$ is a fibration of R-algebras. Denote by $I(B)$ the fibre of $\eta_{\tilde{B}}$ so that there is a homotopy fibre sequence of B-bimodules:

$$I(B) \longrightarrow B \xrightarrow{\eta_B} A .$$

Consider the R-module $A \vee I(B)$. It is clearly an object in $\mathcal{C}_{A/A}$ and, moreover, the obvious map $A \vee I(B) \longrightarrow B$ is a weak equivalence. Then for a based space X the R-algebra B^X, the cotensor of B and X in $\mathcal{C}_{A/A}$ is weakly equivalent to $A \vee I(B)^X$ where $I(B)^X = F_S(\Sigma^\infty X, I(B))$ is the usual cotensor (function spectrum) of X and the R-module $I(B)$.

Now we will describe tensors in $\mathcal{C}_{A/A}$. Let B be a cofibrant object in $\mathcal{C}_{A/A}$ and X be a based space that consists of $n + 1$ points, one of them being the base point. Then the tensor $X \hat{\otimes} B$ is just the n-fold free product of B over A: $B^{\amalg_A n}$. For a general based space X we replace it with the simplicial set X_* whose realization is equivalent to X, and tensor it component-wise with B. The realization of the obtained simplicial object in $\mathcal{C}_{A/A}$ will be homotopy equivalent to $X \hat{\otimes} B$. For example if we take $X = S^1$, the circle and by X_* its simplicial model having two nondegenerate simplices in dimensions 0 and 1 then the resulting simplicial object $X_* \hat{\otimes} B$ is just the bar-construction $\beta_*(A, B, A)$. We have

$$\beta_i(A, B, A) = B^{\amalg_A i}.$$

The face maps are induced in the usual way by the canonical folding map $B \amalg_A B \longrightarrow B$ and the degeneracies are induced by the map ζ_B.

Recall that the functor $\Omega^{ass}_{B|A}$ is defined from the fibre sequence

$$\Omega^{ass}_{B|A} \longrightarrow B \wedge_A B \xrightarrow{m'} B .$$

We will assume that $\Omega^{ass}_{B|A}$ is a cofibrant B-bimodule or else replace it by its cofibrant approximation. By definition,

$$\Omega^{ass}_{A|A}(B) := A \wedge_B \Omega^{ass}_{B|A} \wedge_B A.$$

Both I and $\Omega^{ass}_{A|A}(B)$ considered as functors of B take their values in the category of A-bimodules.

Theorem 5.3. *Let A and B be connective R-algebras. Then $I^{st}(B) \simeq \Omega^{ass}_{A|A}(B)$*

To see this consider the 'universal derivation' map $B \longrightarrow A \vee \Omega^{ass}_{A|A}(B)$ and the composition

$$f_B : I(B) \longrightarrow B \longrightarrow A \vee \Omega^{ass}_{A|A}(B) \longrightarrow \Omega^{ass}_{A|A}(B).$$

It is easy to see that there is a weak equivalence of R-modules

$$\Omega^{ass}_{A|A}(B) \simeq A \wedge_B I(B)$$

and the map f_B could be represented as

$$I(B) \cong B \wedge_B I(B) \xrightarrow{\eta_B \wedge id} A \wedge_B I(B).$$

Note that if the $\eta_B : B \longrightarrow A$ is n-connected then the map f_B is $2n + 1$-connected. Further if $I(B)$ is m-connected then $I(S^n \hat{\otimes} B)$ is $n + m$-connected and the map

$$f_{S^n \hat{\otimes} B} : I(S^n \hat{\otimes} B) \longrightarrow \Omega^{ass}_{A|A}(S^n \hat{\otimes} B)$$

is $2(m + n) + 1$-connected. It follows that the map

$$\Omega^n f_{S^n \hat{\otimes} B} : \Omega^n I(S^n \hat{\otimes} B) \longrightarrow \Omega^n \Omega^{ass}_{A|A}(S^n \hat{\otimes} B)$$

is $2m+1$-connected and therefore $I^{st}(B) = [\Omega^{ass}_{A|A}]^{st}(B)$. The results of Section 2 imply that the functor $B \to [\Omega^{ass}_{A|A}]^{st}(B)$ preserves tensors and therefore

$$\Omega^n \Omega^{ass}_{A|A}(S^n \hat{\otimes} B) \simeq \Omega^n \Sigma^n \Omega^{ass}_{A|A}(B) \simeq \Omega^{ass}_{A|A}(B).$$

which implies that $[\Omega^{ass}_{A|A}]^{st}(B) \simeq \Omega^{ass}_{A|A}(B)$. This finishes our sketch of the proof of Theorem 5.3.

What is the relation of the functor $\Omega^{ass}_{A|A}(B)$ to the topological derivations of A? Let $B = A \coprod A$. Define ζ_B to be the map $A = A \coprod R \xrightarrow{id \coprod 1} A \coprod A$ and $\eta_B : A \coprod A \longrightarrow A$ to be the folding map. Then B becomes an object in $\mathcal{C}^{ass}_{A/A}$. Moreover we have a natural isomorphism for any $C \in \mathcal{C}^{ass}_{A/A}$:

$$h\mathcal{C}^{ass}_{A/A}(B, C) \cong h\mathcal{C}_{R/A}(A, C).$$

Therefore for an A-bimodule M we have

$$Der^0_R(A, M) = h\mathcal{C}_{R/A}(A, A \vee M) \cong h\mathcal{C}^{ass}_{A/A}(B, A \vee M)$$
$$\cong h\mathcal{M}_{A-A}(\Omega^{ass}_{A/A}, M)$$
$$\cong h\mathcal{M}_{A-A}(I^{st}(B), M).$$

We can also obtain in the usual way the enriched version of the above equivalence:

$$\mathbf{Der}_R(A, A \vee M) \simeq F_{A \wedge A^{op}}(I^{st}(A \coprod A), M).$$

Theorem 5.3 has an analogue in the context of commutative R-algebras. The corresponding result was first established by M. Basterra and M. Mandell (cf. also [3]) for TAQ and our proof was modelled on theirs. One should also mention that the original definition of TAQ due to I. Kriz [16] was in terms of stabilization. In contrast with the associative case this interpretation is extremely helpful in concrete calculations.

Let us now describe the result of Basterra and Mandell. A is now a commutative cofibrant R-algebra. Consider the category $\mathcal{C}_{A/A}^{comm}$ of commutative A-algebras over A. Then $I(B) \in \mathcal{M}_A$ is defined from the fibre sequence

$$ I(B) \longrightarrow \tilde{B} \xrightarrow{\eta_{\tilde{B}}} A $$

where $\tilde{B}$ is the fibrant replacement of B in $\mathcal{C}_{A/A}^{comm}$. Recall the definition of the *indecomposables functor* $B \to Q(B)$ from Section 1. It turns out that if A and B are connected then $I^{st}(B)$, the stabilization of the functor I evaluated at B, is weakly equivalent to $Q \circ I(B)$. The proof is very similar to the one given in the associative case. To show that the map

$$ I(S^n \hat{\otimes} B) \longrightarrow Q \circ I(S^n \hat{\otimes} B) $$

is sufficiently highly connected one uses Proposition 2.2. To relate stabilization with TAQ consider $A \wedge A$ as an A-algebra over A. The augmentation $A \wedge A \longrightarrow A$ is just the multiplication map and the A-algebra structure on $A \wedge A$ is defined via the action of A on the left smash factor. Then for an A-module M we have the following isomorphisms in $h\mathcal{M}_A$:

$$ \mathbf{TAQ}_R(A, M) \simeq F_A(Q \circ I(A \wedge A), M) \simeq F_A(I^{st}(A \wedge A), M). $$

In fact the author was informed by M. Basterra that the last isomorphism holds even without the hypothesis that A is connective. That suggests that Theorem 5.3 might also hold without any connectivity assumptions.

6. Calculations

Up until now we considered only formal properties of TAQ and topological derivations and saw that they are quite similar. The difference appears in their calculational aspect. Topological André-Quillen cohomology is far harder to compute and presently very few explicit calculations have been carried out. By contrast, topological derivations are relatively approachable. In most cases their computation reduces to that of topological Hochschild cohomology thanks to the homotopy fibre sequence (3.4). There is an extensive literature dedicated to the computation of THH of different Eilenberg-MacLane spectra of which we mention [5], [9], [20]. Topological Hochschild (co)homology of some spectra other than Eilenberg-MacLane are computed in [25] and [4]. We do not intend to give a detailed review of this literature and restrict ourselves with giving a few examples related to obstruction theory. In this section the symbol '$\wedge$' stands for '$\wedge_S$'.

Our first example is the computation of topological derivations of $H\mathbb{F}_p$.

Theorem 6.1. (1) *There is an isomorphism of graded $\mathbb{F}_p$-vector spaces*

$$THH^*_S(H\mathbb{F}_p, H\mathbb{F}_p) \cong (\mathbb{F}_p[x])^* = \mathrm{Hom}_{\mathbb{F}_p}(\mathbb{F}_p[x], \mathbb{F}_p).$$

where x has degree 2.

(2) *There is an isomorphism of graded $\mathbb{F}_p$-vector spaces*

$$Der^{*-1}_S(H\mathbb{F}_p, H\mathbb{F}_p) \cong (\mathbb{F}_p[x])^*/(\mathbb{F}_p)$$

Note that the statement (2) of the theorem follows from statement (1) because the homotopy fibre sequence (3.4) in our case splits giving the short exact sequence

$$\mathbb{F}_p \longrightarrow THH^*_S(H\mathbb{F}_p, H\mathbb{F}_p) \longrightarrow Der^{*-1}_S(\mathbb{F}_p, \mathbb{F}_p) \ .$$

To obtain the claim about $THH^*_S(H\mathbb{F}_p, H\mathbb{F}_p)$ it is enough to compute topological Hochschild *homology* $THH^S_*(H\mathbb{F}_p, H\mathbb{F}_p)$ because

$$THH^*_S(H\mathbb{F}_p, H\mathbb{F}_p) \cong \mathrm{Hom}_{\mathbb{F}_p}(THH^S_*(H\mathbb{F}_p, H\mathbb{F}_p), \mathbb{F}_p),$$

see Remark 3.3. The computation of $THH^S_*(H\mathbb{F}_p, H\mathbb{F}_p)$ is due to Bökstedt, [5]. Briefly, one uses the spectral sequence

$$E^2_{**} = \mathrm{Tor}_{H\mathbb{F}_p \wedge H\mathbb{F}_{p*}}(\mathbb{F}_p, \mathbb{F}_p) = \mathrm{Tor}_{\mathcal{A}^p_*}(\mathbb{F}_p, \mathbb{F}_p) \Longrightarrow THH^S_*(H\mathbb{F}_p, H\mathbb{F}_p)$$

where $\mathcal{A}^p_* = H\mathbb{F}_p \wedge H\mathbb{F}_{p*}$ is the dual Steenrod algebra mod p. The E_2-term of this spectral sequence is easy to compute. It turns out that for $p = 2$ there are no further differentials and for an odd p one uses the Dyer-Lashof operations to compute the only nontrivial differential d^{p-1}. The same operations also solve the extension problem. The final result is that, regardless of p,

(6.1) $$THH^S_*(H\mathbb{F}_p, H\mathbb{F}_p) \cong \mathbb{F}_p[x]$$

with x in homological degree 2. This finishes our sketch of Theorem 6.1.

Remark 6.2. In fact, Bökstedt's calculation shows that the isomorphism of (6.1) is multiplicative. This fact will be used in the computation of $TAQ^*_S(H\mathbb{F}_p, H\mathbb{F}_p)$.

Our next example is the calculation of $Der^*_{MU}(H\mathbb{F}_p, H\mathbb{F}_p)$ where MU is the complex cobordism spectrum. Recall that the coefficient ring of MU is the polynomial algebra:

$$MU_* = \mathbb{Z}[x_1, x_2, \ldots]$$

where $|x_i| = 2i$. Since MU is known to be a commutative connective S-algebra the spectrum $H\mathbb{F}_p$ is naturally an MU-algebra and it makes sense to consider topological derivations and topological Hochschild cohomology of $H\mathbb{F}_p$ as an MU-algebra. The relevant result is

Theorem 6.3.   (1) *There is an isomorphism of graded $\mathbb{F}_p$-vector spaces*

$$THH^*_{MU}(H\mathbb{F}_p, H\mathbb{F}_p) = \mathbb{F}_p[y_1, y_2, \ldots]$$

where y_i has cohomological degree $2i$.
(2) *There is an isomorphism of graded $\mathbb{F}_p$-vector spaces*

$$Der^{*-1}_{MU}(H\mathbb{F}_p, H\mathbb{F}_p) = \mathbb{F}_p[y_1, y_2, \ldots]/(\mathbb{F}_p).$$

Again the second claim is a consequence of the first one. To verify the first claim we need to compute $\pi_* H\mathbb{F}_p \wedge_{MU} H\mathbb{F}_p$. Using the spectral sequence

$$\operatorname{Tor}^{MU_*}_{**}(\mathbb{F}_p, \mathbb{F}_p) = \operatorname{Tor}^{\mathbb{Z}[x_1, x_2, \ldots]}_{**}(\mathbb{F}_p, \mathbb{F}_p)$$
$$\Rightarrow \pi_* H\mathbb{F}_p \wedge_{MU} H\mathbb{F}_p$$

we see that the ring $\pi_* H\mathbb{F}_p \wedge_{MU} H\mathbb{F}_p$ is isomorphic to $\Lambda_{\mathbb{F}_p}(z_1, z_2, \ldots)$, the exterior algebra over $\mathbb{F}_p$ on generators z_i where $|z_i| = 2i - 1$.

Finally from the spectral sequence

$$\operatorname{Ext}^{**}_{\pi_* H\mathbb{F}_p \wedge_{MU} H\mathbb{F}_p}(\mathbb{F}_p, \mathbb{F}_p) = \operatorname{Ext}^{**}_{\Lambda_{\mathbb{F}_p}(z_1, z_2, \ldots)}(\mathbb{F}_p, \mathbb{F}_p)$$
$$= \mathbb{F}_p[y_1, y_2, \ldots] \Rightarrow THH^*_{MU}(H\mathbb{F}_p, H\mathbb{F}_p)$$

we obtain the desired isomorphism.

Next we would like to discuss the forgetful maps

$$l^*_{MU} : Der^*_{MU}(H\mathbb{F}_p, H\mathbb{F}_p) \longrightarrow [H\mathbb{F}_p, H\mathbb{F}_p]^*_{MU-mod}$$

and

$$l^*_S : Der^*_S(H\mathbb{F}_p, H\mathbb{F}_p) \longrightarrow [H\mathbb{F}_p, H\mathbb{F}_p]^*_{S-mod} = \mathcal{A}^*_p.$$

We already saw that the algebra of cooperations in MU-theory $\pi_* H\mathbb{F}_p \wedge_{MU} H\mathbb{F}_p$ is an exterior algebra on generators z_i. Just as the usual dual Steenrod algebra it is a Hopf algebra and it is easy to see that z_i's are primitive elements in it. Therefore the MU-Steenrod algebra $[H\mathbb{F}_p, H\mathbb{F}_p]^*_{MU-mod}$ is itself an exterior algebra $\Lambda_{\mathbb{F}_p}(z_1^*, z_2^*, \ldots)$ where z_i^*'s are dual to z_i.

Then the map l^*_{MU} sends the derivation $y_i \in Der^{2i-1}_{MU}(H\mathbb{F}_p, H\mathbb{F}_p)$ to the operation $z_i^* \in [H\mathbb{F}_p, H\mathbb{F}_p]^{2i-1}_{MU-mod}$.

Any map of MU-modules $H\mathbb{F}_p \longrightarrow \Sigma^* H\mathbb{F}_p$ can also be considered as a map of S-modules. In other words there is a map from the MU-Steenrod algebra into $\mathcal{A}^*_p$:

$$[H\mathbb{F}_p, H\mathbb{F}_p]^*_{MU-mod} \longrightarrow [H\mathbb{F}_p, H\mathbb{F}_p]^*_{S-mod} = \mathcal{A}^*_p.$$

The latter map is a map of Hopf algebras and should therefore respect primitive elements. We know that the odd degree primitive elements are the Milnor Bocksteins $Q_i \in \mathcal{A}^*_p$. Since $|Q_i| = 2p^i - 1$ only the primitives $z^*_{2p^i-1}$ could have a nonzero image in $\mathcal{A}^*_p$. To see that $z^*_{2p^i-1}$ does indeed have a nonzero image notice that it is the first nonzero Postnikov invariant of the connective

Morava K-theory spectrum $k(n)$. This invariant is nontrivial both in the category of S-modules and MU-modules which shows that the image of $z^*_{2p^i-1}$ is Q_i up to an invertible scalar factor.

This allows one to calculate the image of the map l^*_S. On the one hand its image should be contained in the subspace spanned by the primitive odd degree elements in $\mathcal{A}^*_p$, that is, the elements Q_i. On the other hand any operation Q_i can be lifted to a topological derivation in $Der^*_S(H\mathbb{F}_p, H\mathbb{F}_p)$ since it even lifts to an MU-derivation in $Der^*_{MU}(H\mathbb{F}_p, H\mathbb{F}_p)$. Therefore the image of l_S is the whole subspace of $\mathcal{A}^*_p$ spanned by Q_i's.

We now turn to the calculation of $TAQ^*_S(H\mathbb{F}_p, H\mathbb{F}_p)$ which was done by Kriz, [16]. The method we use is due to Basterra and Mandell. We have

$$TAQ^*_S(H\mathbb{F}_p, H\mathbb{F}_p) \simeq [\operatorname{hocolim}_{n\to\infty} \Omega^n I(S^n \hat{\otimes}(H\mathbb{F}_p \wedge H\mathbb{F}_p)), H\mathbb{F}_p]^*_{H\mathbb{F}_p-mod}$$

$$\simeq [\operatorname{hocolim}_{n\to\infty} \Sigma^{-n} I(S^{n-2} \hat{\otimes}[S^2 \hat{\otimes}(H\mathbb{F}_p \wedge H\mathbb{F}_p)]), H\mathbb{F}_p]^*_{H\mathbb{F}_p-mod}$$

$$\simeq [\Sigma^{-2} \operatorname{hocolim}_{n\to\infty} \Sigma^{-n+2} I(S^{n-2} \hat{\otimes}[S^2 \hat{\otimes}(H\mathbb{F}_p \wedge H\mathbb{F}_p)]), H\mathbb{F}_p]^*_{H\mathbb{F}_p-mod}$$

$$\simeq \Sigma^{-2} TAQ^*_{H\mathbb{F}_p}(S^2 \hat{\otimes}(H\mathbb{F}_p \wedge H\mathbb{F}_p), H\mathbb{F}_p).$$

Lemma 6.4. *There is a weak equivalence of commutative $H\mathbb{F}_p$-algebras*

$$(6.2) \qquad S^2 \hat{\otimes}(H\mathbb{F}_p \wedge H\mathbb{F}_p) \cong H\mathbb{F}_p \vee \Sigma^3 H\mathbb{F}_p.$$

Here the right hand side of (6.2) is the square-zero extension of $H\mathbb{F}_p$ by $\Sigma^3 H\mathbb{F}_p$.

To see this note first that

$$S^1 \hat{\otimes}(H\mathbb{F}_p \wedge H\mathbb{F}_p) \simeq \mathbf{THH}^S(H\mathbb{F}_p, H\mathbb{F}_p).$$

Therefore

$$\pi_* S^1 \hat{\otimes}(H\mathbb{F}_p \wedge H\mathbb{F}_p) = \mathbb{F}_p[x]$$

with $|x| = 2$, see Remark 6.2. Denote the commutative S-algebra $S^1 \hat{\otimes}(H\mathbb{F}_p \wedge H\mathbb{F}_p)$ by X. We have the following weak equivalences of commutative S-algebras:

$$S^2 \hat{\otimes}(H\mathbb{F}_p \wedge H\mathbb{F}_p) \simeq S^1 \hat{\otimes} X \simeq H\mathbb{F}_p \wedge_X H\mathbb{F}_p.$$

From the spectral sequence

$$\operatorname{Tor}^{X_*}_{**}(\mathbb{F}_p, \mathbb{F}_p) = \Lambda_{\mathbb{F}_p}(y) \Rightarrow \pi_* H\mathbb{F}_p \wedge_X H\mathbb{F}_p$$

we see that

$$\pi_* H\mathbb{F}_p \wedge_X H\mathbb{F}_p = \pi_* S^2 \hat{\otimes}(H\mathbb{F}_p \wedge H\mathbb{F}_p) = \Lambda_{\mathbb{F}_p}(y).$$

Here y has degree 3. It is not hard to see, using Proposition 2.2 and Theorem 2.4 that any augmented $H\mathbb{F}_p$-algebra whose coefficient ring is an exterior algebra on one generator y in positive degree is in fact weakly equivalent to the square-zero extension $H\mathbb{F}_p \vee \Sigma^{|y|} H\mathbb{F}_p$. This finishes our sketch of the proof of Lemma 6.4.

It follows that

$$TAQ^*_S(H\mathbb{F}_p, H\mathbb{F}_p) = \Sigma^{-2} TAQ^*_{H\mathbb{F}_p}(H\mathbb{F}_p \vee \Sigma^3 H\mathbb{F}_p, H\mathbb{F}_p).$$

So we need to compute the topological Andrè-Quillen cohomology of the 'Eilenberg-MacLane object' $H\mathbb{F}_p \vee \Sigma^3 H\mathbb{F}_p$ in the category of commutative augmented $H\mathbb{F}_p$-algebras. There is an obvious element ('fundamental class') $x \in TAQ^3_{H\mathbb{F}_p}(H\mathbb{F}_p \vee \Sigma^3 H\mathbb{F}_p, H\mathbb{F}_p)$ corresponding to the identity map on $H\mathbb{F}_p \vee \Sigma^3 H\mathbb{F}_p$. Further one introduces the action of Steenrod operations on TAQ. Let $p = 2$. Then there exist operations

$$Sq^i : TAQ^q_{H\mathbb{F}_p}(-, H\mathbb{F}_p) \longrightarrow TAQ^{q+i}_{H\mathbb{F}_p}(-, H\mathbb{F}_p)$$

defined for $q < i$ and satisfying the usual Adem relations. Applying Sq^i to $y = \Sigma^{-2}x \in TAQ^1_{H\mathbb{F}_p}(H\mathbb{F}_p, H\mathbb{F}_p)$ one obtains new elements in $TAQ^*_{H\mathbb{F}_p}(H\mathbb{F}_p, H\mathbb{F}_p)$. The final result is (for $p = 2$):

Theorem 6.5. $TAQ^*_S(H\mathbb{F}_p, H\mathbb{F}_p)$ *is spanned by the elements* $Sq^{s_1}Sq^{s_2}\cdots$ $Sq^{s_r}y$ *where the sequence* $(s_1, s_2, \ldots, s_r)$ *is Steenrod admissible and* $s_r > 3$.

There is a similar result for odd primes which uses the odd primary operations in TAQ instead of Steenrod squares.

To conclude this section note that the forgetful map

$$TAQ^*_S(H\mathbb{F}_p, H\mathbb{F}_p) \longrightarrow [H\mathbb{F}_p, H\mathbb{F}_p]^*_{S-mod} = \mathcal{A}^*_p$$

is not as interesting as in the associative case. It is easy to see that its image contains (multiples of) the Bockstein homomorphism β corresponding to the topological singular extension

$$H\mathbb{F}_p \longrightarrow H\mathbb{F}_{p^2} \longrightarrow H\mathbb{F}_p$$

and nothing else. Indeed, if it contained, e.g. the operation Q_i for some $i > 0$ then the corresponding singular extension $H\mathbb{F}_p \longrightarrow ? \longrightarrow \Sigma^{2(p^i-1)}H\mathbb{F}_p$ would realize the 2-stage Postnikov tower of $k(n)$ as a commutative S-algebra which is impossible.

7. Existence of S-algebra structures

Topological Andrè-Quillen cohomology was originally introduced in [16] with the purpose of proving that BP, the Brown-Peterson spectrum supports a structure of an E_∞-ring spectra (or, equivalently, a commutative S-algebra structure), a long-standing problem posed by P. May. Unfortunately, this problem is still open, however the problem concerning *associative* structures turned out to be much more manageable. The first result in this direction is the seminal paper of A. Robinson [33] where he proved that there is an A_∞-structure on Morava K-theory spectra at odd primes. Working in the category of MU-modules it is possible to obtain considerably stronger results. Namely, in [18] and [10] it was proved that a broad class of complex-oriented cohomology theories admit structures of S-algebras. These structures are typically non-unique and in the last section we say something about the corresponding moduli spaces. Here we describe the relevant results following [18].

Let us assume that the coefficient ring R_* of our 'ground' commutative S-algebra R is concentrated in even degrees. For an element $x \in R_*$ we will denote by R/x the cofibre of the map $R \xrightarrow{x} R$. Let I_* be a graded ideal generated by a (possibly infinite) regular sequence of homogeneous elements $(u_1, u_2, \ldots) \in R_*$. We assume in addition that each u_k is a nonzero divisor in the ring R_*. Then we can form the R-module R/I as the infinite smash product of R/u_k. It is known by work of Strickland [35] that there is a structure of an R-ring spectrum on R/I. Clearly the coefficient ring of R/I is isomorphic to R_*/I_* where R_*/I_* is understood to be the direct limit of $R_*/(u_1, u_2, \ldots, u_k)$. Let us denote the R-algebra R/I by A. Our standing assumption is that A has a structure of an R-algebra (i.e. strictly associative). This may seem a rather strong condition but, as we see shortly such a situation is quite typical. In fact P. Goerss proved in [10] that any spectrum obtained by killing a regular sequence in MU, the complex cobordism S-algebra, has a structure of an MU-algebra.

The construction we are about to describe allows one to construct new R-algebras by 'adjoining' the indeterminates u_k to the R-algebra A. Let us introduce the notation $A(u_k^l)_*$ for the R_*-algebra

$$\lim_{n \to \infty} R_*/(u_1, u_2, \ldots, u_{k-1}, u_k^l, u_{k+1}, \ldots, u_n).$$

For each l reduction modulo u_k^l determines a map of R_*-algebras

$$A(u_k^{l+1})_* \to A(u_k^l)_*.$$

Now we can formulate our main theorem in this section.

Theorem 7.1. *For each l there exist R-algebras $A(u_k^l)$ with coefficient rings $A(u_k^l)_*$ and R-algebra maps*

$$R_{l,k} : A(u_k^{l+1}) \to A(u_k^l)$$

which give the reductions mod u_k^l on the level of coefficient rings.

Let us explain the idea of the proof. One uses induction on l. The case $l = 1$ is just our original assumption that A is an R-algebra. Suppose that the R-algebras $A(u_k^l)$ with the required properties were constructed for $l \leq i$. One then shows that R/u_k^{i+1} possesses an R-ring structure and there exists a homotopy cofibre sequence of R-modules

$$(7.1) \qquad R/u_k^{i+1} \longrightarrow R/u_k^i \longrightarrow \Sigma^{|u_k|i+1} R/u_k$$

where the first arrow is an R-ring spectrum map realizing the reduction mod u^i in homotopy and the second arrow is an appropriate Bockstein operation. Taking the smash product of (7.1) with

$$R/u_1 \wedge R/u_2 \wedge \ldots \wedge R/u_{k-1} \wedge R/u_{k+1} \wedge \ldots$$

we get the following homotopy cofibre sequence of R-modules:

$$(7.2) \qquad A(u_k^{i+1}) \longrightarrow A(u_k^i) \longrightarrow \Sigma^{|u_k|i+1}A.$$

Finally one proves that the sequence (7.2) could be improved to a topological singular extension by computing $Der_R^*(A(u_k^i), A)$ and analyzing the forgetful map

$$Der_R^*(A(u_k^i), A) \longrightarrow [A(u_k^i), A]_{R-mod}^* .$$

Theorem 7.1 leads to a host of examples of S-algebras. Let $R = MU_{(p)}$. the complex cobordism spectrum localized at p and $I_* = (p, x_1, x_2, \ldots)$ be the maximal ideal in $MU_{(p)*} = \mathbb{Z}_{(p)}[x_1, x_2, \ldots]$. Then $MU/I \simeq H\mathbb{F}_p$ is an MU-algebra. For any polynomial generator $x_k \in MU_{(p)}$ and any i we can construct an $MU_{(p)}$-algebra realizing in homotopy the $MU_{(p)*}$-module $\mathbb{F}_p[x_k^i]$. Taking the homotopy inverse limit of these we get an $MU_{(p)}$-algebra whose coefficient ring is $\mathbb{F}_p[x_k]$. Iterating this procedure we can realize topologically the $MU_{(p)*}$-algebra $\mathbb{F}_p[x_{k_1}, x_{k_2}, \ldots]$ for any sequence (possibly infinite) of polynomial generators $x_{k_1}, x_{k_2}, \ldots$. We could also set $I_* = (x_1, x_2, \ldots)$ to obtain integral versions of these $MU_{(p)}$-algebras.

Recall that the Brown-Peterson spectrum BP is obtained from $MU_{(p)}$ by killing all polynomial generators in $MU_{(p)*}$ except for $v_i = x_{2(p^i-1)}$ provided that we use Hazewinkel generators for $MU_{(p)*}$. Therefore $BP_* = \mathbb{Z}_{(p)}[v_1, v_2, \ldots]$. It follows that BP possesses an $MU_{(p)}$-algebra structure. We define

$$BP\langle n \rangle = BP/(v_i, i > n)$$
$$P(n) = BP/(v_i, i < n)$$
$$B(n) = v_n^{-1}P(n)$$
$$k(n) = BP/(v_i, i \neq n)$$
$$K(n) = v_n^{-1}BP/(v_i, i \neq n)$$
$$E(n) = v_n^{-1}BP/(v_i, i > n).$$

Note that inverting an element in the coefficient ring of an $MU_{(p)}$-ring spectrum is an instance of Bousfield localization in the homotopy category of $MU_{(p)}$-modules. Further Bousfield localization preserves algebra structures we conclude that all spectra listed above admit structures of $MU_{(p)}$-algebras (and therefore MU-algebras).

Finally we need to mention the result of Hopkins and Miller, cf. [29], which states that a suitable completion E_n of the spectrum $E(n)$ (which came to be popularly known as the Morava E-theory) admits a unique structure of an S-algebra. A generalization of this theorem is discussed in [18]. Furthermore, in [11] Hopkins and Goerss proved using a more advanced technology that E_n in fact admits a unique structure of a *commutative* S-algebra. This result was also obtained in the recent paper by Richter and Robinson [31].

8. SPACES OF MULTIPLICATIVE MAPS

In this section we will consider the problem of computing the homotopy type of the mapping space $\mathrm{Map}_{\mathcal{C}_R}(A, B)$ where A is a cofibrant R-algebra. Unexpectedly, it turns out that the space of based loops on $\mathrm{Map}_{\mathcal{C}_R}(A, B)$ is weakly equivalent to an infinite loop space. More precisely, in [19] the following theorem is proved.

Theorem 8.1. *For a cofibrant algebra A and a map of R-algebras $f : A \to B$ the space of d-fold based loops $\Omega^d(\mathrm{Map}_{\mathcal{C}_R}(A, B), f)$ is weakly equivalent to the space $\Omega^\infty \mathrm{Der}_R(A, \Sigma^{-d} B)$ for all $d \geq 1$.*

The main ingredient in the proof of this theorem is the following

Lemma 8.2. *There is a weak equivalence of R-algebras A^{S^d} and the square-zero-extension $A \vee \Sigma^{-d} A$ for all $d \geq 1$.*

Note that, obviously, there is a weak equivalence of R-modules

$$(8.1) \qquad A^{S^d} \longrightarrow A \vee \Sigma^{-d} A$$

The lemma is proved by computing the topological derivations of A^{S^d} and showing that the map of (8.1) could be improved to a topological derivation of A^{S^d} with values in $\Sigma^{-d} A$.

The proof of Theorem 8.1 is actually very easy. We have the following commutative diagram of spaces where both rows are homotopy fibre sequences:

$$\begin{array}{ccccc}
\mathcal{T}op_*(S^d, \mathrm{Map}_{\mathcal{C}_R}(A, B)) & \longrightarrow & \mathcal{T}op(S^d, \mathrm{Map}_{\mathcal{C}_R}(A, B)) & \longrightarrow & \mathrm{Map}_{\mathcal{C}_R}(A, B) \\
\downarrow & & \downarrow & & \| \\
? & \longrightarrow & \mathrm{Map}_{\mathcal{C}_R}(A, B^{S^d}) & \longrightarrow & \mathrm{Map}_{\mathcal{C}_R}(A, B)
\end{array}$$

Here the horizontal rightmost arrows are both induced by the inclusion of the base point into S^d. Since the right and the middle vertical arrows are weak equivalences (even isomorphisms) it follows that the map $\mathcal{T}op_*(S^d, \mathrm{Map}_{\mathcal{C}_R}(A, B)) \to ?$ is a weak equivalence. But Lemma 8.2 tells us that the R-algebra B^{S^d} is weakly equivalent as an R-algebra to $B \vee \Sigma^{-d} B$. In other words the term ? is weakly equivalent to the topological space of maps $A \to B \vee \Sigma^{-d} B$ which yield f when composed with the projection $B \vee \Sigma^{-d} B \longrightarrow B$. Therefore ? is weakly equivalent to $\Omega^\infty \mathrm{Der}_R(A, \Sigma^{-d} B)$ and our theorem is proved.

Corollary 8.3. *For a cofibrant algebra A and a map of R-algebras $f : A \to B$ there is a bijection between sets $\pi_d(\mathrm{Map}_{\mathcal{C}_R}(A, B), f)$ and $\mathrm{Der}_R^{-d}(A, B)$ for $d \geq 1$. If $d \geq 2$ then this bijection is an isomorphism of abelian groups.*

One might wonder whether Theorem 8.1 remains true in the context of commutative S-algebras. The answer is no. The crucial point is the weak equivalence of S-algebras $S \vee S^{-1}$ and S^{S^1}. It is clear that $\pi_0 S \wedge_{S \vee S^{-1}} S$

is the divided power ring. However N. Kuhn and M. Mandell proved that $\pi_0 S \wedge_{S^{S^1}} S$ is the ring of numeric polynomials. Therefore $S \vee S^{-1}$ and S^{S^1} cannot be weakly equivalent as commutative S-algebras.

Another point of view is afforded by Mandell's theorem [22] which states that the homotopy category of connected p-complete nilpotent spaces of finite p-type can be embedded into the category of E_∞ algebras over $\bar{\mathbb{F}}_p$ as a full subcategory. Here $\bar{\mathbb{F}}_p$ is the algebraic closure of the field $\mathbb{F}_p$. This embedding is via the singular cochains functor $C^*(-, \mathbb{F}_p)$. The category of E_∞ algebras over $\bar{\mathbb{F}}_p$ is Quillen equivalent to the category of commutative $H\bar{\mathbb{F}}_p$-algebras. Consider an arbitrary space X (connected, p-complete, nilpotent and of finite p-type) and the corresponding $H\bar{\mathbb{F}}_p$-algebra $\tilde{X}$ which we assume to be cofibrant. Then the space of maps in $\mathcal{C}^{comm}_{H\bar{\mathbb{F}}_p}$ from $\tilde{X}$ into $H\bar{\mathbb{F}}_p$ will be weakly equivalent to X. This shows that the space of commutative R-algebra maps could have an essentially arbitrary homotopy type.

However, there is a context in which the analogue of Theorem 8.1 holds for commutative R-algebras. Let us suppose that our ground S-algebra R is rational, that is, local with respect to $H\mathbb{Q}$. Let A and B be commutative R-algebras where A is cofibrant as before. Then it is easy to see that A^{S^1} and $A \vee \Sigma^{-1}A$ are weakly equivalent as *commutative* R-algebras. Indeed, $A \vee \Sigma^{-1}A$ is weakly equivalent as a commutative R-algebra to $A \wedge R[S^{-1}]$ where $R[S^{-1}]$ is the free commutative R-algebra on the R-module S^{-1}. We have $\pi_* A^{S^1} = \Lambda_{A_*}(x)$ where the exterior generator x has degree -1. The element x determines a map of R-modules $S^1 \longrightarrow A^{S^1}$ and therefore a map of R-algebras $A \vee \Sigma^{-1}A \simeq A \wedge R[S^{-1}] \longrightarrow A^{S^1}$ which is a weak equivalence. Further there are the following equivalences of commutative R-algebras:

$$A^{S^d} \cong A^{S^1 \wedge S^{d-1}} \cong (A^{S^1})^{S^{d-1}} \simeq (A \vee \Sigma^{-1}A)^{S^{d-1}} \simeq A \vee \Sigma^{-d}A.$$

So we in this case we have the exact analogue of Lemma 8.2. The remainder of the proof is the same and we obtain

Theorem 8.4. *Let R be a rational commutative S-algebra, Then for a cofibrant R-algebra A and a map of R-algebras $f : A \to B$ the space of d-fold based loops $\Omega^d(\mathrm{Map}_{\mathcal{C}^{comm}_R}, f)$ is weakly equivalent to the space $\Omega^\infty \mathbf{TAQ}_R(A, \Sigma^{-d}B)$. Furthermore there is a bijection between the sets $\pi_d(\mathrm{Map}_{\mathcal{C}^{comm}_R}(A, B), f)$ and $TAQ_R^{-d}(A, B)$ for $d \geq 1$. If $d \geq 2$ then this bijection is an isomorphism of abelian groups.*

For $R = H\mathbb{Q}$, the Eilenberg-MacLane spectrum the category of commutative R-algebras is Quillen equivalent to the category of commutative differential graded algebras over $\mathbb{Q}$. Via this equivalence Theorem 8.4 translates into a statement in rational homotopy theory which does not seem to have appeared in the literature.

9. MODULI SPACES

For an R-module X it is natural to look at the set of all non-isomorphic R-algebra structures on X. (We assume that X supports at least one R-algebra structure.) This set is actually π_0 of the Dwyer-Kan *classification space*, cf. [7] which we will now describe.

Consider the subcategory W of $\mathcal{C}_R$ consisting of those R-algebras which are weakly equivalent to X as R-modules. The morphisms of W are the weak equivalences of R-algebras. Then the *moduli space* $\mathcal{M}(X)$ is by definition the nerve of W.

For a cofibrant R-algebra A we will denote by $haut(A)$ the topological monoid of homotopy auto-equivalences of A. For a non-cofibrant A we take for $haut(A)$ the monoid of homotopy auto-equivalences of its cofibrant approximation. Thus $haut(A)$ is well defined up to homotopy. We can form its classifying space $Bhaut(A)$. Then the results of Dwyer and Kan imply the following

Theorem 9.1. *There is a weak equivalence of spaces*

$$\mathcal{M}(X) \simeq \coprod Bhaut(A)$$

where the disjoint union is taken over the set of connected components of W.

We assumed from the beginning the existence of at least one R-algebra structure on X, that gives the space $\mathcal{M}(X)$ a base point. Let us denote the corresponding R-algebra by A. The connected component of A in $\mathcal{M}(X)$ is, therefore, weakly equivalent to $Bhaut(A)$.

Note that it is possible to consider other moduli spaces for X. For example, we could insist that the homotopy X_* has a fixed ring structure, or that X itself has a fixed R-ring spectrum structure up to homotopy. Each of these choices affects π_0 of our moduli space but not the connected components themselves, and therefore, not the higher homotopy groups.

Assume, without loss of generality, that A is a cofibrant R-algebra. Clearly $haut(A)$ is a disjoint union summand of the space $\mathrm{Map}_{\mathcal{C}_R}(A, A)$ of all multiplicative self-maps of A. Taking the identity map $A \longrightarrow A$ as the base point in both $haut(A)$ and $\mathrm{Map}_{\mathcal{C}_R}(A, A)$ we see that

$$\Omega haut(A) \simeq \Omega \mathrm{Map}_{\mathcal{C}_R}(A, A).$$

But we know from the previous section that there is a weak equivalence of spaces

$$\Omega \mathrm{Map}_{\mathcal{C}_R}(A, A) \simeq \Omega\Omega^\infty \mathbf{Der}_R(A, A).$$

This gives the following

Theorem 9.2. *The space $\Omega^2 \mathcal{M}(X)$ of two-fold loops on the moduli space $\mathcal{M}(X)$ is weakly equivalent to $\Omega\Omega^\infty \mathbf{Der}_R(A, A)$. In particular it is an infinite loop space.*

In particular, we have an isomorphism

$$\pi_i \mathcal{M}(X) \cong Der_R^{1-i}(A, A)$$

for $i \geq 2$. The computation of π_1 and π_0 is, of course, a completely different story.

What about the moduli space of *commutative* R-algebra structures? Theorem 9.1 has an obvious analogue in the context of commutative R-algebras. However the analogue of Theorem 9.2 falls through because for a commutative R-algebra A the space of self-maps of A in $\mathcal{C}_R^{comm}$ is not in any obvious way related to $\mathbf{TAQ}_R(A, A)$. However Hopkins and Goerss proved in [11] that E_n, the Morava E-theory spectrum mentioned at the end of Section 7 admits a unique structure of a commutative S-algebra and the space of its self-maps is homotopically discrete with the set of connected components being equal to S_n, the Morava stabilizer group. In our interpretation it means that the appropriate moduli space of E_n is weakly equivalent to BS_n, the classifying space of S_n.

REFERENCES

[1] M. André, *Méthode simpliciale en algèbre homologique et algèbre commutative.* Lecture Notes in Mathematics, Vol. 32 Springer-Verlag, Berlin-New York 1967.

[2] M. Basterra, *André-Quillen cohomology of commutative S-algebras.* J. Pure Appl. Algebra 144 (1999), no. 2, 111–143.

[3] M. Basterra & R. McCarthy, *Γ-homology, topological André-Quillen homology and stabilization.* Topology Appl. 121 (2002), no.3, 551–566.

[4] A. Baker & A. Lazarev, *Topological Hochschild cohomology and generalized Morita equivalence,* preprint, math.AT/0209003.

[5] M. Bökstedt, *Topological Hochschild homology of $\mathbb{Z}$ and $\mathbb{Z}_p$,* preprint.

[6] J. Cuntz & D. Quillen, *Algebra extensions and nonsingularity.* J. Amer. Math. Soc. 8 (1995), no. 2, 251–289.

[7] W. Dwyer & D. Kan, *A classification theorem for diagrams of simplicial sets.* Topology 23 (1984), no. 2, 139–155.

[8] A. Elmendorf, I. Kriz, M. Mandell & J.P. May, *Rings, modules, and algebras in stable homotopy theory.* Mathematical Surveys and Monographs **47** (1997).

[9] V. Franjou & J. Lannes, L. Schwartz, *Autour de la cohomologie de Mac Lane des corps finis.* Invent. Math. 115 (1994), no. 3, 513–538.

[10] P. Goerss, *Associative MU-algebras,* preprint 2001.

[11] P. Goerss & M. Hopkins *Moduli spaces of commutative ring spectra,* this volume.

[12] T. Goodwillie, *Calculus I. The first derivative of pseudoisotopy theory.* K-Theory 4 (1990), no. 1, 1–27.

[13] T. Goodwillie, *Calculus 3. The Taylor series of homotopy functors,* Geometry & Topology 7 (2003), 645–711.

[14] M. Hovey, *Model categories.* Mathematical Surveys and Monographs, 63. American Mathematical Society, Providence, RI, 1999.

[15] M. Hovey, B. Shipley & J. Smith, *Symmetric spectra.* J. Amer. Math. Soc. 13 (2000), no. 1, 149–208.

[16] I. Kriz, *Towers of E_∞-ring spectra with applications to BP,* preprint, 1993.

[17] A. Lazarev, *Homotopy theory of A_∞ ring spectra and applications to MU-modules.* K-Theory 24 (2001), no. 3, 243–281.

[18] A. Lazarev, *Towers of MU-algebras and a generalized Hopkins-Miller theorem*, to appear in the Proc. of the London Math. Soc.

[19] A.Lazarev, *Spaces of multiplicative maps between highly structured ring spectra*, Proceedings of the Isle of Skye Conference on Algebraic Topology, to appear.

[20] A. Lindenstrauss & I. Madsen, *Topological Hochschild homology of number rings.* Trans. Amer. Math. Soc. 352 (2000), no. 5, 2179–2204.

[21] I. Madsen, *Algebraic K-theory and traces.* Current developments in mathematics, 1995 (Cambridge, MA), 191–321, Internat. Press, Cambridge, MA, 1994.

[22] M. Mandell, E_∞ *algebras and p-adic homotopy theory.* Topology 40 (2001), no. 1, 43–94.

[23] J.P. May, E_∞ *ring spaces and* E_∞ *ring spectra.* With contributions by F. Quinn, N. Ray, and J. Tornehave. Lecture Notes in Mathematics, Vol. 577. Springer-Verlag, Berlin-New York, 1977.

[24] J.P. May, *The geometry of iterated loop spaces.* Lectures Notes in Mathematics, Vol. 271. Springer-Verlag, Berlin-New York, 1972.

[25] J. McClure & R. Staffeldt, *On the topological Hochschild homology of* bu. I. Amer. J. Math. 115 (1993), no. 1, 1–45.

[26] T. Pirashvili & B. Richter, *Robinson-Whitehouse complex and stable homotopy.* Topology 39 (2000), no. 3, 525–530.

[27] D. Quillen, *Homotopical algebra.* Lecture Notes in Mathematics, No. 43. Springer-Verlag, Berlin-New York 1967.

[28] D. Quillen, On the (co-)homology of commutative rings. Applications of Categorical Algebra (Proc. Sympos. Pure Math., Vol. XVII, New York, 1968) Amer. Math. Soc., Providence, R.I, (1970), 65–87.

[29] C. Rezk, Notes on the Hopkins-Miller Theorem. *Homotopy theory via algebraic geometry and group representations* (Evanston, IL, 1997), Contemp. Math., 220, Amer. Math. Soc., Providence, RI, (1998), 313–36.

[30] B. Richter, *An Atiyah-Hirzebruch spectral sequence for topological André-Quillen homology.* J. Pure Appl. Algebra 171 (2002), no. 1, 59–66.

[31] B. Richter, A. Robinson, *Gamma homology of polynomial algebras and of group algebras,* Proceedings of the March 2002 Northwestern University. Conference on Algebraic Topology, to appear.

[32] A. Robinson & S. Whitehouse, *Operads and* Γ-*homology of commutative rings.* Math. Proc. Cambridge Philos. Soc. 132 (2002), no. 2, 197–234.

[33] A. Robinson, *Obstruction theory and the strict associativity of Morava K-theories.* Advances in homotopy theory, London Math. Soc. Lecture Note Ser., 139, Cambridge Univ. Press, Cambridge, (1989), 143–152.

[34] A. Robinson, *Gamma homology, Lie representations and* E_∞ *multiplications*, Inv. Math., 152 (2003), 331–348.

[35] N.P. Strickland, *Products on MU-modules.* Trans. Amer. Math. Soc. 351 (1999), no. 7, 2569–2606.

MATHEMATICS DEPARTMENT, BRISTOL UNIVERSITY, BRISTOL, BS8 1TW, ENGLAND.

email-address: a.lazarev@bristol.ac.uk

INDEX

Printed in the United States
by Baker & Taylor Publisher Services